AF777980

CLIMATE CHANGE AND ITS CAUSES, EFFECTS AND PREDICTION

ASSESSING CLIMATE CHANGE IMPACTS ON THE UNITED STATES

CLIMATE CHANGE AND ITS CAUSES, EFFECTS AND PREDICTION

Additional books in this series can be found on Nova's website at:

https://www.novapublishers.com/catalog/index.php?cPath=23_29&seriesp=Climate+Change+and+its+Causes%2C+Effects+and+Prediction

Additional E-books in this series can be found on Nova's website at:

https://www.novapublishers.com/catalog/index.php?cPath=23_29&seriespe=Climate+Change+and+its+Causes%2C+Effects+and+Prediction

CLIMATE CHANGE AND ITS CAUSES, EFFECTS AND PREDICTION

ASSESSING CLIMATE CHANGE IMPACTS ON THE UNITED STATES

ALEX B. MCNEILL
EDITOR

Nova Science Publishers, Inc.

New York

Copyright © 2010 by Nova Science Publishers, Inc.

All rights reserved. No part of this book may be reproduced, stored in a retrieval system or transmitted in any form or by any means: electronic, electrostatic, magnetic, tape, mechanical photocopying, recording or otherwise without the written permission of the Publisher.

For permission to use material from this book please contact us:
Telephone 631-231-7269; Fax 631-231-8175
Web Site: http://www.novapublishers.com

NOTICE TO THE READER

The Publisher has taken reasonable care in the preparation of this book, but makes no expressed or implied warranty of any kind and assumes no responsibility for any errors or omissions. No liability is assumed for incidental or consequential damages in connection with or arising out of information contained in this book. The Publisher shall not be liable for any special, consequential, or exemplary damages resulting, in whole or in part, from the readers' use of, or reliance upon, this material. Any parts of this book based on government reports are so indicated and copyright is claimed for those parts to the extent applicable to compilations of such works.

Independent verification should be sought for any data, advice or recommendations contained in this book. In addition, no responsibility is assumed by the publisher for any injury and/or damage to persons or property arising from any methods, products, instructions, ideas or otherwise contained in this publication.

This publication is designed to provide accurate and authoritative information with regard to the subject matter covered herein. It is sold with the clear understanding that the Publisher is not engaged in rendering legal or any other professional services. If legal or any other expert assistance is required, the services of a competent person should be sought. FROM A DECLARATION OF PARTICIPANTS JOINTLY ADOPTED BY A COMMITTEE OF THE AMERICAN BAR ASSOCIATION AND A COMMITTEE OF PUBLISHERS.

Library of Congress Cataloging-in-Publication Data

Available Upon Request

ISBN: 978-1-60876-160-9

Published by Nova Science Publishers, Inc. + New York

CONTENTS

Preface

Human activities are yielding rising atmospheric concentrations of greenhouse gases and other gases and particulates which are greatly altering the Earth's land cover. A scientific consensus has emerged that those activities, if allowed to continue unabated, will have extensive, highly uncertain, but potentially serious and costly impacts on regional climates and ocean conditions throughout the world. This book presents an overview of the current understanding of the impacts of climate change in the U.S., emphasizing the wide range of uncertainty about the magnitude and timing of those impacts and the implications of that uncertainty for the formulation of effective policy responses.

Chapter 1 - The climate is changing, and these changes are affecting the world around us. In order to deal with the changes that are taking place now and to prepare for those that are likely to happen in the future, decisionmakers need information about global change and its effects on the Nation and the world we live in.

This national scientific assessment integrates, evaluates, and interprets the findings of the U.S. Climate Change Science Program (CCSP) and draws from and synthesizes findings from previous assessments of the science, including reports and products by the Intergovernmental Panel on Climate Change (IPCC). It analyzes current trends in global change, both natural and human-induced, and it projects major trends for the future. It analyzes the effects of these changes on the natural environment, agriculture, water resources, social systems, energy production and use, transportation, and human health. It is intended to help inform discussion of the relevant issues by decisionmakers, stakeholders, and the public. As such, this report addresses the requirements for assessment in the Global Change Research Act of 1990.[1]

This assessment addresses not only climate change, but also other change in the global environment—including water resources, oceans, atmospheric chemistry, land productivity, and ecological systems—that may alter the capacity of Earth to sustain life. This broader set of changes is referred to as 'global change,' as defined in the Global Change Research Act.

Over the past several years, our understanding of climate variability and change and our ability to estimate their future effects has improved significantly. The conclusions in this assessment build on the vast body of observations, modeling, decision support, and other types of activities conducted under the auspices of CCSP and from previous assessments of the science, including reports and products by the IPCC, CCSP, and others. This assessment and the underlying assessments have been subjected to and improved through rigorous peer reviews.

Chapter 2 - Human activities around the world—primarily fossil fuel use, forestry, and agriculture—are producing growing quantities of emissions of greenhouse gases, other gases, and particulates and are also greatly altering the Earth's vegetative cover. A strong consensus has developed in the expert community that if allowed to continue unabated, the accumulation of those substances in the atmosphere and oceans, coupled with widespread changes in patterns of land use, will have extensive, highly uncertain, but potentially serious and costly impacts on regional climate and ocean conditions throughout the world.

This paper summarizes the current state of scientific understanding of the potential effects of projected changes in climate and related developments. The paper describes the wide range of potential impacts, including changes in: seasonal weather patterns; the amount and type of precipitation; storms and sea level; regular climate fluctuations; ocean acidity; ecosystems and biodiversity; agriculture, forestry, and fishing; water supply and other infrastructure; and human health. The discussion focuses mainly on projections of impacts in the United States but also refers to impacts elsewhere that could be particularly severe and could indirectly affect the United States. The paper draws from various primary and secondary sources, particularly the *Fourth Assessment Report of the Intergovernmental Panel on Climate Change* (IPCC), a major national assessment recently released by the National Science and Technology Council in the Executive Office of the President, and a number of surveys published by the U.S. Climate Change Science Program, which integrates executive-branch research on climate and global change.[1]

The paper emphasizes the extensive uncertainty about the future growth of greenhouse-gas emissions and resulting climate-related developments, and the implications of that uncertainty for climate policy. Uncertainty arises from several sources, including limitations in current data, imperfect understanding of physical processes, and the inherent unpredictability of economic activity, technological innovation, and many aspects of the interacting components (land, air, water and ice, and life) that make up the Earth's climate system.

Uncertainty does not imply that nothing is known about future developments, but rather that projections of future changes in climate and of the resulting impacts should be considered in terms of ranges or probability distributions.[2] For example, some recent research suggests that the median increase in average global temperature during the 21st century will be in the vicinity of 9° Fahrenheit (F) if no actions are taken to reduce the growth of greenhouse-gas emissions. However, warming could be much less or much greater than that median level, depending on the growth of emissions and the response of the climate system to those emissions.

The uncertainties surrounding the extent of climate change and its effects on human welfare complicate policy choices because any given emissions scenario could result in much more or much less dramatic changes in climate (and consequent impacts) than are expected. A full evaluation of any policy requires an assessment not only of its most likely effect on climate impacts but also of how it might shift the probabilities of impacts that are unlikely but would be very severe if they occurred. Because of that uncertainty, policymakers might wish to consider taking more action as a hedge against those severe outcomes than they would just to address the expected or most likely outcome.

In: Assessing Climate Change Impacts on the United States
Editor: Alex B. McNeill
ISBN: 978-1-60876-160-9
© 2010 Nova Science Publishers, Inc.

Chapter 1

SCIENTIFIC ASSESSMENT OF THE EFFECTS OF GLOBAL CHANGE ON THE UNITED STATES*

Committee on Environment and Natural Resources,
National Science and Technology Council

SECTION I. EXECUTIVE SUMMARY

I.1. Introduction

The climate is changing, and these changes are affecting the world around us. In order to deal with the changes that are taking place now and to prepare for those that are likely to happen in the future, decisionmakers need information about global change and its effects on the Nation and the world we live in.

This national scientific assessment integrates, evaluates, and interprets the findings of the U.S. Climate Change Science Program (CCSP) and draws from and synthesizes findings from previous assessments of the science, including reports and products by the Intergovernmental Panel on Climate Change (IPCC). It analyzes current trends in global change, both natural and human-induced, and it projects major trends for the future. It analyzes the effects of these changes on the natural environment, agriculture, water resources, social systems, energy production and use, transportation, and human health. It is intended to help inform discussion of the relevant issues by decisionmakers, stakeholders, and the public. As such, this report addresses the requirements for assessment in the Global Change Research Act of 1990.[1]

This assessment addresses not only climate change, but also other change in the global environment—including water resources, oceans, atmospheric chemistry, land productivity, and ecological systems—that may alter the capacity of Earth to sustain life. This broader set of changes is referred to as 'global change,' as defined in the Global Change Research Act.

* This is an edited, reformatted and augmented version of a National Science and Technology Council publication dated May 2008.

Over the past several years, our understanding of climate variability and change and our ability to estimate their future effects has improved significantly. The conclusions in this assessment build on the vast body of observations, modeling, decision support, and other types of activities conducted under the auspices of CCSP and from previous assessments of the science, including reports and products by the IPCC, CCSP, and others. This assessment and the underlying assessments have been subjected to and improved through rigorous peer reviews.

I.2. Causes of Climate Change

Our understanding of climate change continues to grow, enabling scientists to draw increasingly certain conclusions about its causes and impacts. For example, in their most recent assessment of climate change science, the IPCC concluded that it is unequivocal that the average temperature of Earth's surface has warmed recently and it is *very likely* (greater than 90% probability)[2] that most of this global warming is due to increased concentrations of human- generated greenhouse gases. Several lines of evidence, including those outlined in the following sections, point to a strong human influence on climate. Although these individual lines of evidence vary in their degrees of certainty, when considered together they provide a compelling and scientifically sound explanation of the changes to Earth's climate—including changes in surface temperature, ice extent, and sea level rise—observed at global and continental scales over the past few decades. [IV.1 and IV.2][3]

Several different types of gases in the atmosphere warm the planet by trapping energy that would otherwise be emitted to space. These 'greenhouse gases' include water vapor, carbon dioxide, methane, ozone, nitrous oxide, and several fluorine- and chlorine-containing gases. Of the greenhouse gases that are increasing in atmospheric concentration as a direct result of human activities, carbon dioxide is contributing most to the recent warming. The globally averaged concentration of carbon dioxide in the atmosphere has increased from about 280 parts per million (ppm) in the 18th century to 383 ppm in 2007. Emissions of carbon dioxide from fossil fuel use and from the effects of land use change are the primary sources of this increase. The current atmospheric concentration of carbon dioxide greatly exceeds the natural range of the last 650,000 years (180 to 300 ppm) as determined from ice cores. Indeed, the atmospheric levels of all major greenhouse gases have also increased significantly in the past century. [IV. 1 .a]

An increasing body of scientific research supports the conclusion that, while greenhouse gases are but one of many factors that affect climate, they are *very likely* the single largest cause of the recent warming. Other factors that affect climate may contribute either a warming or cooling influence. For example, some types of tiny particles in the air introduce a cooling influence. These particles, known as aerosols, include sulfate, organic carbon, nitrate, and mineral dust. Black carbon aerosols introduce a warming influence. Deposition of black carbon on snow and ice also contributes a warming influence on the climate by decreasing surface reflectivity that would otherwise deflect more solar energy back into space. Aerosols have been observed to increase cloudiness and cloud reflectivity, both of which have a cooling influence. Other changes to surface reflectivity, as well as variations in solar irradiance, can have either warming or cooling influences. [IV. 1 .a]

Studies that rigorously quantify the effect of different external influences on observed changes (attribution studies) conclude that most of the recent global warming is *very likely* due to human- generated increases in greenhouse gas concentrations. A large number of climate model simulations show that natural factors alone cannot explain the observed warming in the second half of the 20th century of Earth's land masses and oceans, or that of the North American continent. On the other hand, simulations that include human factors are able to reproduce important large-scale features of the recent changes. Several studies indicate that natural factors played an important role in the warming of the first half of the 20th century. Attribution studies show that it is *likely* (greater than 66% probability) that there has been a substantial human contribution to the surface temperature increase in North America.

According to the IPCC, model simulations of regional (sub-continental-scale) climate have improved, and a limited number of regional climate features are consistently captured in multiple climate model projections. Regional changes in the United States that are consistently projected in climate models are discussed throughout the report, and include some significant regional changes in temperature and precipitation. There are limits to how effectively climate change and its impacts can be projected on small scales, such as at the scale of a single city or county, a single state, or even a group of states, especially over the next decade or two. Climate changes on small scales are difficult to discern, in part, because climate variability averaged over small regions is greater than the variability averaged over large regions. For these and other reasons, attribution of the drivers of long-term temperature changes on time scales of less than 50 years and at regional scales (e.g., county, state, or multiple states, as opposed to continental), with limited exceptions, has not yet been established. [IV.2.]

In addition to average temperatures, recent work shows that human activities have also *likely* influenced extremes in temperature. Many indicators of climate extremes—including the annual numbers of frost days, warm and cold days, and warm and cold nights—show changes that are consistent with warming. For example, there is evidence that human-induced warming may[4] have substantially increased the risk of extremely warm summer conditions in some regions. Discernible human influences extend to additional aspects of climate, including the recent decreases in Arctic sea ice extent, patterns of sea level pressure and winds, and the global-scale pattern of land precipitation. [IV.2.a]

According to CCSP Synthesis and Assessment Product (SAP) 3.3, it is *very likely* that the human-induced increase in greenhouse gases has contributed to the increase in sea surface temperatures in the hurricane formation regions. There is a strong statistical connection between tropical Atlantic sea surface temperatures and Atlantic hurricane activity as measured by an index that accounts for storm intensity, frequency, and duration on decadal timescales over the past 50 years. This evidence suggests a substantial human contribution to recent hurricane activity. However, a confident assessment of human influence on hurricanes will require further studies using models and observations. [IV.2.a]

I.3. Trends and Projections of Physical Changes

This section describes observed historical trends and model-based projections of future changes to physical and chemical components of the environment.

I.3.a. Temperature

The climate system is warming, as is now evident from direct observations of increases in global average air and ocean temperatures and inferences from widespread melting of snow and ice, rising global average sea level, and other indicators. As determined by the IPCC, the globally averaged temperature rise over the last 100 years (1906–2005) is 1.33 ± 0.32 °F when estimated by a linear trend. The rate of global warming over the last 50 years (0.23 ± 0.05 °F per decade) is almost double that for the past 100 years (0.13 ± 0.04 °F per decade). In addition, as assessed by the IPCC, it is *very likely* that average Northern Hemisphere temperatures during the second half of the 20th century were warmer than any other 50-year period in the last 500 years, and it is *likely* that this 50-year period was the warmest Northern Hemisphere period in the last 1,300 years. Land regions have warmed faster than the oceans—about double the ocean rate after 1979 (more than 0.49 °F per decade). The greatest warming is at high northern latitudes during spring and winter. [IV.1.b]

Like global average temperatures, U.S. average temperatures also increased during the 20th and into the 21st century, according to federal statistics. The last decade is the warmest in more than a century of direct observations in the United States. Average temperatures for the contiguous United States have risen at a rate near 0.6 °F per decade in the past few decades. But warming is not distributed evenly across space or time. [IV.1.b]

The number of U.S. heat waves has been increasing since 1950, though it should be noted that the heat waves associated with the severe drought of the 1930s remain the most severe in the U.S. historical record. There have also been fewer unusually cold days during the last few decades, and the last 10 years have seen fewer severe cold waves than for any other 10-year period in the historical record, which dates back to 1895. [IV.1.b]

The IPCC concluded that continued greenhouse gas emissions at or above current rates are expected to cause further warming and to induce many changes during the 21st century that will *very likely* be larger than those of the 20th century. For the next 20 years, a globally averaged warming of about 0.3 to 0.4 °F per decade is projected for a range of scenarios of greenhouse gas emissions[5]. Through about 2030, there is little difference in the warming rate projected using a variety of emissions scenarios. Possible future variations in natural factors, such as a large volcanic eruption, could introduce variations to this long-term warming projection. Even if atmospheric greenhouse gas levels remained constant, the globally averaged warming would continue to be nearly 0.2 °F per decade during the next two decades because of the time it takes for the climate system, particularly the oceans, to reach equilibrium. [IV.3.b]

By the mid-21st century, the effect of the choice of emission scenario becomes more important in terms of the magnitude of the projected warming, with model projections of increases in globally averaged temperature of approximately 2 to 3 °F for several of the IPCC scenarios. According to the IPCC, all of North America is *very likely* to warm during this century, and to warm more than the global average increase in most areas. Nearly all the models assessed by the IPCC project that the average warming in the United States will exceed 3.6 °F, with 5 out of 21 models projecting that average warming will exceed 7.2 °F by the end of the century. The largest warming in the United States is projected to occur in winter over northern parts of Alaska. In regions near the coasts, the projected warming during the 21st century is expected to be less than the national average. According to CCSP SAP 3.3, abnormally hot days and nights and heat waves are *very likely* to become more frequent, and cold days and cold nights are *very likely* to become much less frequent over North America. [IV.3.b]

I.3.b. Precipitation, runoff, and drought

According to historical records, the total annual precipitation over the contiguous United States has increased at an average rate of 6% per century from 1901 to 2005, with significant variability over time and by region. The greatest increases in precipitation were in the northern Midwest and the South. The contiguous United States has had statistically significant increases in heavy precipitation, primarily during the last three decades of the 20th century and over the eastern parts of the country. [IV.1.c]

In keeping with the overall precipitation trends, most of the United States has experienced decreases in drought severity and duration during the second half of the 20th century. However, a severe drought has affected the southwestern United States from 1999 through 2007. The southeastern United States has also recently experienced severe drought. On a longer time scale, reconstructions of droughts using tree rings and geological evidence show that much more severe droughts have occurred over the last 2,000 years than those that have been observed in the instrumental record, notably, the Dust Bowl drought of the 1 930s and extensive drought in the 1950s. [IV.1.d]

Streamflow in the eastern United States has increased 25% in the last 60 years. However, it has decreased by about 2% per decade in the central Rocky Mountain region over the past century. The annual peak of streamflow in snowmelt-dominated western mountains is now generally occurring at least a week earlier than in the middle of the 20th century. Winter stream flow is increasing in basins with seasonal snow cover. The fraction of annual precipitation falling as rain (rather than snow) increased in the last half century. [IV.1.d]

Most climate models project an increase in winter precipitation in the northern tier of states and a decrease in portions of the Southwest during the 21st century. Summer precipitation is projected to decrease in the Northwest of the contiguous United States and increase in Alaska; it is uncertain whether summer precipitation will increase or decrease over large portions of the interior United States. In northern regions of North America, the magnitude of precipitation increase is projected to be greatest in autumn, whereas winter precipitation is projected to increase by the largest fraction relative to its present amount. A majority of climate models generally show winter increases in northern regions and summer decreases in western and southern regions. [IV.3.b]

In the 21st century, precipitation over North America is projected to be less frequent but more intense. This increase in storminess is projected to be accompanied by greater extreme wave heights along the coasts. [IV.3.b]

I.3.c. Ice and snow

Observations indicate that annual average Arctic sea ice extent decreased by 2.7 ± 0.6% per decade between 1978 and 2005. Larger decreases were observed in summer (7.4 ± 2.4%

per decade). In 2007, Arctic sea ice extent was approximately 23% below the previous all-time minimum observed in 2005. In addition to the decreased extent, the average sea ice thickness in the central Arctic *very likely* decreased by up to approximately 3 feet from 1987 to 1997, according to the Arctic Climate Impacts Assessment and the IPCC. Along the Alaskan coast, reductions in the thickness and spatial extent of sea ice are creating more open water, allowing winds to generate stronger waves, which increase shoreline erosion. Antarctic sea ice extent shows no statistically significant average trend. [IV. 1 .d]

The snow-covered area of North America increased in the November to January season from 1915 to 2004 due to increases in precipitation. However, spring snow cover in mountainous regions of the western United States generally decreased during the latter half of the 20th century. The IPCC determined that this latter trend is *very likely* due to long-term warming, with potential influence from decadal-scale natural variability. In Alaska, where the warming has been particularly pronounced, the permafrost base has been thawing at a rate of up to 1.6 inches per year since 1992. [IV.1.d]

The date that rivers and lakes freeze over has become later (average rate of 5.8 ± 1.6 days per century) and the ice breakup date has happened earlier (average rate of 6.5 ± 1.2 days per century), according to an analysis of 150 years of available data for the Northern Hemisphere. In addition to these changes in seasonal ice and snow, glaciers have been losing mass in the northwestern United States and Alaska, with losses especially rapid in Alaska after the mid-1990s. [IV.1.d]

Snow cover is projected to continue to decrease as the climate warms. According to the IPCC, results from multiple model simulations indicate that an Arctic Ocean free of summer ice is *likely* by the end of the century, with some models suggesting that this could occur as soon as 2040. Glaciers and terrestrial ice sheets are projected to continue to lose mass as increases in summertime melting outweigh increases in wintertime precipitation. This will contribute to sea level rise. Widespread increases in thaw depth are projected over most permafrost regions. [IV.3.b]

I.3.d. Sea level

There is strong evidence that global average sea level gradually rose during the 20th century, after a period of little change between A.D. 0 and A.D. 1900, and is currently rising at an increased rate. The global average rate of sea level rise from 1993 to 2003 was 0.12 ± 0.03 inches per year, significantly higher than the 20th century average rate of 0.067 ± 0.02 inches per year. Two major processes lead to changes in global mean sea level on decadal and longer time scales, and each currently account for about half of the observed sea level rise: expansion of the ocean volume due to warming, and the exchange of water between the oceans and land reservoirs of water, including glaciers and land ice sheets. [IV. 1 .d]

U.S. sea level data from at least as far back as the early 20th century show that along most of the U.S. Atlantic and Gulf Coasts, sea level has been rising 0.8 to 1.2 inches per decade. The rate of relative sea level rise varies from a few inches per decade along the Louisiana Coast (due to sinking land) to a drop of a few inches per decade in parts of Alaska (due to rising land). [IV.1.d]

Along with increases in global ocean temperatures, the IPCC projects that global sea level will rise between 7 and 23 inches by the end of the century (2090–2099) relative to the base period (1980–1999). According to the IPCC, the average rate of sea level rise during the 21st century is *very likely* to exceed the 1961–2003 average rate. Storm surge levels are expected to increase due to projected sea level rise. Combined with non-tropical storms, rising sea level extends the zone of impact from storm surge and waves farther inland, and will *likely* result in increasingly greater coastal erosion and damage according to the IPCC. Potential accelerations in ice flow of the kind recently observed in some Greenland outlet glaciers and West Antarctic ice streams could substantially increase the contribution from the ice sheets to sea level, a possibility not reflected in the aforementioned projections. Understanding of these processes is limited and there is no consensus on their magnitude and thus on the upper bound of sea level rise rates. [IV.3.b]

I.3.e. Atlantic hurricanes

As recognized in recent assessments, detection of long-term trends in tropical cyclone activity is complicated by multi-decadal variability and the quality of the tropical cyclone records prior to routine satellite observations that began in about 1970. Even taking these factors into account, CCSP SAP 3.3 concluded that it is *likely* that the annual numbers of tropical storms, hurricanes, and major hurricanes in the North Atlantic have increased over the past 100 years, a time in which Atlantic sea surface temperatures also increased. Additionally, there is evidence for an increase in extreme wave height characteristics over the past couple of decades, associated with more frequent and more intense hurricanes. [IV. 1 .c]

It is *likely* that hurricane rainfall and wind speeds will increase in response to human-caused warming, according to CCSP SAP 3.3. There is less confidence in the projected changes in the number of tropical cyclones. The apparent increase in the proportion of very intense storms since 1970 in some regions is much larger than simulated by current models for that period, highlighting the uncertainty associated with this issue. [IV.3.b]

Trends in other extreme weather events that occur at small spatial scales—such as tornadoes, hail, lightning, and dust storms—cannot be determined at the present time due to insufficient evidence. [IV.3 .b]

I.3.f. Abrupt Climate Change

An abrupt climate change occurs when the climate system crosses a threshold, which triggers a transition into a new state that may have large and widespread consequences. Over at least the last 100,000 years, abrupt regional warming (up to 29 °F within decades over Greenland) and cooling events occurred repeatedly over the North Atlantic region. Greenhouse warming and other human alterations of the Earth system may increase the possibility of abrupt climate change. One such possible change is a rapid decrease in the rate of flow of the major deep ocean circulation pattern in the North Atlantic that affects North American and European climate. The IPCC reported that an abrupt slowdown of this circulation during the 21st century is *very unlikely* (less than 10% probability), but longer-term changes cannot currently be assessed with confidence. Other possible abrupt changes are rapid disintegrations of the Greenland Ice Sheet and the West Antarctic Ice Sheet, both of which could raise sea level by several feet. Although models suggest the complete melting of the Greenland Ice Sheet (leading to at least 20 feet of sea level rise) would only require sustained warming in the range of 3.4 to 8.3 °F (relative to preindustrial temperatures), it is expected to be a slow process that would take many centuries to complete. There is presently no consensus on the long-term future of the West Antarctic Ice Sheet or its contribution to sea level rise. [IV. 3 .b]

I.3.g. Stratospheric ozone

Ozone at very high altitudes in the atmosphere plays an important role in shielding life on Earth from harmful ultraviolet radiation. Springtime polar ozone depletion continues to be severe when winter temperatures in the polar stratosphere (above approximately 6 miles) are particularly cold. According to the World Meteorological Organization, the average concentration of atmospheric ozone outside of polar regions is no longer declining, as it was in the 1990s. Measurements from some stations in relatively unpolluted locations indicate that ultraviolet radiation levels have been decreasing since the late 1990s, in accordance with observed ozone increases. However, ultraviolet radiation levels are still increasing at some Northern Hemisphere stations as a consequence of long-term changes in other factors such as clouds and atmospheric particulates that also affect ultraviolet radiation. [IV. 1 .a]

According to the World Meteorological Organization, it is *unlikely* that total ozone averaged over the region 60° S to 60° N will decrease significantly below the low values of the 1990s, because the abundances of ozone-depleting substances have peaked and are in decline. The current best estimate is that ozone between 60° S and 60° N will return to pre-1980 levels around the middle of the 21st century. Model simulations suggest that changes in climate, specifically the cooling of the stratosphere associated with increases in the abundance of carbon dioxide, may hasten the return of global column ozone to pre-1980 values by up to 15 years. [IV.3.d]

I.4. Effects of Global Change on the Natural Environment and Human Systems

According to CCSP SAP 4.3, it is *very likely* that temperature increases, increasing carbon dioxide levels, and altered patterns of precipitation are already affecting U.S. water resources, agriculture, land resources, biodiversity, and human health, among other things. SAP 4.3 also concluded that it is *very likely* that climate change will continue to have significant effects on these resources over the next few decades and beyond. [V.]

I.4.a. Biological diversity, ecosystem composition, and the natural environment

Ecosystems provide society with a number of goods (e.g., food, fiber, fuel, pharmaceutical products) and services (e.g., cycling of water and nutrients, regulation of weather and climate, removal of waste products, recreational and spiritual opportunities), and are essential to human health and well-being. Biodiversity (i.e., the variety of all forms of life, from genes to species to ecosystems) is a fundamental building block of many of the services that ecosystems provide. It is intrinsically important both because of its contribution to the functioning of ecosystems and because it is difficult or impossible to recover or replace once it has eroded. The effects of climate change on U.S. ecosystems include changes in the timing and length of the growing season, primary production, and species distributions. According to CCSP SAP 4.3, it is *very likely* that climate change will increase in importance as a driver for changes in biodiversity over the next several decades, although for most ecosystems it is not currently the largest driver of change. Other elements of global change (e.g., land management and use and nitrogen deposition) will continue to play a significant role in determining the future fate of ecosystems. Adaptive responses will as well. [V.1]

Some general findings from CCSP SAP 4.3, the IPCC, and other assessments, are provided below:

- The consistency of observed significant changes in physical and biological systems and observed significant warming across the globe *likely* cannot be explained entirely by natural variability or other confounding non-climate factors. [V.1]
- There has been a significant lengthening of the growing season and increase in net primary productivity in the higher latitudes of North America. Over the last 19 years, global satellite data indicate that the onset of spring is taking place 10 to 14 days earlier across the temperate latitudes. Net primary productivity is projected to increase at high latitudes due to extended growing seasons and carbon dioxide fertilization. Projections for temperate latitudes are unclear due to uncertainty in whether precipitation increases will be great enough to offset evapotranspiration increases. [V.1 .a]
- In an analysis of 866 peer-reviewed papers exploring the ecological consequences of climate change worldwide, nearly 60% of the 1,598 species studied exhibited shifts in their distributions and/or timing of their annual cycles that correspond to recent large-scale climate change patterns. [V.1]
- The resilience of many ecosystems is *likely* to be exceeded this century by an unprecedented combination of climate change, associated disturbances (e.g., flooding, drought, wildfire, insects, ocean acidification), and other global change drivers (e.g., land use change, pollution). [V.1 .a]
- In North America, warming has generally resulted in and is expected to continue to result in shifts of species ranges poleward and to higher altitudes. However, species that require higher-elevation habitat, such as alpine ecosystems, may have nowhere to migrate. [V.1 .a]
- Over the course of this century, net carbon uptake by terrestrial ecosystems is projected to peak before mid-century and then weaken or even reverse, thus amplifying climate change. [V.1.a]
- Changes in temperature and precipitation will *very likely* decrease the cover of vegetation that protects the ground surface from wind and water erosion. [V.1 .e]
- Many effects of climate change on U.S. ecosystems and wildlife may emerge most strongly through potential changes in the intensity and frequency of extreme events such as hurricanes and disturbances such as wildfires. In the near term, fire effects are expected to generally exceed direct climate effects on ecosystems. [V.1 .c]
- While there will always be uncertainties associated with the future extent of climate change, the response of ecosystems to climate impacts, and the effects of management, it is both possible and essential for management practices to help protect climate-sensitive ecosystems. [V.1.f]

Some regional and biome-specific findings from CCSP SAP 4.3 and the IPCC are provided below:

- The rapid rates of warming that have been seen in the Arctic in recent decades (and are projected for at least the next century) are dramatically reducing the snow and ice covers that provide denning and foraging habitat for polar bears and other ice-dependent species. [V.1 .e]

- The Alaskan tree line is expected to move northward and to higher elevations. Forests will replace significant amounts of existing tundra, and tundra vegetation will move into polar deserts. These changes are projected to increase carbon uptake, which would tend to offset warming effects. However, the reduced reflectivity associated with the vegetation land cover is expected to outweigh this, causing further warming. [V.1 .e]
- Where adequate water is available, nitrogen deposition and warmer temperatures have *very likely* increased forest growth and will continue to do so in the near future. However, it is difficult to separate the role of climate from other factors. Rising carbon dioxide levels will *very likely* increase photosynthesis in forests, but this increase will *likely* only enhance wood production in young forests on fertile soils. [V.1 .e]
- In the last three decades, the wildfire season in the western United States has lengthened and burn durations have increased. Climate change has also *very likely* increased the size and number of insect outbreaks and tree mortality that help to fuel wildfires in the interior West, the Southwest, and Alaska. These trends are *very likely* to continue. [V.1 .c]
- Many plants and animals in arid ecosystems are near their physiological limits for tolerating temperature and water stress and even slight changes in stress will have significant consequences. Climate change in arid regions is *very likely* to be detrimental to river and riparian ecosystems, increase erosion, and promote invasion of exotic grass species in arid lands. Climate change in arid regions is also *likely* to create physical conditions conducive to wildfire. In arid regions where ecosystems have not co-evolved with a fire cycle, the loss of iconic megaflora, such as saguaro cacti and Joshua trees, is *very likely*. [V.1 .a]
- On small oceanic islands with cloud forests or high-elevation ecosystems, such as the Hawaiian Islands, human-induced climate change, land use changes, and invasive species are *likely* to have synergistic effects that drive several species (e.g., some endemic birds) to extinction. [V.1 .c]
- Erosion and ecosystem loss is affecting many parts of the U.S. coastline, but it is unclear to what extent these losses result from temperature and precipitation changes, sea level rise, and other human drivers. Coastal wetland loss is occurring where these ecosystems are squeezed between natural and artificial landward boundaries and rising sea levels. To date, more than 50% of the original salt marsh habitat in the United States has been lost. Approximately 20% of the remaining coastal wetlands in the U.S. mid-Atlantic region are potentially at risk of inundation between 2000 and 2100. Salt marsh biodiversity is projected to decrease in northeastern marshes through expansion of non-native species such as *Spartina alterniflora*. Erosion of barrier islands has increased the height of waves that reach the shorelines of coastal bays. [V.1 .c]
- The increasing carbon dioxide level in the atmosphere has made the oceans more acidic. This acidification is expected to have negative impacts on marine shell-forming organisms and consequently large portions of the marine food chain. Corals

in many tropical regions are experiencing substantial mortality from increasing water temperatures and increasing storm intensity in some regions, on top of a host of other ongoing challenges from development and tourism, fishing, and pollution. Increasing ocean acidification is expected to exacerbate these effects. [V.1 .a]

- Land-based ecosystems in the northeastern and southeastern United States will *likely* become carbon sources, while the western United States will *likely* remain a carbon sink. [V.1 .a]

I.4.b. Agriculture and food production

The many U.S. crops and livestock (valued at about $200 billion in 2002) are strongly affected by weather and climate factors (such as temperature, precipitation, carbon dioxide concentrations, and water availability). Vulnerability of this sector to climate change is a function of many interacting factors, including pre-existing climatic and soil conditions, changes in pest competition, water availability, and the sector's capacity to cope and adapt through management practices, seed and cultivar technology, and changes in economic competition among regions [V.2]. The following findings are based on CCSP SAP 4.3 and the IPCC:

- With increased carbon dioxide levels and temperature, the lifecycle of grain and oilseed crops will *likely* progress more rapidly. But, as temperatures rise, these crops will increasingly begin to fail, especially if climate variability increases and precipitation lessens or becomes more variable. [V.2]
- The marketable yield of many horticultural crops (e.g., tomatoes, onions, and fruits) is *very likely* to be more sensitive to climate change than grain and oilseed crops. [V.2.c]
- Climate change is *likely* to lead to northward migration of weed species. Increasing carbon dioxide levels are *likely* to help many weeds, particularly some types of invasive weeds, more than most cash crops. Recent research also suggests that glyphosate, the most widely used herbicide in the United States, loses its efficacy on weeds grown at the increased carbon dioxide levels that are projected for the coming decades. [V.2.d]
- Disease pressure on crops and domestic animals will *likely* increase with earlier springs and warmer winters, which will allow proliferation and higher survival rates of pathogens and parasites. Regional variation in warming and changes in rainfall will also affect spatial and temporal distribution of disease. [V.2.e]
- Projected increases in temperature and a longer growing season will *likely* extend forage production into late fall and early spring, thereby decreasing need for winter season forage reserves. However, these benefits will *very likely* be affected by regional variations in water availability. [V.2.f]
- Climate-change-induced shifts in plant species are already underway in rangelands. Establishment of perennial herbaceous species is reducing the availability of soil moisture early in the growing season. [V.3.e]
- Shifts in the productivity and type of plants will *likely* also have significant impact on livestock operations. Higher temperatures will *very likely* reduce livestock production during the summer season, but these losses will *very likely* be partially offset by

warmer temperatures during the winter season. For ruminants, current management systems generally do not provide shelter to buffer the adverse effects of changing climate. Such protection is more frequently available for non-ruminants (e.g., swine and poultry). [V.2.f]

- Cold freshwater fisheries are *likely* to be negatively affected. Warm freshwater fisheries will generally benefit. The results for cool-water fisheries will be mixed, with gains in the northern and losses in the southern portions of ranges. Effects of increasing temperature on marine fisheries are already occurring, with rapid poleward shifts in some regions. These shifts are expected to continue in the future. [V.2.g]

I.4.c. Water resources

Plants, animals, natural and managed ecosystems, and human settlements are susceptible to variations in the storage, fluxes, and quality of water, all of which are sensitive to climate change. The effects of climate on the Nation's water storage capabilities and hydrologic functions will have significant implications for water management and planning as natural processes become more variable. Other factors affecting water resources include water pollution, damming of rivers, wetland drainage, reduced stream flow, and lowering of the groundwater table (e.g., due to irrigation). Although climate-related changes have been small compared to these other pressures to date, climate change is expected to result in increasing effects in the future. [V.4]

Although U.S. water management practices are generally quite advanced, particularly in the West, the reliance on past conditions as the foundation for current and future planning and practice will no longer be tenable as climate change and variability increasingly create conditions that are well outside of historical parameters, eroding predictability. The findings below, reported by the IPCC, CCSP SAP 4.3, and other sources, are based on the ongoing and projected water cycle changes described earlier. [V.4]

- In some mountain areas, earlier snowmelt peaks result in reduced low flows in the summer and fall. Continuing shifts in this direction are *very likely* and may substantially affect the performance of reservoir systems through changes in the seasonality of streamflow. [V.4.a]
- Water quality is sensitive to increased water temperatures, changes in precipitation, and other climate-related factors. However, most water quality changes observed so far across the continental United States are *likely* attributable to causes other than climate change. Higher temperatures and nutrient loads will tend to reduce the oxygen content of water with potential negative impacts on aquatic organisms. Increases in intense rain events will tend to result in the introduction of more sediment, nutrients, pathogens, and toxics into water bodies from non-point sources. The intrusion of saline water into groundwater supplies is *likely* to adversely affect water quality in coastal regions around the United States. These water quality changes could impose enormous costs on water treatment infrastructure. [V.4.c]
- Stream temperatures are *likely* to increase as the climate warms and are *very likely* to have effects on aquatic ecosystems and water quality. Changes in temperature will be most evident during low flow periods, when they are of greatest concern. [V.4.d]

- Projections suggest that efforts to offset the declines in available surface water by increasing withdrawal of groundwater will be hampered by decreases in groundwater recharge in some water-stressed regions, such as the southwestern United States. [V.4.b]
- Less reliable supplies of water are expected to create challenges for managing urban water systems as well as for industries that depend on large volumes of water. Across North America, vulnerability to extended drought is increasing as population growth and economic development create more demands from agricultural, municipal, and industrial uses, resulting in frequent over-allocation of water resources. Examples of vulnerable U.S. regions include: the heavily-used water systems of the West that rely on capturing snowmelt runoff, such as the Columbia and Colorado River systems; portions of California; the New York area, as a consequence of greater water supply variability; and many islands such as the U.S. territories of Puerto Rico and U.S. Virgin Islands. [V.4.e]
- Trends toward more efficient water use are *likely* to continue in the coming decades. Pressures for reallocation of water will be greatest in areas of highest population growth, such as the Southwest. Declining per capita (and, in some cases, total) water consumption will help mitigate the impacts of climate change on water resources. [V.4.d]

I.4.d. Social systems and settlements

Human systems include social, economic, and institutional structures and processes. These systems are influenced by multiple factors and stresses (e.g., access to financial resources, urbanization, and shifts in demographics). U.S. settlements will feel the effects of climate change as they interact with these other factors. Climate change effects could push stressed systems beyond sustainable thresholds. Climate sensitivity varies across settlements and industrial sectors. While it may appear that industrialized countries like the United States are well equipped to cope with gradual climate change at a national level, at a local level there may be substantial variability in climate effects and capacities to adapt. On the other hand, some U.S. settlements may find opportunities in climate change. [V.5]

According to CCSP SAP 4.6, some of the key aspects of human settlements that are affected by climate change include: human health; water and other urban infrastructures; energy requirements; urban metabolism (e.g., the welfare and activities of urban communities); economic competitiveness, opportunities, and risks; and social and political structures. Several of these challenges are discussed in other sections of the Executive Summary (and the body of the report).

Globally, the most vulnerable industries, settlements, and societies are generally those in coastal and river flood plains, those whose economies are closely linked with climate-sensitive resources, and those in areas prone to extreme weather events, especially in places that are being rapidly urbanized. Poor communities can be especially vulnerable, particularly those concentrated in high-risk areas. According to the IPCC, the most vulnerable areas in the United States are *likely* to be Alaska (e.g., indigenous communities dependent on hunting climate- sensitive species), coastal and river basin locations that are susceptible to flooding, arid areas where water scarcity is a pressing issue, and areas whose economic bases are climate-sensitive. It is possible that regions exposed to risks from climate change will see movement of population and economic activity to other locations. One reason is public perceptions of risk, but a more powerful driving force may be the availability of insurance. [V.5]

Some of the key findings from CCSP SAP 4.6 and the IPCC related to settlements are outlined below.

- Population growth is generally shifting toward areas (e.g., coastal regions) more likely to be vulnerable to the effects of climate change. Demand for waterfront property and land for building in the United States continues to grow, increasing the value of property at risk. [V.5.b]
- Coastal population increases together with *likely* increases in hurricane rainfall and wind speeds and greater storm surge due to sea level rise will continue to increase coastal vulnerabilities in the Southeast and Gulf Coast. Urban centers that were once assumed to have a high adaptive capacity remain vulnerable to extreme events such as hurricanes. [V.5.b]
- The degradation of coastal ecosystems, especially wetlands and coral reefs, can have serious implications for the well-being of societies that depend on them. The costs of adaptation for vulnerable coasts are generally much less than the costs of not acting. [V.5 .c]
- For small islands, particularly in the Pacific, some studies suggest that sea level rise could reduce island size, raising concerns for parts of Hawaii and other U.S. territories. [V.5 .c]
- Climate and extreme events can have substantial effects on local economies. For example, tourism could be affected by drought-influenced water levels in rivers and reservoirs, cleanup following multiple storm outbreaks, and changes in the length of the tourist season (e.g., ski season and beach season). [V.5.d]
- As discussed elsewhere, wildfires have increased in extent and severity in recent years and are *very likely* to intensify in a warmer future. At the same time, the population has been expanding into fire-prone areas, increasing society's vulnerability to wildfire. [V.5.d]

I.4.e. Human health

Climate variability and change can affect health directly and indirectly. The heat stress associated with a warmer environment can directly affect the body. Climate change can also make it possible for animal-, water-, and food-borne diseases to spread or emerge in areas where they had been limited or had not existed, or it can make it possible for such diseases to disappear by making areas less hospitable to the disease carrier or pathogen. Climate can also affect the incidence of diseases associated with air pollutants and aeroallergens.

Climate impacts on health are complex and will be influenced by multiple factors, including demographics; population and regional vulnerabilities; the social, economic, and cultural context; availability of resources and technological options; built and natural environments; public health infrastructure; and the availability and quality of health and social services. CCSP SAP 4.6 concluded that climate change is *very likely* to accentuate the disparities already evident in the American health care system. As discussed in CCSP SAP 4.6 and the IPCC, the extent to which communities are prepared and have the capacity to adapt will determine the severity of the following climate change impacts. Conclusions from those reports are discussed below. [V.6]

Temperature-related conclusions include the following:

- It is *very likely* that heat-related morbidity and mortality will increase over the coming decades, however net changes in mortality are difficult to estimate because, in part, much depends on complexities in the relationship between mortality and global change. High temperatures tend to exacerbate chronic health conditions. An increased frequency and severity of heat waves is expected, leading to more illness and death, particularly among the young, elderly, frail, and poor. [V.6]
- In many cases, the urban heat island effect may increase heat-related mortality. High temperatures and high air pollution can interact to result in additional health impacts. [V.6]
- Climate change is projected to lead to fewer deaths from cold exposure. [V.6.a]

Conclusions related to animal-, food-, and water-borne diseases include the following:

- Climate change is *likely* to increase the risk and geographic spread of vector-borne infectious diseases, including Lyme disease and West Nile virus. [V.6.c]
- There will *likely* be an increase in the spread of several food and water-borne pathogens among susceptible populations depending on the pathogens' survival, persistence, habitat range and transmission under changing climate and environmental conditions. However, major human epidemics of these diseases in the United States are *unlikely* if the public health infrastructure is maintained and improved as needed. [V.6 and V.6.c]
- Federal and state laws and regulatory programs protect much of the U.S. population from waterborne disease. However, if climate variability increases, current and future deficiencies in watershed protection, infrastructure, and storm drainage systems will tend to increase the risk of contamination events. [V.6.c]

Conclusions related to air quality include the following:

- As the climate becomes warmer and more variable, air quality is *likely* to be affected. [V.6.e]
- Climate change can be expected to influence the concentration and distribution of air pollutants through a variety of processes, including the modification of biogenic emissions, the change of chemical reaction rates, wash-out of pollutants by precipitation, and modification of weather patterns that influence pollutant buildup. [V.6.e]
- In studies holding pollution emissions constant, climate change was found to lead to increases in regional ground-level ozone pollution in the United States and other countries. It is well-documented that breathing air containing ozone can reduce lung function, increase susceptibility to respiratory infection, and contribute to premature death in people with heart and lung disease. [V.6.e] (The health effects of stratospheric ozone depletion on the United States have not been assessed here.)

- Climate change and changes in carbon dioxide concentration could increase the production and allergenicity of airborne allergens and affect the growth and distribution of weeds, grasses, and trees that produce them, which may increase the incidence of allergic rhinitis. [V.6.e]
- Uncertainties in climate models make the direction and degree of change in air quality and aeroallergens somewhat speculative. [V.6.d]

Conclusions related to extreme events include the following:

- Increases in extreme weather (e.g., storms, flooding) and accompanying events (e.g., wildfire resulting from prolonged drought) may lead to increases in deaths, injuries, infectious diseases, interruptions of medical care for chronic disease treatment, and stress-related disorders and other adverse effects associated with social disruption and migration. [V.6.d]
- Extreme climate events may also have substantial mental health impacts (e.g., post-traumatic stress disorder and depression). [V.6]
- High-density populations in low-lying coastal regions, such as the U.S. Gulf of Mexico, experience a high health burden from weather disasters, particularly among lower income groups. [V.6]
- Wildfires pose significant direct health threats. They can also have substantial effects through increased eye and respiratory illnesses due to fire-related air pollution and mental health impacts from evacuations, lost property, and damage to resources. Wildfires, with their associated decrements to air quality and pulmonary effects, are likely to increase in frequency, severity, distribution, and duration in the Southeast, the Intermountain West and the West. [V.6 and V.6.b]
- Morbidity and mortality due to an event will tend to increase with the intensity and duration of the event, but will tend to decrease with advance warning and preparation. [V.6.a]

I.4.f. Energy production, use, and distribution

To date, most discussions on energy and climate change have focused on mitigating human effects on climate. However, along with this role as a *driver* of climate change, the energy sector will be subject to the *effects* of climate change. As discussed in the IPCC and CCSP SAP 4.5, climate change is *likely* to affect the use and production of energy in the United States. It is *likely* to affect physical and institutional infrastructures and is *likely* to interact with and possibly exacerbate ongoing environmental change and environmental and population-related pressures in settlements. Concerns about climate change impacts could change perceptions and valuations of energy technology alternatives. Responses to climate change may lead to the following changes in energy demand according to the IPCC and CCSP SAP 4.5: [V.7]

- decreases in the amount of energy consumed in residential, commercial, and industrial buildings for space heating and water heating; [V.7.a]
- increases in energy used in residential, commercial, and industrial buildings for space cooling; [V.7.a]

- increases in energy consumed for residential and commercial refrigeration and industrial process cooling (e.g., in thermal power plants or steel mills); [V.7.a]
- increases in peak demand for electricity in most regions of the country, except in the Pacific Northwest; [V.7.a]
- increases in energy used to supply other resources for climate-sensitive processes, such as pumping water for irrigated agriculture and municipal uses; [V.7.a]
- changes in the balance of energy use among delivery forms and fuel types, as between electricity used for air conditioning and natural gas used for heating; [V.7] and
- changes in energy consumption in key climate-sensitive sectors of the economy, such as transportation, construction, agriculture, and others. [V.7]

According to CCSP SAP 4.5, in the absence of energy efficiency measures directed at space cooling, climate change is expected to cause a significant increase in the demand for electricity in the United States, which would require the building of additional electricity production facilities (and probably transmission facilities) at an estimated cost of many billions of dollars. [V.7.a]

Climate change could affect production, supply, and transmission of energy in the United States in the following ways according to the IPCC and CCSP SAP 4.5:

- direct impacts from increased intensity of extreme weather events, [V.7.b]
- reduced water supplies in regions dependent on water resources for hydropower and/or thermal power plant cooling, [V.7.b]
- facility siting decisions affected by changing conditions, [V.7.b] and
- positive or negative impacts on production of biomass, wind power, or solar energy where climate conditions change. [V.7.b]

Significant uncertainty exists about the potential impacts of climate change on energy production and distribution, in part because the timing and magnitude of climate impacts are uncertain. Nonetheless, every existing source of energy in the United States has some vulnerability to climate variability. Although effects on the existing infrastructure might be categorized as modest, local and industry-specific impacts could be large, especially in areas that may be prone to disproportional warming (e.g., Alaska) or weather disruptions (e.g., the Gulf Coast and Gulf of Mexico). [V.7.b]

I.4.g. Transportation

Increasing global temperatures, rising sea levels, and changing weather patterns pose significant challenges to our country's transportation venues including: roadways, railways, transit systems, marine transportation systems, airports, and pipeline systems. The U.S. transportation network is vital to the Nation's economy, safety, and quality of life. The following findings, derived from the IPCC and CCSP SAP 4.7 (focused on Gulf Coast transportation impacts), relate to the relationship between climate change and transportation in North America: [V.8]

- Warmer or less snowy winters are *likely* to reduce delays, improve ground and air transportation reliability, and decrease the need for winter road maintenance. However, more intense winter storms could increase risks for traveler safety and require increased localized snow removal. [V.8]
- Increasing frequency, intensity, or duration of heat spells could cause railroad tracks to buckle or kink and could affect roads through softening and traffic-related rutting. [V.8]
- Coastal and riverine flooding and landslides are *very likely* to cause negative impacts on roads, rails, and ports. The crucial connectivity of the transportation system means that the services of the network can be threatened even if small segments are wiped out. [V.8]
- Transportation infrastructure is *likely* to be particularly affected in more northerly latitudes. Permafrost degradation reduces surface load-bearing capacity and potentially triggers landslides. While the season for transport by barge is *likely* to be extended, the season for ice roads that utilize waterways is *likely* to be compressed. [V.8]
- Climate change may worsen the vulnerability of Gulf Coast and eastern transportation systems to hurricanes and tropical storms as warming seas give rise to more energetic storms and as sea level rises. [V.8.d]
- An increase in the frequency of extreme precipitation events may contribute to increased accident rates; more frequent short-term flooding and bridge scour, as well as more culvert washouts; exceeding the capacity of storm drain infrastructure; and more frequent landslides, requiring increased maintenance. [V.8]
- Increases in precipitation and the frequency of severe weather events could negatively affect aviation. Higher temperatures affect aircraft performance and increase the necessary runway lengths. [V.8]
- Some of these risks are expected to be offset by improvements in technology and information systems. [V.8]

SECTION II. INTRODUCTION

II.1. Objectives

Earth's climate and other key aspects of the natural environment are changing, and these changes are significantly affecting society and nature. The U.S. Climate Change Science Program (CCSP) and its international partners are making a concerted scientific effort to better understand the characteristics and causes of these changes, their past and future course, and their implications. By communicating the findings of these efforts, CCSP is making important strides toward its overarching vision:

> A nation and the global community empowered with the science-based knowledge to manage the risks and opportunities of change in the climate and related environmental systems.

The specific objectives of this assessment are to integrate, evaluate, and interpret the findings of CCSP to support informed discussion about the relevant issues by decisionmakers, stakeholders, the media, and the public. This assessment reviews a wide range of observed and projected vulnerabilities, risks, and impacts associated with global change, focusing particularly on the effects of climate change. It also discusses the effects of changing ozone concentrations; ecosystem shifts; and changes in air quality, land and water use, and ocean chemistry. This assessment focuses primarily on the United States. However, given the global nature of the issues, it also discusses international impacts.

Sidebar

Global Change: Changes in the global environment (including alterations in climate, land productivity, oceans or other water resources, atmospheric chemistry, and ecological systems) that may alter the capacity of the Earth to sustain life (from the *Global Change Research Act of 1990*).

Climate Change: A statistically significant variation in the mean state of the climate and/or in its variability, persisting for an extended period (typically decades or longer), due to either natural or human causes (adapted from the *CCSP Strategic Plan*).

This report is pursuant to the Global Change Research Act of 1990,[6] which requires the U.S. Global Change Research Program[7] to "combine and interpret information readily usable by policymakers attempting to formulate effective strategies for preventing, mitigating, and adapting to the effects of global change." The specific assessment requirement is as follows:

Scientific Assessment

> On a periodic basis (not less frequently than every 4 years), the Council, through the Committee,[8] shall prepare and submit to the President and the Congress an assessment which –
>
> (1) integrates, evaluates, and interprets the findings of the Program and discusses the scientific uncertainties associated with such findings;
>
> (2) analyzes the effects of global change on the natural environment, agriculture, energy production and use, land and water resources, transportation, human health and welfare, human social systems, and biological diversity; and
>
> (3) analyzes current trends in global change, both human-induced and natural, and projects major trends for the subsequent 25 to 100 years.[9]

This scientific assessment addresses each of the three components of this requirement.

The findings contained in this assessment build upon the vast body of observations, process studies, modeling, and other types of basic and applied research conducted under the auspices of CCSP. In so doing, it integrates, evaluates, and interprets findings of the program, as called for in the first paragraph of the assessment requirement of the Global Change Research Act of 1990. The following sections of the introduction describe the manner in which uncertainty is characterized in this assessment, its relationship to other assessments, and how the assessment is structured.

II.2. Relationship to Other Assessments and Primary Scientific Literature

This scientific assessment integrates, evaluates, and interprets the findings of CCSP and other scientific investigations of global change. It analyzes the trends and effects of global environmental changes, with a particular focus on climate change. This scientific assessment primarily draws from and synthesizes findings from previous assessments of the state of the science. All of these underlying assessments involved leading experts in the relevant disciplines who analyzed the existing body of primary, peer-reviewed scientific literature, and observations. These underlying assessments were all subjected to and improved through rigorous peer reviews of their own.

In some cases, this assessment cites the primary literature, particularly when the underlying assessment derived a particular conclusion from a single or small number of studies. In cases where the authors of the underlying assessments drew from a large number of studies or where they exercised their expert judgment on a particular issue, the assessment itself is cited.

This scientific assessment draws significantly from and extends the findings of the Intergovernmental Panel on Climate Change (IPCC) Fourth Assessment Reports that were released in 2007.[10] Within the IPCC reports, this assessment draws most heavily from the North America chapter of the IPCC Working Group II report, as well as several chapters from the IPCC Working Group I report. The IPCC reports, which represent the international scientific consensus on climate change, are written and reviewed by hundreds of the world's leading environmental and social scientists studying climate change. The United States was a leading contributor to the IPCC:

- U.S. scientists actively participated, serving as authors and reviewers;
- CCSP sponsored key science that underpinned the IPCC reports; and
- the United States hosted the Technical Support Unit that facilitated production of the IPCC report on the physical science of climate change.

This scientific assessment also incorporates findings from several CCSP synthesis and assessment products that integrate and analyze CCSP results related to specific sectors, disciplines, and cross-cutting issues.[11] The CCSP synthesis and assessment products focus on issues of importance, with a particular emphasis on North America.

Other key assessments used in this report include the United Nations Environment Programme's *Scientific Assessment of Ozone Depletion: 2006* (WMO, 2007), the *Arctic Climate Impact Assessment* (ACIA, 2005), and several reports from the National Research Council. Nearly all of the major assessment reports cited in this document were released within the past few years. The *National Assessment of the Potential Consequences of Climate Variability and Change* (NAST, 2001) is also cited and served as a foundation for a significant portion of the research carried out under CCSP that is drawn from in the assessments cited in this document.

II.3. Characterization of Uncertainty

In characterizing uncertainty, this assessment uses an approach similar to the one used in the IPCC Fourth Assessment Report,[12] which was also generally used in the CCSP synthesis and assessment reports that provided input to this assessment. The terms used to describe the levels of confidence and likelihood associated with individual findings in the IPCC Fourth Assessment Report and CCSP synthesis and assessment products have been adopted in this assessment.

II.3.a. Description of confidence

On the basis of a comprehensive reading of the literature, their analysis of current knowledge, and their expert judgment, the authors of the underlying assessments have assigned confidence levels to major statements. These statements of confidence (printed in italics) use the following definitions:

Very high confidence	At least a 9 out of 10 chance of being correct
High confidence	About an 8 out of 10 chance
Medium confidence	About a 5 out of 10 chance
Low confidence	About a 2 out of 10 chance
Very low confidence	Less than a 1 out of 10 chance

II.3.b. Description of likelihood

Likelihood refers to an assessment of the probability, based on expert judgement or quantitative analysis (when available), that some well-defined outcome has occurred or will occur in the future. The descriptions of likelihood (printed in italics) use the following definitions:

Virtually certain	greater than 99% probability of occurrence
Very likely	90 to 99% probability
Likely	66 to 90% probability
More likely than not	greater than 50% probability
About as likely as not	33 to 66% probability
Unlikely	10 to 33% probability
Very unlikely	1 to 10% probability
Exceptionally unlikely	less than 1% probability

Non-italicized terms and phrases such as "may," "are expected," and "are projected" are used to indicate the possibility of the occurrence of an event or phenomenon without assignment of a formal level of likelihood.

II.4. Report Structure

The body of the report is divided into three main sections:

- **Evaluation of Overall Progress by CCSP**. This section provides illustrations of the program's integrated efforts toward its major objectives and interprets some of the major advances that have been made.

- **Trends and Projections of Global Environmental Change**. This section describes key observations, causes, and future projections of global change, both nationally and globally. This section provides much of the underpinning for the subsequent section.

- **Analysis of the Effects of Global Change on the Natural Environment and Human Systems.** This section describes the effects of climate and other environmental drivers on key U.S. assets and sectors, including natural and managed ecosystems, water resources, human health, welfare, settlements, energy supply and demand, and transportation.

Section III: Evaluation of Overall Progress by CCSP

The Climate Change Science Program (CCSP) articulated five overarching goals in its strategic plan (CCSP, 2003):

1. Improve knowledge of the Earth's past and present climate and environment, including its natural variability, and improve understanding of the causes of observed variability and change.
2. Improve quantification of the forces bringing about changes in the Earth's climate and related systems.
3. Reduce uncertainty in projections of how the Earth's climate and related systems may change in the future.
4. Understand the sensitivity and adaptability of different natural and managed ecosystems and human systems to climate and related global changes.
5. Explore the uses and identify the limits of evolving knowledge to manage risks and opportunities related to climate variability and change.

This section reports key progress toward these goals since the completion of CCSP's Strategic Plan in 2003. In providing an assessment that includes consideration of the accomplishments of CCSP over the last several years, it is important to differentiate between different kinds of accomplishments. A conceptual framework for addressing the program's accomplishments and developing performance metrics by which its progress could be evaluated was provided by the National Research Council (NRC) in its report, *Thinking Strategically: The Appropriate Use of Metrics for the Climate Change Science Program* (NRC, 2005b). In this report, the NRC identified five different types of metrics: process metrics, input metrics, output metrics, outcome metrics, and impact metrics.

The progress that CCSP has made over the past several years comes from a variety of efforts. In implementing its program, CCSP uses four approaches of scientific research, observations, decision support, and communications (CCSP, 2003). In particular, as new observational capabilities become available, they can be thought of as providing an input metric (e.g., launch of a new satellite) that leads to outputs (e.g., large data sets that are made available to the scientific research community through a data archive). Scientists use the observations (together with relevant models of varying degrees of complexity) to enhance the state of scientific knowledge, and thereby create outcomes that can be evaluated using a variety of metrics.

Similarly, complex modeling and assimilation systems that are developed to describe prior and/or predict future Earth system evolution, support assessments, and produce consistent data sets that can be used widely by the scientific community may be thought of as inputs, while the results from them can be considered as outputs, with corresponding metrics associated with them as their capabilities and availability increase.

The following are among the output metrics identified by the NRC (2005b):

1. The research has engendered significant new avenues of discovery.
2. The program has led to the identification of uncertainties, increased understanding of uncertainties, or reduced uncertainties that support decision-making or facilitate the advance of other areas of science.
3. The program has yielded improved understanding, such as (a) more consistent and reliable predictions or forecasts, (b) increased confidence in our ability to simulate and predict climate change and variability, and (c) broadly accepted conclusions about key issues or relationships.

Impact metrics would be applicable once the scientific results can be used for policymaking, resource management, and decision support. Indeed, the NRC has defined three classes of impact metrics (NRC, 2005b):

1. The results of the program have informed policy and improved decision making.
2. The program has benefited society in terms of enhancing economic vitality, promoting environmental stewardship, protecting life and property, and reducing vulnerability to the impacts of climate change.
3. Public understanding of climate issues has increased.

A number of examples of progress that CCSP has made toward meeting these goals are outlined in this section. Many of these are best considered as outcome goals, as they represent the knowledge, understanding, and predictive capability that come from the program. Some, especially those associated with Goal 5, can be thought of as impact goals because of how that goal is defined. However, given the large investment made by CCSP agencies in observational and modeling capability, and the very significant increase in capability that has become available since the publication of the *CCSP Strategic Plan* in 2003, it is important that the input and output metrics most closely associated with these advances also be addressed. Thus, this section begins with a number of CCSP accomplishments that can best be thought of as input metrics and concludes with a number of cross-cutting accomplishments that are primarily output metrics. Program outcomes and impacts have been assessed in a number of NRC reports and elsewhere.

CCSP provides key large-scale observational efforts that underpin our Nation's climate research, including satellite observations, surface-based measurement networks, and field campaigns. However, many of the observing systems used by CCSP were developed primarily for purposes other than climate research and are not included in CCSP's budget. During the approximately five years since the release of the strategic plan, several important advances in observational capability have come to fruition through enhanced deployment of surface-based networks, launch and/or releases of global data sets from satellites, and the conduct of field campaigns that integrate surface-based, airborne, and satellite-based measurements. The data from these investments are made broadly available to scientific and applications communities and they have had an appreciable impact on climate research in the United States.

This section briefly summarizes some of these accomplishments. These accomplishments, which are drawn largely from the program's annual reports (*Our Changing Planet*, OCP) for FY 2004 through FY 2009, are not intended to comprehensively describe the program's inputs and outputs. This section is not a comprehensive assessment of the program's capability to deliver the knowledge necessary to address the key scientific and societal issues related to the effects of global change. Rather, this section is intended to provide illustrative examples of the broad and integrated range of CCSP's activities.

III.1. Program Inputs

III.1.a. Surface-based observing networks

Ocean Monitoring: CCSP agencies have been contributing to a significant growth in worldwide ocean *in situ* monitoring over the past several years. An initial ocean observing system for climate has reached 59% completion towards its planned completion in 2013, with the United States currently supporting nearly 50% of the ocean-based observing platforms [OCP 2009[13]]. A network of tropical moored buoys has been extended into the Indian Ocean, with the number of sites increasing from 6 [OCP 2007] to 8 [OCP 2008] to a planned total of 12 by the end of 2008 [OCP 2009]. The Argo network of profiling arrays is now globally complete, with some 3,000 instruments deployed around the world [OCP 2009], with the United States responsible for approximately half of them [OCP 2008].

Aerosol and Radiation Monitoring: The Aerosol Robotic Network (AERONET) continues to expand, with some 230 automated sites now instrumented with sun photometers around the world. Sites added most recently include the Tibetan plateau and the Ganges floodplain of India. Scientific advances implemented recently into the network include more accurate measures of particle size and improved determination of aerosol light absorption [OCP 2009]. Three- dimensional information about aerosol distributions is being obtained from the Micro-Pulse Lidar Network (MPL Net), for which 13 stations are now operating around the world, with 5 more to be put into place in the near future. Most MPL Net stations are co-located with Aerosol Robotic Network stations so that the column data obtained with passive techniques can be inter- compared with profile data obtained with active approaches. Data products produced by MPL Net include aerosol optical depth, absorption, and size distribution; aerosol and cloud heights; and planetary boundary layer structure and evolution [OCP 2009]. The Baseline Surface Radiation Network has been expanded to 38 sites and has data records up to 15 years long at its original 6 stations [OCP 2008].

Atmospheric Radiation Measurement (ARM) Program: The ARM program has deployed a mobile facility that can be temporarily relocated around the world to provide the same type of comprehensive surface-based measurements of atmospheric radiation that are made at the ARM permanent sites. The mobile facility was first deployed in Point Reyes, California, as part of a joint study by the Department of Energy and the Department of Defense [OCP 2007], and was then deployed to Niamey, Niger, in 2006 [OCP 2008]. During this deployment, the mobile facility was able to make detailed observations of a large Saharan dust storm that took place over North Africa in March 2006 [OCP 2008]. In 2007 it was deployed to the Black Forest region of Germany, and a deployment to China is planned for 2008 [OCP 2009]. The results from the mobile facility can be compared to those obtained from the ARM fixed sites in the Arctic, tropics, and mid-latitudes.

AmeriFlux: Based on systematic eddy-covariance measurements, AmeriFlux Network sites provide data on exchange of carbon dioxide (CO_2), water vapor, and energy between air and plant canopies for a range of different types of ecosystems. Time averages of these continuous measurements are expressed as net ecosystem exchange (NEE) for any selected time period, and the annual CO_2 NEE, for example, is often termed net ecosystem production (NEP), or the quantity of carbon gained or lost by the system. Some AmeriFlux data products also include measures of respiration, which when combined with NEE enable estimates of gross primary production (GPP). Both NEP and GPP are important for calculating terrestrial carbon budgets and for carbon cycle analysis and modeling. AmeriFlux data products also include a number of corollary biometric and micrometeorological measurements that are used to understand carbon cycle processes. (See <public>.)

Ground-Water Climate Response Network: The federal government, in cooperation with hundreds of federal, state, and local agencies, collects nationally consistent information about the Nation's groundwater resources and helps define and manage those resources. The U.S. Geological Survey (USGS) maintains a network of wells to monitor the effects of droughts and other climate variability on groundwater levels. The Ground-Water Climate Response Network consists of a national network of about 140 wells monitored as part of the USGS Ground-Water Resources Program, supplemented by more than 200 wells monitored as part of the Cooperative Water Program that meet the same network criteria.

Soil Climate Analysis Network (SCAN) and Snowpack Telemetry (SNOTEL): SCAN is an *in situ* real-time observation network. The information from SCAN aids in data assimilation for drought assessments; flood response; integrated pest management; land productivity in relation to soil moisture and temperature changes; help in predicting shifts in wetlands, crops, and other ecosystems; and would serve as ground-truth for satellite soil moisture information. SCAN data are being examined for use in modeling carbon fluxes. At each SCAN site, minimum measurements include precipitation, relative humidity, wind speed, solar radiation, and barometric pressure above ground, and soil moisture and soil temperature at several depths below the surface. SNOTEL site measurements include snow depth and moisture content. (See <www.wcc.nrcs.usda.gov>.)

U.S. Historical Climatology Network (USHCN): The USHCN is a high-quality data set of daily and monthly records of basic meteorological variables from over 1,000 observing stations across the 48 contiguous United States. Daily data include observations of maximum and minimum temperature, precipitation amount, snowfall amount, and snow depth from 1,062 stations; monthly data consist of monthly averaged maximum, minimum, and mean temperature and total monthly precipitation from 1,221 stations. Most of these stations are U.S. Cooperative Observing Network stations located generally in rural locations, while some are National Weather Service First Order stations that are often located in more urbanized environments. The USHCN has been developed over the years to assist in the detection of regional climate change and to analyze U.S. climate.

III.1.b. Satellite observations

GRACE: The Gravity Recovery and Climate Experiment consists of two satellites flying in formation, from which spatial and temporal variability in Earth's gravitational field can be measured and used to determine information about the distribution of mass in the Earth. Not only can GRACE provide information about the relatively time-independent variability of the gravitational field associated with Earth's solid matter, but it can provide information about shorter-term variability associated with changes in the distribution of water (both frozen and liquid). This latter variability can then be used to provide information about distributions of groundwater, mass of ice sheets, and ocean bottom pressure (from which deep ocean currents can be evaluated) [OCP 2007]. The ability of GRACE to make usefully accurate surface-water storage estimates over seasonal time scales demonstrates its potential for applied use, and the availability of comparisons between observed and modeled water storage estimates provides increased confidence in the global annual patterns and seasonal cycle of water storage. The satellite observations have shown some low-latitude biases in the models, however, and the availability of the satellite data should contribute to future improvement of models of water storage [OCP 2008].

ICESat: The Ice, Cloud, and Land Elevation Satellite was launched in January 2003. While it measures surface elevations of ice and land, vertical distributions of clouds and aerosols, and vegetation canopy heights, all with unprecedented accuracy, its primary emphasis has been on acquiring time series of ice-sheet elevation changes that can be used to determine the mass balance of ice sheets, most notably those in Greenland and Antarctica. Additional ice-related capabilities have included the ability to characterize detailed topographic features of ice sheets, ice shelves, and ice streams and the ability to gather new observations of sea-ice thickness distribution [OCP 2007].

SORCE: The Solar Radiation and Climate Experiment satellite was launched in 2003. This satellite contains four instruments measuring not only the total solar irradiance (TSI), but also spectrally resolved solar irradiance (SSI) over a very broad range of wavelengths. The SORCE satellite overlapped other U.S. satellites measuring TSI and SSI, allowing for improved capability for determining trends, which can be very challenging in the absence of overlap. This is due to the challenge to metrology that is associated with space-based solar irradiance measurement, for which the variation in TSI over a solar cycle is of the order of only 0.3% (peak-to-peak). Besides supporting long-term trend observations, SORCE has contributed to studies of shorter-term solar variability. In 2004, SORCE was able to document both changes in solar luminosity associated with the transit of Venus across the Sun and the largest solar X-ray flare ever recorded [OCP 2006]. The four-plus years of SORCE observations are providing previously unavailable information concerning the full wavelength dependence of solar irradiance over a significant fraction of the solar cycle, including at visible and infrared wavelengths where such data were previously only available over very brief periods (e.g., the length of a Space Shuttle mission) [OCP 2008]. Continuity of TSI measurements is planned with the Glory satellite to be launched in 2009. Additional TSI and SSI measurements are being considered as part of the plan for re-manifestation of the climate sensors that were lost from the National Polar-Orbiting Operational Environmental Satellite System as part of the Nunn- McCurdy certification process completed in 2006.

Aura: The Aura spacecraft, designed to study the trace constituent and aerosol composition of the troposphere and stratosphere, was launched on 15 July 2004. Aura contains four instruments whose measurement goals include ozone depletion, global air quality, and the radiative forcing of climate. It is an international mission, with instruments or components contributed by the United Kingdom, the Netherlands, and Finland. While most of Aura's instruments provide markedly enhanced capabilities over those flown previously, Aura can also provide continuity in measurements of ozone column and profile, aerosol distributions, and trace constituent profiles that were obtained from other platforms, most notably the Upper Atmosphere Research Satellite, which took data from 1991 through 2005, and the Total Ozone Mapping Spectrometer series of satellites [OCP 2006]. The Aura satellite flies as part of an international constellation ('A-Train') of polar-orbiting spacecraft whose equatorial crossing times are all within approximately 15 minutes of each other in the early afternoon (~1 :30 p.m. at the locality that they fly over). The near simultaneity of the measurements from the different spacecraft and instruments will allow for synergistic studies that would not be possible without the coordinated flights. Aura results are affecting a number of areas, including studies of the relationship between upper tropospheric water vapor and ice that demonstrate the existence of a positive feedback associated with cloud- induced transport of moisture to the upper troposphere [OCP 2007] and those of how ozone, water vapor, and ozone-depleting trace gases are transported between the troposphere and stratosphere, which will allow for improvements in the models used to represent the response of the ozone layer to future concentrations of greenhouse gases [OCP 2008].

COSMIC: The Constellation Observing System for Meteorology, Ionosphere, and Climate consists of six satellites containing radio-occultation instruments based on the Global Positioning System technique for determining temperature and moisture profiles of the atmosphere. The system was launched on 14 April 2006. It is a joint U.S.–Taiwanese effort, and represents the first focused effort to use the radio-occultation technique to obtain a spatially distributed set of profiles (~2500 every 24 hours) that can be used for both climate studies and weather forecasting. The high vertical resolution of the radio-occultation technique and the absolute nature of the measurement will make the COSMIC data useful for climate studies [OCP 2007].

CloudSat and CALIPSO: The Cloud Satellite and Cloud-Aerosol Lidar and Infrared Pathfinder Satellite Observation satellites were launched in June 2006 to use radar and lidar observations, respectively, for determination of the three-dimensional distribution of clouds and aerosols in Earth's atmosphere. These satellites were launched into the A-train constellation (see Aura section above), and the active remote-sensing techniques used on these spacecraft complement the passive techniques used on the other members of the constellation. Both CloudSat and CALIPSO represent partnerships—CloudSat with the Canadian Space Agency and the U.S. Air Force, and CALIPSO with the French space agency, CNES. Among the scientific issues being addressed by CloudSat and CALIPSO are the relationship between aerosols, clouds, and precipitation and the distribution of clouds over polar regions in winter [OCP 2009].

III.1.c. Field campaigns

TC4: CCSP scientists completed the Tropical Composition, Cloud, and Climate Coupling (TC4) experiment in Costa Rica during the summer of 2007. This field campaign, which included three large aircraft (ER-2, WB-57, DC-8) flying coordinated trajectories, focused on identifying and quantifying the chemical and dynamical processes that take place in Earth's tropical tropopause layer. A particular emphasis was the study of the composition, formation, and radiative properties of clouds (cirrus and sub-visible cirrus) in this region. Besides the aircraft, the campaign made use of data from the A-Train constellation of satellites, as well as three ground- based and balloon sonde stations [OCP 2009].

CLASIC: The Cloud and Land Surface Interaction Field Campaign was carried out in June 2007 in the region surrounding the ARM network's Southern Great Plains site in Oklahoma. Three 'supersites' were heavily instrumented to obtain ground-based measurements linking observed carbon and moisture fluxes to atmospheric structure, and several research aircraft were used as well. This campaign was coordinated with the Mid-Continent Initiative study of the North American Carbon Program to help understand the influence of the land surface on atmospheric concentrations of aerosols, gases, and other constituents [OCP 2009].

ICARTT: The International Consortium for Atmospheric Research on Transport and Transformation was carried out in the summer of 2004 to study the factors involved in the intercontinental transport of pollution and its effects on radiation balance over North America and the North Atlantic. This campaign involved scientists from the United States and five other countries, working together with aircraft-, ship-, satellite-, and surface-based observations [OCP 2006].

MILAGRO: The Megacity Initiative: Local and Global Research Observations was carried out in the spring of 2006, bringing together several planned field experiments that were addressing the impacts of large sources of air pollution in North America (most notably the Mexico City area) on the tropospheric trace constituent and aerosol distributions over a broader geographic region. This effort brought together measurement initiatives by CCSP agencies, and allowed for synergistic observations using surface-based, airborne, and satellite-based platforms. Results showed higher than expected formation of organic aerosols within the atmosphere, and should help improve future calculations of aerosol influences on climate [OCP 2006, 2007, 2008].

III.1.d. Data management

Giovanni: A new tool to facilitate the use of space-based remote-sensing data by eliminating a number of tedious steps that investigators would normally have to take before they could begin data analysis has been developed and implemented at the National Aeronautics and Space Administration's (NASA) Goddard Space Flight Center. This tool is known as Giovanni (Goddard Earth Sciences Data and Information Services Center Interactive Online Visualization and Analysis Infrastructure), and will make it easier for scientists and other users to access data from NASA's Earth Observing System. Among the steps previously required that will be eliminated through use of Giovanni are searching for appropriate data files, requesting them from a central archive, transferring the files to the user's own computing system, and extracting the relevant data from potentially unfamiliar data formats. This online tool will allow scientists to spend a larger fraction of their time working with the actual data and thus better comprehend regional events and the interconnected nature of global environmental processes [OCP 2008]. (See <daac.gsfc.nasa.gov/techlab/giovanni/>.)

Carbon Dioxide Information and Analysis Center (CDIAC): The Department of Energy (DOE) maintains CDIAC to provide comprehensive, long-term data management support, analysis, and information services in support of DOE climate change research, the global climate research community, and the general public. CDIAC data support research on both the present-day carbon budget and temporal changes in carbon sources and sinks. Particular issues that can be addressed with CDIAC data include emission rates and atmospheric and oceanic concentrations of CO_2, as well as information on fluxes of CO_2 within the atmosphere and between the land surface and the atmosphere [OCP 2007]. In 2008, CDIAC will begin to offer CO_2 measurements from buoys, research cruises, and volunteer observing ships, as well as syntheses of data from the Atlantic and Pacific regions [OCP 2009]. (See <cdiac.ornl.gov/>.)

Global Change Master Directory (GCMD): The GCMD provides descriptive and spatial information about data sets relevant to global change research. It contains over 18,200 metadata descriptions of data sets coming from some 2,800 government agencies, research institutions, archives, and universities. Besides facilitating research on climate change, the GCMD contains data and metadata that support work in numerous Earth science disciplines, with a particular emphasis on supporting multidisciplinary research. Software upgrades are made periodically in response to user needs and to capitalize on new technology [OCP 2008]. (See <gcmd.nasa.gov/>.)

Global Observing System Information Center (GOSIC): GOSIC is an operational data facility implemented through the National Oceanic and Atmospheric Administration's (NOAA) National Climate Data Center on behalf of the global observing community. GO SIC provides information and facilitates easier access to data produced by the Global Climate Observing System, the Global Ocean Observing System, the Global Terrestrial Observing System, and other observing programs. While the data themselves may be highly distributed, the existence of a single entry point will facilitate the use of these data by a wide variety of users. GOSIC includes search capability across a broad range of international data centers to support full utilization of data sets produced by multiple nations [OCP 2009]. (See <www.gosic.org/>.)

Earth Observing System Data and Information System (EOSDIS): EOSDIS provides convenient mechanisms for locating and accessing products from NASA's Earth Observing System. Data may be obtained by users either electronically or on physical media. EOSDIS provides access to a very large set of data through the eight Distributed Active Archive Centers and several additional data sources. These centers process data products from instrument data, archive and distribute the resulting data, and provide a full range of user support. More than 2,100 distinct data products are archived and distributed through the Distributed Active Archive Centers and associated centers [OCP 2007]. (See <www.esdis.eosdis.nasa.gov/>.)

Earth Science Research, Education, and Applications Solutions Network (REASoN): NASA's REASoN program created 40 cooperative agreements that are designed to support the development of higher-level science products based on data from NASA's satellite programs, to use these products to advance Earth system research, to develop and demonstrate new technologies for data management and distribution, and to contribute to interagency efforts to improve the maintenance and accessibility of data and information systems. The REASoN tasks were initiated in 2004 and most completed their work in 2007 [OCP 2007]. A follow-on program, Making Earth System Data Records for Use in Research Environments (MEaSUREs), was initiated in 2007. (See <reason-projects.gsfc.nasa.gov/>.)

Scientific Data Stewardship (SDS): The SDS is a paradigm for data management implemented at NOAA to use an integrated suite of functions for preserving and exploiting the full scientific value of environmental data. The functions include careful monitoring of system performance for long-term applications, generating authoritative long-term records from multiple observing platforms, assessing the state of Earth system components, and proper archiving of and timely access to data and metadata [OCP 2006].

National Data Centers: The Comprehensive Large-Array Data Stewardship System is NOAA's data system for providing data from current and future polar-orbiting and geostationary operational satellite systems. The system will provide permanent, secure storage and safe, efficient access between data centers and customers. The system will be sized to handle a significant increase in data rates that will occur as the next generation of observing systems is implemented. Initial accomplishments include implementation of free Internet-based customer access to data from NOAA's Geostationary Operational Environmental Satellite [OCP 2006]. (See <www.class.ngdc.noaa.gov/>.)

III.1.e. Modeling

IPCC Model Analyses and Evaluation: Three U.S. groups carried out a number of model runs to support the development of the IPCC Fourth Assessment Report. These groups were those at the National Center for Atmospheric Research, the NOAA Geophysical Fluid Dynamics Laboratory, and the NASA Goddard Institute for Space Studies. All models are consistent in projecting global warming in the 21st century in response to anthropogenic forcing. They include improved representation of the El Niño–Southern Oscillation (ENSO) compared to previous models, although they do not fully represent many aspects of ENSO including its effects on climate in North America and elsewhere [OCP 2008]. Data from these model runs are available from the Program for Climate Model Diagnosis and Intercomparison (PCMDI; <www-pcmdi.llnl.gov/>) at the Lawrence Livermore National Laboratory. PCMDI, together with its university and government laboratory partners, is applying its collective expertise to support modeling studies initiated by the IPCC. It is providing facilities for the storage and distribution of terascale data sets from multiple coupled atmosphere–ocean general circulation model (GCM) simulations of present-day climate as well as climate changes resulting from large transient increases in carbon dioxide [OCP 2006].

North American Regional Climate Change Assessment Program (NARCCAP): NARCCAP is an effort funded by multiple agencies to meet the climate scenario needs of the United States and Canada. It aims to develop an ensemble of regional climate change scenarios for North America and develop and apply statistical methods to systematically investigate the uncertainties in future climate projections at the regional level. NARCCAP uses a multi-model approach that includes four global atmosphere–ocean coupled models to provide global climate change simulations; six regional climate models to downscale the global simulations; and two global atmosphere models in time-slice experiments to provide climate change scenarios at the same horizontal resolution as the regional climate models. These regional and global simulations contributed to the IPCC Fourth Assessment Report. The NARCCAP model archive is accessible from the Earth System Grid. (See <www.narccap.ucar.edu>.)

III.2. Program Outputs

This section describes CCSP progress toward its overarching goals since 2003, when the program's initial Strategic Plan was released.

III.2.a. Goal 1: Improve knowledge of the Earth's past and present climate and environment, including its natural variability, and improve understanding of the causes of observed variability and change

CCSP research requires a solid foundation of observations, monitoring data, and analyses of these data to provide a better understanding of Earth system processes, to understand the magnitude and extent of climate variations, and to test and to improve models. Analyses of collected observations and monitoring data underpin all aspects of climate system study. In the past few years, key analyses of collected data have provided important insights into understanding the nature and variability of the Earth system.

One of the most notable advances has been the increased certainty ascribed by the IPCC to the influence of humans on climate. The 2001 IPCC Working Group I report (IPCC, 2001) concluded that "...most of the observed warming over the last 50 years is likely to have been due to the increase in greenhouse gas concentrations." By 2007, the IPCC (2007a) had strengthened that conclusion to "very likely." This change was made in part on the basis of work carried out under the auspices of CCSP, including expanded observational records, improvements in the simulation of many aspects of climate, and a series of new attribution studies that evaluated whether observed changes are quantitatively consistent with the expected response to external forcings and inconsistent with alternative physically plausible explanations.

Another example of a body of work with profound impact is the progress made in understanding temperature and moisture changes at continental scales, and in understanding the magnitude of climate change at high latitudes, in the Arctic and Antarctic regions of Earth. Analysis of temperature and moisture records together with satellite images and ground-based measurements for North America and Europe show that both continents are experiencing earlier transitions from winter to summer. Warmer spring temperatures are causing earlier spring green-up of vegetation, together with longer growing seasons overall. Observations in the western United States also indicate that the annual peak in spring river runoff is occurring earlier in the season and is supplying less water during the growing season (Mote et al., 2005). At high latitudes, satellite, airborne, and ground-based observations suggest that significant changes are occurring in the mass balance of the Greenland and Antarctic Ice Sheets. These changes are inferred to be caused by warming at high latitudes. New satellite-based observations of the polar regions indicate significant reductions in the volume of the Greenland Ice Sheet (Velicogna and Wahr, 2005), declining Arctic sea ice cover, and loss of ice mass in Antarctica despite no measurable change in snowfall over the last 50 years (Velicogna and Wahr, 2006). Analyses of observations together with climate model simulations suggest that the pattern of high-latitude temperature change is more readily explained by human activity and natural climate forcing than by internal variability alone.

Another key temperature finding comes from the body of analytical work reported in the CCSP synthesis and assessment report, *Temperature Trends in the Lower Atmosphere: Steps for Understanding and Reconciling Differences* (Karl et al., 2006). Previously reported data showing discrepancies between the amounts of warming near the surface and higher in the atmosphere have been used to challenge the reliability of climate models and the reality of human-induced global warming. In these earlier observations, surface data showed substantial global-average warming, while early versions of satellite and radiosonde data showed little or no warming above the surface. Now that errors in the satellite and radiosonde data have been identified and corrected, the discrepancy no longer exists. New data sets have also been developed that do not show such discrepancies. For recent decades, all current atmospheric data sets now show global-average warming that is similar to the surface warming. A next step is to improve climate-modeling results: while these observations are consistent with the results from climate models at the global scale, discrepancies in the tropics remain to be resolved. In the meantime, these observations and modeling results have provided increased scientific confidence in our understanding of observed climatic changes and their causes.

It is critical to observe and understand variations in the average state of particular climate parameters like temperature and precipitation, but it is as important to understand natural variability and changes in the frequency or intensity of extreme events like drought and unusually wet conditions. CCSP has made important progress in understanding the climate system's natural variability. One of the most well known variability events is the El Niño phenomenon, which recurs on a time scale of approximately two to seven years and involves a warming of the eastern tropical Pacific in combination with changes in atmospheric circulation. Recent research links decadal changes in this pattern to droughts and wet conditions over North America and suggests that a portion of such decadal changes may be predictable (Seager et al., 2005). Additional recent accomplishments on the topic of drought used the geologic record to reconstruct drought history over the past millennium, using proxy records derived from tree rings and sediment cores (Cook et al., 2004). This research indicates that recent U.S. droughts, while severe enough to cause substantial impacts to society, are relatively minor in comparison to naturally occurring droughts over the past thousand years. Interpreting changes in the characteristics of extreme events remains one of CCSP's ongoing research frontiers.

Additional work on atmospheric conditions provided an improved understanding of climate influences on ozone distribution. Using satellite measurements corroborated by surface measurements, an important study found increases in ozone in the Antarctic middle stratosphere during Southern Hemisphere summer (December). Model simulations showed that these increases were caused by the delayed transition from dynamic spring conditions to more stable summer conditions due to the spring ozone hole. The lengthening of less-stable spring dynamics forces the descent of ozone-rich air from higher levels of the atmosphere to the mid-stratosphere (about 30 km altitude). The same study also found that future greenhouse gas increases would produce similar ozone increases. Another study found that doubling of CO_2 caused a strengthening of the atmospheric circulation responsible for the global distribution of ozone. The results of this study indicate that total ozone will increase at high latitudes of the Northern and Southern Hemispheres and decrease in the tropics.

Oceans, too, are experiencing change. Observations of global sea level increases are consistent with the declining volume of land ice as well as observations of ocean warming, which contributes to sea level rise by expanding ocean volume. Observations of the North Atlantic indicate a reduction in salinity (Curry and Mauritzen, 2005), which climate system models indicate may lead to a slowdown of the large-scale ocean circulation that transports heat to high- latitude regions (Stouffer et al., 2006a).

Ocean heat storage is the largest component of Earth's climate system for storing the energy imbalance between the sources and sinks of thermal energy, thus trends in this storage rate are important to quantify. Global-scale observations of ocean temperature indicate an overall pattern of warming (e.g., Levitus et al., 2001). Such trends are generally consistent with climate model projections that include greenhouse gas forcings (e.g., Barnett et al., 2005). However, observation-based estimates of ocean heat storage show significant regional interannual and decadal-scale variations. Recent studies indicate that the Southern Hemisphere warming may have been underestimated (Gille, 2008), and that correction of the systematic instrumentation errors results in diminished decadal variability and a larger, steadier long-term trend (Wijffels et al., 2008).

III.2.b. Goal 2: Improve quantification of the forces bringing about changes in the Earth's climate and related systems

It is essential to understand the factors responsible for global environmental change in order to make long-term climate projections. These forcing factors include changing levels of greenhouse gases, land cover changes, airborne dust and other particles (aerosols), and solar variability. CCSP research gives considerable attention to identifying and quantifying the effects of these forcing factors, and to understanding the ways in which factors cause feedbacks among Earth systems. As in the previous goal, the following examples of progress toward CCSP Goal 2 result from the integrated focus of multiple CCSP research elements.

Climate and the global carbon cycle are a tightly coupled system where changes in climate affect the transfer of atmospheric CO_2 to the terrestrial biosphere and the ocean, and vice versa. An important conclusion of recent carbon cycle research is that future warming is likely to lead to a further decrease in the efficiencies of land and ocean in absorbing excess CO_2 (i.e., a positive feedback) (Fung et al., 2005). This assessment is based on advances in U.S. and global carbon observations and improvements in carbon cycle models. Controlled experiments on carbon uptake and release in ecosystems are one means of improving our understanding of carbon cycle dynamics, which can contribute to corresponding carbon cycle model improvements. For example, Free-Air Carbon Dioxide Enrichment experiments, in which CO_2 is purposely injected into the air around a small plot of land, have led to the conclusion that the mass of carbon in ecosystems initially tends to increase when exposed to increased levels of CO_2 (Jastrow et al., 2005; Norby et al., 2005). This increase may be limited by the availability of nutrients, although a comprehensive meta-analysis indicates that nitrogen supply generally keeps pace with plant demands in natural systems (Luo et al., 2006). Other controlled experiments in which ecosystems are purposely warmed generally indicate greater ecosystem CO_2 release with higher temperatures. However, there are still significant uncertainties associated with the biospheric response to climate change, particularly with respect to the complex and dynamic nature of ecosystems and their interactions with climate and the hydrologic cycle.

CCSP has made significant advances in understanding the processes responsible for the production and destruction of other greenhouse gases, including methane and nitrous oxide. For example, recent analyses estimate that approximately 60% of all methane emissions from wetlands occur in the tropics (Melack et al., 2004). In polar regions, recent studies have elucidated processes by which carbon, currently trapped either as organic matter or methane hydrates in the permafrost (frozen soil), is released to the atmosphere (Zimov et al., 2006). Warming increases these releases and can create an amplifying feedback loop. Another example of progress on non-carbon greenhouse gases is the work that has improved understanding of interactions between climate variability and near-surface ozone, which is a health hazard.

One of the largest uncertainties in projections of potential future climate change is the role of aerosols. Recent research has reduced some of this uncertainty, in part through efforts made possible by the Climate Change Research Initiative. In the first phase of preparing the synthesis and assessment report that deals with aerosol properties and their impacts on climate, a comprehensive paper has been published that reviews recent progress in characterizing aerosols and assessing the direct effect of aerosols on climate change (Yu et al., 2006). This work served as a major resource in the preparation of the IPCC Fourth Assessment Report.

Another important recent advance is improved estimates of the amount of carbon being sequestered in North America and globally, and in particular, how the rate of carbon uptake is changing in all ecosystems. These estimates are made through the innovative combination of carbon cycle models and observations of carbon concentrations and isotopes (Fung et al., 2005). A key goal of the North American Carbon Program is to further improve estimates of carbon sources and sinks. Work of this nature is vital for assessing the efficacy of natural carbon uptake, as well as the potential for purposeful carbon capture in managed ecosystems. Observations of ocean carbon are important for addressing uncertainties associated with the global carbon budget. New global-scale ocean carbon analyses indicate increasing carbon concentrations in ocean water. In addition to confirming the oceans as a significant carbon sink, this information is also being used to estimate the increase in ocean acidity caused by increasing amounts of dissolved CO_2 and the potentially deleterious consequences for marine ecosystems (Orr et al., 2005). Recent measurements of carbon sedimentation along continental shelves have shown these regions to be responsible for a significant fraction of oceanic carbon uptake (Muller-Karger et al., 2005).

Recent climate warming has been particularly intense in boreal and Arctic regions, leading to concern that increasing air temperature in these ecosystems may lead to an increase in the incidence of wildfires. Longer growing seasons lead to increased fire fuel loads, and increased temperatures can lead to drier conditions. In turn, wildfire provides rapid release of carbon and produces aerosols that affect atmospheric conditions. Beyond the emission of CO_2 and other greenhouse gases, understanding the consequences of large-scale wildfires for climate is challenging due to the many additional ways in which they influence the lower atmosphere and surface of Earth. A recent study in Alaska found that there was intensification in the climatc warming in the first year after a major fire but a slight decrease in the local climate warming when averaged over the 80 years of the study. The long-term result, which was primarily due to plant regrowth increasing the summer reflectivity of the burned surface, appeared to be more significant than the greenhouse gases emitted by the fire (Randerson et al., 2006). The study results suggest that future increases in wildfire in some parts of the boreal zone of Alaska may have different feedbacks to global warming than previously thought.

Scientists are also studying the possibility that increased permafrost thawing due to warming in Arctic regions could cause the release of substantial amounts of carbon long held in the frozen tundra. There appear to be two potential mechanisms for the carbon to reach the atmosphere: drainage of the carbon-rich river flow into the Arctic Ocean with subsequent emission, and direct respiration or recycling of the newly thawed carbon. Measurements made in the Yukon River Basin in northern Canada have shown that the latter process predominates (Guo and MacDonald, 2006). These results have significance for understanding both the movement of carbon through Arctic landscapes, and the potential effects of that carbon on ecosystems and the atmosphere.

CCSP's interdisciplinary research on the carbon cycle has also produced a set of analyses using long-term observations of several young and mature forests. Results from this work show that forest carbon storage has been increasing in these ecosystems and is not in balance with the carbon lost by respiration and decay. This result is contrary to the contemporary concept of near balance of carbon sources and sinks in mature forests (Zhou et al., 2006). The gain in forest carbon is typical of findings from U.S.-based large-scale networks, as well as observations made in mature forests in China. Therefore, evidence is mounting that these sinks for atmospheric CO_2 offer significant potential for modulating the rate of atmospheric CO_2 increase (Urbanski et al., 2007).

In western states, large changes in land cover and land use have occurred over the past century, with rapidly expanding urbanization along the Pacific coast and extensive agricultural development inland. Researchers exploring the effects of urbanization and agriculture on regional climate have found that irrigated agriculture in California tended to lower average and maximum near-surface air temperatures, while conversion of natural vegetation to urban areas tended to increase near-surface air temperatures. The surface temperature changes and their associated effects on the atmosphere also caused changes in the regional airflow. Overall, it was found that conversion of natural vegetation to irrigated agriculture has likely had a larger effect on the climate of California than urban growth, but that increased conversion of irrigated land to urban/suburban development could alter this conclusion (Kueppers et al., 2007).

III.2.c. Goal 3: Reduce uncertainty in projections of how the Earth's climate and related systems may change in the future

Uncertainty provides a measure of the reliability of forecasts of future climate and related Earth systems. Reducing uncertainty is crucial to providing decisionmakers with tools for assessing strategies for adaptation, mitigation, and other forms of risk reduction. However, the wording of the goal is incomplete—*reducing uncertainty* is only part of the story. Improving the projections themselves and understanding both the nature and implications of uncertainties are equally important, as noted in the NRC metrics report (NRC, 2005b). The thrust of this goal is, therefore, to improve projections, and to characterize their uncertainty, in order to improve the utility of projections of how Earth's climate and related systems may change in the future. CCSP has significantly advanced the ability to estimate future Earth system conditions at time scales ranging from months to centuries and at spatial scales ranging from regional to global. The primary tools for Earth system prediction and projection are computer models that reflect the best available knowledge of Earth system processes. Reducing uncertainty requires continual integration of observations and modeling across the full range of climate and Earth systems research.

When recent model simulations of the climate of the past 100 years are compared to observations at coarse temporal and spatial scales, the results generally indicate improvements over previous generations of models, including the ability to represent weather systems, climate variability (e.g., monsoons and El Niño), ocean processes (e.g., the Gulf Stream), surface hydrology, and other Earth system processes, components, and dynamics (Collins et al., 2006; Schmidt et al., 2006). One of the ways in which these models have advanced is through improvements in the representation of the processes responsible for key Earth system feedbacks such as those associated with water vapor, clouds, sea ice, and the carbon cycle (Delworth et al., 2006; Gnanadesikan et al., 2006; Wittenberg et al., 2006).

For a model to produce a realistic climate projection, it must include realistic representations of physical processes such as cloudiness, precipitation, and solar energy. Understanding the influence and feedbacks of clouds on climate has proven to be a challenging task for scientists worldwide, but progress is being made. Recent innovative studies using newly developed, detailed models of cloud processes that are coupled with a global climate model provide results that are significantly more consistent with observations than traditional cloud modeling techniques. The incorporation of improved cloud representation in climate models is expected to reduce the uncertainty in predictions of the global and regional water cycle and surface climate.

Energy from the Sun not reflected back to space provides the driving energy to Earth's weather and climate systems. Clouds are a major component in the global reflectance of sunlight. Year- to-year variability in the global reflectance is dominated by the variability of cloudiness in the tropics. On the other hand, scientists have recently found little change in the year-to-year variability of reflectance at middle and high latitudes despite decreases in the highly reflective snow and sea ice cover. This result appears to be due to the compensating increase in cloud cover balancing the decreasing surface-level reflectance. Clouds continue to provide the largest source of uncertainty in model estimates of climate sensitivity, although a recent study finds evidence that, in most climate models used in the IPCC Fourth Assessment Report, clouds provide a positive feedback.

The magnitude of future warming will be strongly influenced by the extent to which atmospheric water vapor concentration increases in response to an initial warming caused by increases in CO_2 and other greenhouse gases. An accurate representation of this feedback in climate models is critical for making long-term climate projections. Recent innovative analyses have shown that water vapor increases in the upper atmosphere measured by satellites and balloon-borne sensors are generally consistent with state-of-the-art climate model simulations, lending credence to the ability of current models to represent the water vapor feedback (Cess, 2005; Soden et al., 2005).

Analyses of climate model simulations generated for the IPCC Fourth Assessment Report have identified several additional characteristics of climate change projections common to all of the models (Held and Soden, 2006). Examples of these robust model projections include strong sub-tropical drying, weakening of large-scale tropical atmospheric motions, and expansion of the poleward upper atmospheric wind pattern known as the Hadley circulation. In another study, several models were used to investigate the effects of the freshwater input from melting ice and glaciers on the currents in the North Atlantic (Stouffer et al., 2006a). These currents are important due to their large-scale transport of heat. The study concluded that, in response to expected levels of freshwater input in the northern North Atlantic, the average modeled large- scale deep ocean current weakens by about 30% by the end of the century. All models simulate some weakening of this deep circulation, but no model simulates a complete shutdown of it.

CCSP researchers also use the geological record to test and apply climate models, particularly in cases where that knowledge has a bearing on climate change processes relevant to current society. One such analysis involves the largest known extinction in Earth's history, which took place approximately 250 million years ago at the Permian–Triassic boundary when approximately 95% of marine and 75% of terrestrial species were lost. In this study, a climate model simulation indicated that the elevated levels of CO_2 during this period led to climatic conditions inhospitable to both marine and terrestrial life (Kiehl and Shields, 2005). It is hypothesized that a critical level of high-latitude warming was reached where the connection of oxygen-rich surface waters to the deep ocean was dramatically reduced—thus leading to a shutdown of marine biologic activity, which in turn led to increased atmospheric CO_2 and accelerated warming.

The historical record provides a broader set of observations to test and apply climate models to help reduce uncertainty in their projections of the future. A recent study used a simple model to attempt to reproduce paleoclimate reconstructions of Northern Hemisphere temperature over the past seven centuries in response to estimated solar, volcanic, and greenhouse gas forcing during this period (Hegerl et al., 2006). This study suggests that, for the current century, very high climate sensitivities predicted by some models for a doubling of atmospheric greenhouse gas concentrations are less likely than previously thought.

The CCSP modeling strategy utilizes a multi-tiered approach in which new and improved Earth system sub-models (e.g., clouds, ecosystem dynamics, and sea ice) are developed and tested by individual researchers or small research teams. When significant improvements in these sub- models arise, they are integrated as appropriate into high-end Earth system models.

A result of these ongoing efforts is a set of U.S. models that expand beyond earlier atmosphere–ocean models to include relatively sophisticated representations of land surface hydrology, sea ice, ecosystems, and atmospheric chemistry. Several U.S. Earth system modeling centers have used variations of these models to produce ensembles of projections that are providing important new perspectives on potential future climate system change (Meehl et al., 2004a; Stouffer et al., 2006b). These ensembles are also being used to characterize the intrinsic uncertainty associated with potential future climate change.

A set of new high-resolution climate model simulations has been completed for North America that provides information at a scale finer than 100 km x 100 km (Han and Roads, 2004; Leung et al., 2004; Mason, 2004; Wood et al., 2004).The ability of these regional-scale models to represent climate processes is being assessed.[14] These regional and global simulations, based on models developed at U.S. institutions, contributed to the IPCC Fourth Assessment Report.

Because Earth system models are extremely complex and benefit greatly from input and evaluation by multiple research teams, several new efforts have been initiated to enable sharing, testing, and improvement of these models by diverse groups of researchers (Meehl et al., 2004b, 2005). Many of the recent model simulations referred to above are now widely available through a new capability for data archiving and dissemination developed by the PCMDI.[15] Large strides have been made in creating climate model code according to a set of standards that facilitates exchange of sub-models (e.g., the Earth System Model Framework), which enables researchers to readily trace the source of differences between various models and between models and observations. The U.S. Climate Variability and Predictability Program (CLIVAR) is exploring a new approach for bringing together observers, theorists, and high-end modelers to improve key model deficiencies (USCLIVAR, 2002; Bretherton et al., 2004). This approach is attempting to significantly reduce the time lags that often exist between the observation of key climate processes and the integration of these processes into more comprehensive Earth system models. Several high-end Earth system modeling efforts in the United States, which involve many different, independent research teams, are using these types of new collaborative approaches and tools to evaluate, improve, and integrate model components.

The projections made by CCSP research pertain not just to physical climate, but also to other components of the Earth system, including atmospheric chemistry. Continuing research has provided an estimate that the recovery of the Antarctic ozone hole will occur approximately 10 to 20 years later than the previous estimate of 2050 (Newman et al., 2006). As a result of the Montreal Protocol and its amendments, the use of ozone-depleting substances has been greatly reduced. Improved understanding of atmospheric dynamics now gives 2001 as a better estimate of when the peak in ozone-depleting substances occurred in the Antarctic stratosphere. This date is later than had been estimated previously and results in a longer projected time scale for recovery back to pre-1980 (unperturbed) levels of these substances.

III.2.d. Goal 4: Understand the sensitivity and adaptability of different natural and managed ecosystems and human systems to climate and related global changes

Goals 4 and 5 are expected to be areas of growth for CCSP over the coming years, in keeping with the increased availability of results from Goal 1 through 3 as inputs. Progress is reported here for the past several years to the present. Significant advances have been made in understanding the potential impacts of climate change and in the improvement of methodologies as new information has become available. CCSP research typically uses many different sources of information, including analyses of paleoclimate data, direct monitoring and observations, process studies, and model-based projections. Increasingly, research also accounts for the dynamic nature of the response of human and natural systems to climate change.

Research to understand ecosystem and human system sensitivity and adaptability encompasses and integrates a wide range of potential impacts on societal needs such as water, health, and agriculture, as well as potential impacts on natural terrestrial and marine ecosystems. This integration is exemplified by the development of several of CCSP's synthesis and assessment products (SAPs), particularly the seven products under Goal 4. One example is SAP 4.1, which combines census population data, topographic elevation, shore protection, and land use information to study the potential socioeconomic impacts of different scenarios of sea level rise. Other important research into the potential implications of sea level rise also points to the need to account for a wide variety of factors when assessing future impacts. For example, some measures to protect coastlines may carry negative side effects, such as the potential for wetlands loss when inland barriers are constructed, preventing the wetlands from migrating inland in response to rising sea level (Cahoon et al., 2006).

Similarly, the adaptability of complex coastal ecosystems is becoming better understood. For example, studies in a Chesapeake Bay marsh ecosystem showed that rising sea level, increasing CO_2, and high rainfall can interact and improve the growth of a relatively tall bulrush at the expense of a hay-like cordgrass that grows in thick mats (Erickson et al., 2007). The thick cordgrass mats trap sediment and organic material more effectively than do the bulrushes, and so bulrush-dominated marshes are less able to rise in elevation through the addition of sediments. Such changes in species composition, caused by interacting global change factors, can therefore influence the ability of coastal marshes to adapt and keep abreast of sea level rise.

Carbon cycle scientific research and modeling are highly relevant to carbon management needs, as demonstrated by a recent study that estimated the spatial variability of net primary production and potential biomass accumulation over the conterminous United States (Potter et al., 2007). This study's model-based predictions indicate a potential to remove carbon from the atmosphere at a rate of 0.3 Gt of carbon per year through afforestation of low-production crop and rangeland areas. This rate of carbon sequestration could offset about one-fifth of the annual fossil fuel emissions of carbon in the United States.

Observational and modeling studies of terrestrial ecosystems indicate a wide variety of changes in which it appears that climate variations play a significant role. For example, recent evidence indicates a northward expansion of the ranges of many bird and butterfly species in the United States corresponding to warming in the region (Sekercioglu et al., 2004). Declines in Arctic sea ice, observed both *in situ* and by satellites, have been linked to increasing vulnerability of polar bear populations (ACIA, 2005). Satellite and *in situ* observations also indicate a trend toward earlier growth of spring vegetation (Angert et al., 2005). In addition to temperature and hydrologic changes, the increasing level of atmospheric CO_2 is thought to play a role in changing ecosystem distributions and characteristics due to its fertilizing effect. Agricultural yield models account for this effect and project a range of agricultural impacts depending on the magnitude and nature of future climate change, crop types, and the types of adaptive measures that are adopted. Recent research indicates that different strategies may be required to manage insects, weeds, and diseases in agricultural systems (Ziska and Runion, 2006).

Another example of the ecological consequences of climate change involving insects and affecting adaptability is the devastation of millions of acres of western U.S. and Canadian pines by bark beetles during the warmth and drought of 2000 to 2004. Recent modeling and observations revealed that beetles invading the northernmost lodgepole pine trees are now only a few miles from previously pristine jack pine populations (Logan and Powell, 2007). This may create a direct pathway of invasion to valued pine forests in the eastern United States and Canada.

CCSP's integrated approach to understanding the sensitivity and adaptability of natural ecosystems to climate change has also been applied in remote, high-latitude regions. The West Antarctic Peninsula is experiencing some of the largest, most rapid warming on Earth, which is causing loss of sea ice and increased snow precipitation. In turn, these changes are having major contrasting impacts on the adaptability of different penguin species. For example, in the vicinity of Anvers Island near the West Antarctic Peninsula during the last three decades, populations shifted south, so that local abundance of the ice-dependent and snow-intolerant Adelie penguins decreased by 65% (currently about 5,000), while the abundance of Chinstraps and Gentoos increased by 2,730% and 4,600% (currently about 300 and 650), respectively (Ducklow et al., 2007). Climate warming in the Canadian Arctic has caused significant declines in total cover and thickness of sea ice and progressively earlier ice breakup in some areas. These changes have had a substantial effect on polar bear populations, causing them to extend their normal fast for longer periods during the open-water season (Stirling and Parkinson, 2006).

Components of CCSP research funded in part through the U.S. Joint Global Ocean Flux Study Program have explored ecosystem impacts in the open ocean resulting from climate variability and change as well as from changes in ocean chemistry and thermal structure. An example of a chemical impact is the chain of events causing the oceans to become more acidic due to chemical changes resulting from the absorption of increasing concentrations of atmospheric CO_2 (Orr et al., 2005). Ocean warming tends to increase vertical stratification (layering) and thus slow the overturning of nutrient-rich deep-ocean waters (Schmittner, 2005). Recent model projections suggest that increased ocean acidification and increased layering of the upper ocean due to warming are likely to reduce plankton production. These model results are supported by satellite observations indicating significant changes in photosynthetic plankton concentrations, including declines in the North Atlantic and Pacific and increases in the Indian Ocean (Gregg et al., 2003).

In addition to managed ecosystems, CCSP research has expanded understanding of the sensitivity and adaptability of a variety of other societal sectors. One of these is human health, which may be affected by changes in temperature and storm intensity, or through changes in distributions of insects that carry pathogens. An example of research in this area is the effects of climate change on heat waves. Observational and modeling work suggests that the probability of heat waves such as the one that occurred in Europe in 2003 has increased significantly, and that future warming may make heat waves of similar magnitude a normal summer occurrence within several decades (Meehl and Tebaldi, 2004). Recent research on the societal dimensions of climate variations has shown that physical climate analyses, such as the aforementioned study of heat waves, must be assessed within a complex fabric of other social and environmental factors (Poumadere et al., 2005). An example is the general increase in financial losses due to hurricanes over the past century, which is probably attributable more to expanding coastal development than to any changes in hurricane characteristics (Pielke et al., 2005). In regions such as central Africa, where the capacity to adapt to environmental variations is often relatively low, recent research has shown strong correlations between year-to-year climate variations and malaria outbreaks (Thomson et al., 2006).

These are a few examples of CCSP's research examining the sensitivity and adaptability of human and natural systems to climate variability and change. It is clear from this work that climate variations can have both beneficial and adverse effects on environmental and socioeconomic systems. However, future projections indicate that the larger the magnitude and rate of climate change, the more likely it is that adverse effects will dominate (NRC, 2002).

III.2.e. Goal 5: Explore the uses and identify the limits of evolving knowledge to manage risks and opportunities related to climate variability and change

The Nation's basic research on global environmental variability and change has led toward a number of important opportunities for applying that knowledge to help decisionmakers make good choices supported by sound science. Over the past several years, CCSP has taken three main approaches to explore and communicate the potential uses and limits of this knowledge: 1) developing scientific syntheses and assessments; 2) supporting and exploring adaptive management and planning capabilities; and 3) developing methods to support climate change policy inquiries. Some key examples of progress are provided below.

The key focus of the program's synthesis and assessment activities is its current suite of 21 SAPs, which are intended to provide current evaluations of the science foundation for use in informing public debate, policy, and operational decisions and for defining and setting the future direction and priorities of the program. For example, SAP 1.1 (Karl et al., 2006) largely resolved a long-standing debate concerning an apparent discrepancy of global temperature trends at the surface and in the lower atmosphere; SAP 2.1a (Clarke et al., 2007) provided key policy-relevant insights regarding technology options for stabilizing greenhouse gas concentrations; and SAP 4.4 (Julius et al., 2008) frames an important set of considerations for management of ecosystems on federally managed lands. These reports examine various aspects of climate change science and impacts on a national, regional, or sectoral basis as appropriate to each topic. A description of each product and its schedule for completion can be found at <www.climatescience.gov/Library/ sap/sap-summary.php>.

Another important focus for the program's synthesis and assessment activities has been its involvement in the IPCC and other international assessment activities, including providing coordination and support to the IPCC Fourth Assessment Report. The IPCC's major activity is to prepare at regular intervals these comprehensive assessments of policy-relevant scientific, technical, and socioeconomic information appropriate to the understanding of human-induced climate change, potential impacts of climate change, and options for mitigation and adaptation. Approximately 120 U.S. scientists are IPCC authors and 15 are review editors. The United States co-chairs and hosts IPCC Working Group I, which primarily addresses physical science aspects of climate change. The United States has also played significant roles in the World Meteorological Organization (WMO) / United Nations Environment Programme (UNEP) ozone assessments (e.g., WMO, 2007), the Arctic Climate Impact Assessment (ACIA, 2005), and the Millennium Ecosystem Assessment (MEA, 2005), among others.

The second of CCSP's decision-support approaches is the exploration of adaptive management strategies. Activities under this approach develop and evaluate options for adjusting to variability and change in climate and other conditions through 'learning by doing' and integrating knowledge with practice. This area of work grows out of the insight that a key to assessment and decision support is close and ongoing interaction between users and producers of information. While this area is less well developed than some others in CCSP's portfolio, it is a strong and integral component of the missions of several CCSP member agencies, and those agencies provide extensive stakeholder engagement and decision support for adaptation, mitigation, and management of risk to policymakers and land managers at national, regional, state, and local levels.

As one example of the development of methods to support climate change policy inquiries, CCSP scientists developed and documented a 'water supply stress index' that calculates water shortage risks across the conterminous United States. The index is based on models and observations that integrate the effects of climate, land cover, and current water uses by municipalities and industries on water supply (Sun et al., 2006). The water supply stress index and the methods associated with it will be used by local and regional decisionmakers to quantify the likelihood of future water shortages under changing climate, water, and land uses for determining adaptation practices. Incorporation of the subsurface water table into regional climate models is important, since land cover changes produce significant effects on the water table and the hydrologic cycle. Shallow water tables can be either a sink or source of water to the surface soil, depending on the relative balance of infiltration versus evaporation (Fan et al., 2007). Recent studies using detailed observations and regional climate models have found that the fraction of rainfall that either recharges groundwater or ends up as streamflow tends to decrease when the fraction of land devoted to agriculture increases. This result suggests that intensive agriculture can amplify surface water stresses, particularly during drought conditions (Jayawickreme and Hyndman, 2007).

Another example of this work is an ongoing project that brings together researchers who study climate processes and their effects on the U.S. Southwest with individuals and organizations that need climate information to make informed decisions (Jacobs et al., 2005). Numerous tangible benefits from this project have helped a wide variety of decisionmakers, from state and local water planners to farmers to public health officials. For example, the project developed a suite of products that make predictions of water availability months in advance, allowing water managers to adjust reservoir levels accordingly to meet the competing demands for this scarce resource.

Yet another example is the combined use of satellite-based observations of fires and moisture conditions together with seasonal climate forecasts to provide information to fire managers to help them make early and effective decisions about the resources they will need to cope with emerging fires and fire-season dangers. One way in which this information is communicated is through annual workshops targeted separately at eastern and western U.S. fire hazards, which bring together climate scientists and forecasters with fire managers to produce seasonal fire outlooks.[16] There are many other examples of the exploratory use of seasonal-to-interannual climate information for decisionmaking both domestically and internationally.

In addition to CCSP's work on adaptive management at seasonal-to-interannual time scales, the program is also developing valuable information for long-term (decades to centuries) adaptation issues. One example is the program's analyses of ways in which agricultural practices might be adjusted to take advantage of rising CO_2 levels and to cope with potentially warmer temperatures and decreased moisture availability (Boote et al., 2005). Recent work has shown that sufficient variability exists within some crop species to begin selecting for crop varieties that could maintain or increase yields in a future enhanced CO_2 environment.

An example of regional decision support is the work carried out by the Consortium for Atlantic Regional Assessment (CARA), which is providing data and tools to help decisionmakers understand how outcomes of their decisions could be affected by potential changes in both climate and land use. On an interactive, user-friendly Web site, CARA has organized data on climate (historical records and future projections from seven global climate models), land cover, and socioeconomic and environmental variables to help inform local and regional decisionmakers.[17] The CARA tools and tutorials are designed to help decisionmakers understand the issues related to land use and climate change by gathering, organizing, and presenting information for evaluating alternative mitigation strategies.

A workshop involving scientists and managers, co-led by several CCSP agencies under the auspices of the U.S. Coral Reef Task Force, resulted in the publication of *A Reef Manager's Guide to Coral Bleaching* (Marshall and Schuttenberg, 2006). The combined research results among state/territorial, federal, academic, nongovernmental, and international scientists concluded that warming sea surface temperatures are a key factor in mass coral bleaching events. The guide provides managers with strategies to support the natural resilience of coral reefs in the face of climatic change.

CCSP's third decision-support approach is to help inform inquiries related to climate change policy, in part by using comparative analyses of climate change scenarios. One example is a collaborative effort between climate scientists and New York City water infrastructure planners that is using regional-scale hydrologic scenarios to inform the long-lasting investments that are being considered in the modernization of the city's water supply system.[18] Another example is the application of carbon cycle research to assess the potential feasibility, magnitude, and permanence of a variety of different carbon sequestration options (Sarmiento et al., 1999). An initial result from this line of work is the preliminary conclusion that the restoration of inland wetlands could be a particularly efficient means for sequestering carbon in North American prairie lands (Euliss et al., 2006).

CCSP researchers have also developed new metrics for estimating greenhouse gas emissions and carbon sequestration in the agricultural and forestry sectors (Birdsey, 2006). These sectors can reduce atmospheric greenhouse gas concentrations by increasing carbon sequestration in biomass and soils, by reducing fossil fuel emissions through use of biomass fuels, and by substituting agricultural and forestry products that require less energy than other materials to produce. The DOE's National Greenhouse Gas Registry is using the new metrics as the basis for reporting greenhouse gas information from the agricultural and forestry sectors.[19]

Another important way in which CCSP is helping to inform climate change policy inquiries is through integrated assessment modeling, which considers the social and economic factors that may lead to climate change (e.g., greenhouse gas emissions) and the resultant effects of those activities on the Earth system and human welfare. These models are useful for considering the costs and effects of various policy options. One important result of this work suggests that reducing emissions of greenhouse gases other than CO_2 could be an economically efficient first step in reducing the overall atmospheric burden of greenhouse gases (Hansen and Sato, 2004). Another important new set of analyses assesses various policy options while accounting for inherent scientific and economic uncertainty (Webster et al., 2003).

These examples collectively provide a sketch of the progress CCSP has made toward its goals. Additional information on these items and others can be found in CCSP's annual report to Congress, *Our Changing Planet.* While these accomplishments, taken together, represent significant progress in climate change science and constitute a substantial portfolio of work, many questions remain to be answered within each of CCSP's strategic goals. The remainder of the report presents science findings, many of which originated from CCSP-sponsored work, organized by topic.

SECTION IV. TRENDS AND PROJECTIONS OF GLOBAL ENVIRONMENTAL CHANGE

IV.1. Radiative Forcing and Observed Climate Change

Radiative forcing is a change caused by an environmental factor in the balance of solar radiation coming into and the infrared and reflected shortwave radiation going out of the atmosphere. As such, measures of radiative forcing—expressed in watts per square meter (W/m^2)—can be used to weigh the relative importance of various factors, such as greenhouse gases or land cover, in terms of their contribution to climate change. A positive forcing means the factor causes a warming effect, and a negative forcing means the factor causes a cooling effect.

This section describes radiative forcing and the factors that contribute to it, as well as observations of climate change both globally and in the United States. This Scientific Assessment does not delineate specific sectoral contributions to increasing greenhouse gas levels, such as the contributions from the transportation sector described in OFCM (2002).

The radiative forcing values presented here come from the Working Group I contribution to the Intergovernmental Panel on Climate Change (IPCC) Fourth Assessment Report (IPCC, 2007a) and show the change between the pre-industrial time of 1750 and the year 2005. These values represent global averages—they are the result of *global* changes in atmospheric concentrations of greenhouse gases and other factors, and are therefore not specific to the activities of any country in isolation.

IV.1.a. Radiative forcing due to greenhouse gases and other factors

Greenhouse gases have a positive forcing (a warming effect) because they absorb and radiate outgoing infrared radiation that would otherwise escape directly into space. This radiation by greenhouse gases occurs in all directions, with a net effect of warming the air and surface below (in comparison to the absence of greenhouse gases). Water vapor is the most important naturally occurring greenhouse gas. However, direct anthropogenic emissions of water vapor correspond to less than 1% of natural sources and make a negligible contribution to radiative forcing.

The level of scientific understanding of human influences on climate improved significantly between the IPCC's Third Assessment Report (IPCC, 2001) and the Fourth Assessment Report (IPCC, 2007a). In 2007, the IPCC reported the global average net radiative forcing was +1.6 (+0.6 to +2.4) W/m^2. The IPCC concluded with *very high confidence* (at least a 9 out of 10 chance of being correct)[20] that human activities have led to a net warming of the climate.

Figure IV.1 shows the IPCC's (2007d) globally averaged radiative forcing estimates for anthropogenic greenhouse gas emissions and other factors for 2005.

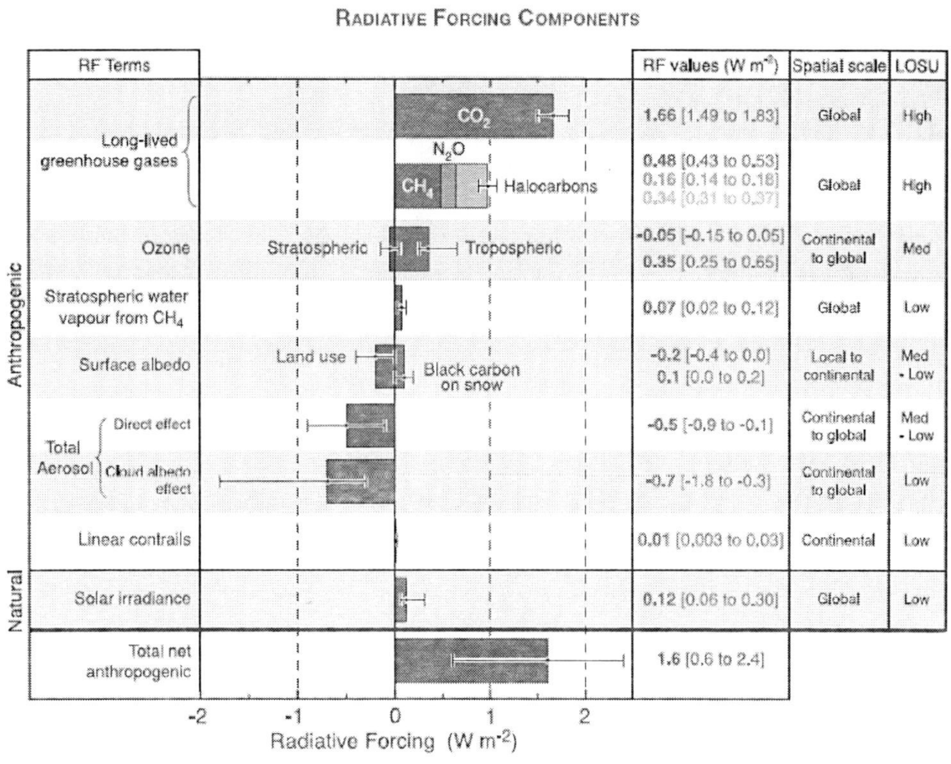

Figure IV.1. Global average radiative forcing (RF) estimates and ranges in 2005 for anthropogenic carbon dioxide (CO_2), methane (CH_4), nitrous oxide (N_2O) and other important factors, together with the typical geographical extent (spatial scale) of the forcing and the assessed level of scientific understanding (LOSU). The net anthropogenic radiative forcing and its range are also shown. These require summing asymmetric uncertainty estimates from the component terms and cannot be obtained by simple addition. Additional forcing factors about which there is a very low level of scientific understanding are not included here. Volcanic aerosols, which also contribute a natural forcing, are not included in this figure due to their episodic nature. The range for linear contrails does not include other possible effects of aviation on cloudiness. Source: IPCC (2007d).

Carbon dioxide, methane, and nitrous oxide

The greenhouse gases carbon dioxide (CO_2), methane (CH_4) and nitrous oxide (N_2O) persist in the atmosphere for decades to centuries to longer periods. As such, their emission has long-term impacts on climate. The combined radiative forcing due to the cumulative increase (from 1750 to 2005) in atmospheric concentrations of CO_2, CH_4, and N_2O is +2.30 W/m^2, with an uncertainty range of +2.07 to +2.53 W/m^2. The observed accumulation of these gases in the atmosphere is primarily from human activities, and as such, the positive radiative forcing from these gases is primarily anthropogenic in origin as well. The IPCC (2007d) stated that the rate of increase in positive radiative forcing due to these three greenhouse gases during the industrial era is *very likely* (greater than 90% probability)[21] to have been unprecedented in more than 10,000 years.

The largest positive radiative forcing is due to CO_2 (+1.66 ± 0.17 W/m^2), and the trend for the radiative forcing from CO_2 is increasing. The period from 1995 to 2005 showed a 20% increase in its radiative forcing—the largest change for any decade in the last 200 years. The IPCC concluded, "Emissions of CO_2 from fossil fuel use and from the effects of land use change on plant and soil carbon are the primary sources of increased atmospheric CO_2" (Solomon et al., 2007).

The next largest components of positive radiative forcing are CH_4 (+0.48 ± 0.05 W/m^2) and N_2O (+0.16 ± 0.02 W/m^2). Current atmospheric CH4 levels are almost triple that of the pre-industrial era. These levels are due to continuing anthropogenic emissions of CH_4, which are greater than natural emissions (Solomon et al., 2007). N_2O increases in the atmosphere since pre-industrial times are also primarily from human activities, in particular land use associated with agriculture.

Fluorocarbons, chlorocarbons, and sulfur hexafluoride

The ozone-depleting substances covered under the Montreal Protocol—chlorofluorocarbons (CFCs), hydrochlorofluorocarbons (HCFCs), and chlorocarbons—are also strong greenhouse gases. As a group, they contributed +0.32 ± 0.03 W/m^2 to anthropogenic radiative forcing in 2005. In the two decades since the Montreal Protocol, the atmospheric concentrations of CFCs peaked and have since begun to decline. Increases in HCFCs have slowed notably in recent years. In response, the radiative forcing from CFCs and HCFCs peaked in 2003 and is now beginning to decline (Forster et al., 2007).

These ozone-depleting substances have reduced global stratospheric ozone levels by about 4% from 1980 levels (IPCC, 2007d). The decreased levels of stratospheric ozone are estimated to have a slightly negative radiative forcing effect of -0.05 ± 0.10 W/m^2. This estimate has a medium level of scientific understanding (Solomon et al., 2007).

As a group, hydrofluorocarbons (HFCs), perfluorocarbons (PFCs), and sulfur hexafluoride (SF_6) had a total radiative forcing in 2005 of +0.0 17 ± 0.002 W/m^2 (Forster et al., 2007). While the current atmospheric concentrations of these gases—all anthropogenic compounds—are small, their concentrations are rapidly increasing (Solomon et al., 2007).

Ozone

Ozone (O_3) in the troposphere (the atmosphere from Earth's surface to 10–16 km altitude) has a positive radiative forcing, with an estimated value of +0.35 (+0.25 to +0.65) W/m^2. Unlike the greenhouse gases mentioned above, tropospheric O_3 is much shorter lived in the atmosphere. Its atmospheric lifetime is of the order of weeks to months (as opposed to decades to centuries for the well-mixed greenhouse gases) and is therefore not as well mixed in the global atmosphere. Tropospheric O_3 is a secondary product of photochemistry in the atmosphere and changes in its abundance are driven by changes in emissions of its precursors, mainly nitrogen oxides, carbon monoxide, and hydrocarbons, which all have significant anthropogenic sources.

According to the World Meteorological Organization (WMO, 2007), spring polar O_3 depletion continues to be severe when winter temperatures in the polar stratosphere (above approximately 10 km) are particularly cold. The average concentration of atmospheric O_3 outside of polar regions is no longer declining, as it was in the 1990s. Measurements from some stations in unpolluted locations indicate that ultraviolet radiation levels have been decreasing since the late 1990s, in accordance with observed ozone increases. However, ultraviolet radiation levels are still increasing at some Northern Hemisphere stations as a consequence of long-term changes in other factors such as clouds and atmospheric particulates that also affect ultraviolet radiation. Trends in tropospheric O_3 vary in sign and magnitude depending on location. It is noted that low latitudes show significant upward trends in tropospheric O_3 (Solomon et al., 2007).

Water vapor

Although water vapor is the most important naturally occurring greenhouse gas, direct anthropogenic emissions of water vapor correspond to less than 1% of natural sources and make a negligible contribution to radiative forcing. However, as temperatures increase, tropospheric water vapor concentrations increase, representing a key feedback but not a forcing of climate change (Solomon et al., 2007). Feedbacks are defined as processes in the climate system (such as a change in water vapor concentrations) that can either amplify or dampen the system's initial response to radiative forcing changes (NRC, 2003). From this perspective, water vapor is excluded as a radiative forcing agent in Figure IV.1. The IPCC (Solomon et al., 2007) concluded that observations demonstrate increases in both lower-tropospheric and upper-tropospheric water vapor since at least the 1980s.

Water vapor in the stratosphere (the atmosphere at 10 to 50 km above Earth's surface) is increasing due to oxidation of CH_4 and is estimated to have a positive radiative forcing of $+0.07 \pm 0.05$ W/m^2 (Solomon et al., 2007). The concentration of any species, including water vapor, changes with altitude, often in relatively predictable ways, such that the concentration profile in relation to altitude will have a predictable 'structure' to it. The contribution of CH_4 to the corresponding change in this vertical structure of the water vapor near the tropopause is uncertain and leads to a low overall level of scientific understanding for this radiative forcing component. The IPCC also noted that other potential human causes of stratospheric water vapor increases that could contribute to radiative forcing are poorly understood (Solomon et al., 2007).

Aerosols

Aerosols are non-gaseous substances, either solid particles or liquid droplets, suspended in the atmosphere. Some aerosols, for example sulfate aerosols that are mainly the result of sulfur dioxide (SO_2) emissions, exert a negative forcing (cooling effect) by reflecting and scattering incoming solar radiation. Whereas, other types of aerosols, for example black carbon, absorb incoming solar radiation, leading to a positive forcing. As such, anthropogenic emissions of aerosols contribute to both positive and negative radiative forcing. These direct interactions of aerosols with radiation are listed under the aerosol direct effect radiative forcing term in Figure IV.1.

In addition to directly reflecting solar radiation, aerosols cause an additional, indirect negative forcing effect by enhancing cloud albedo, which is a measure of reflectivity or brightness. This occurs because aerosols act as particles around which cloud droplets can form. An increase in the number of aerosol particles leads to a greater number of smaller cloud droplets, which leads to enhanced cloud albedo. Aerosols also influence cloud lifetime and precipitation, but these effects are considered part of the climate response and the IPCC made no central estimates of these indirect forcing effects.

The uncertainties in the radiative forcing estimates associated with the direct and indirect effects of aerosols are better understood now than at the time of the IPCC Third Assessment Report (2001a). Nevertheless, they remain the dominant uncertainty in radiative forcing (IPCC, 2007d). In particular, significant changes in the estimates of the direct radiative forcing due to biomass- burning, nitrate, and mineral dust aerosols have occurred since the Third Assessment Report (2001a). The IPCC (2007d) estimated that the net effect of all aerosols (primarily sulfate, organic carbon, black carbon, nitrate, and dust) is a cooling effect, with a total direct radiative forcing of -0.5 (-0.9 to -0.1) W/m^2 and an additional indirect cloud albedo (i.e., enhanced reflectivity) forcing of -0.7 (-1.8 to -0.3) W/m^2. Black carbon aerosols cause yet another forcing effect by decreasing the surface albedo of snow and ice (+0.1 (0.0 to +0.2) W/m^2).

Land cover changes, linear contrails, and solar irradiance

Three last radiative forcing estimates are listed in Figure IV. 1. Changes in surface albedo due to human-induced land cover changes, such as deforestation, exert a forcing of -0.2 (-0.4 to 0.0) W/m^2. Persistent linear contrails from global aviation contribute a small radiative forcing of +0.0 1 (+0.003 to +0.03) W/m^2. Changes in solar irradiance since 1750 are estimated to have caused a radiative forcing of +0.12 (+0.06 to +0.30) W/m^2.

IV.1.b. Observed temperature changes

Observational evidence from all continents and most oceans shows that many natural systems are being affected by regional climate changes, particularly temperature increases (IPCC, 2007b).

Temperature trends

Numerous lines of evidence robustly lead to the conclusion that the climate system is warming. The IPCC (2007d) stated in its Fourth Assessment Report, "Warming of the climate system is unequivocal, as is now evident from observations of increases in global average air and ocean temperatures, widespread melting of snow and ice, and rising global average sea level." Section IV.2 describes the extent to which changes in global and continental temperature and other climate factors have been observed.

Of the measurable climate properties, air temperature is the most easily measured, directly observable, and geographically consistent indicator of climate change. Data from thousands of worldwide observation sites on land and sea are used to calculate surface temperature, with interpolation (approximation of the values based on discrete points) within areas of the globe where no observational data exists. Although biases may exist in surface temperatures (changes in station exposure and instrumentation over land, or changes in measurement techniques by ships and buoys in the ocean), these biases are *likely* (66 to 90% probability) to be largely random and, therefore, will cancel out over large regions such as the tropics or the entire globe (Karl et al., 2006). Likewise, the heat island effects of urban areas are real, but local. They have been accounted for in the analyses of temperature records and have not biased the large-scale trends (Trenberth et al., 2007).

The IPCC (Trenberth et al., 2007) reported the following trends in global surface temperatures:

- Global mean surface temperatures have risen, especially in the past three decades. The temperature rise over the last 100 years (1906–2005) is 0.74 ± 0.18 °C when estimated by a linear trend (purple line in Figure IV.2). The rate of warming over the last 50 years (0.13 ± 0.03 °C per decade; orange line in Figure IV.2) is almost double that for the past 100 years (0.07 ± 0.02 °C per decade). These conclusions are based on analyses of multiple quality- controlled data sets of *in situ* temperature observations described in Trenberth et al. (2007).

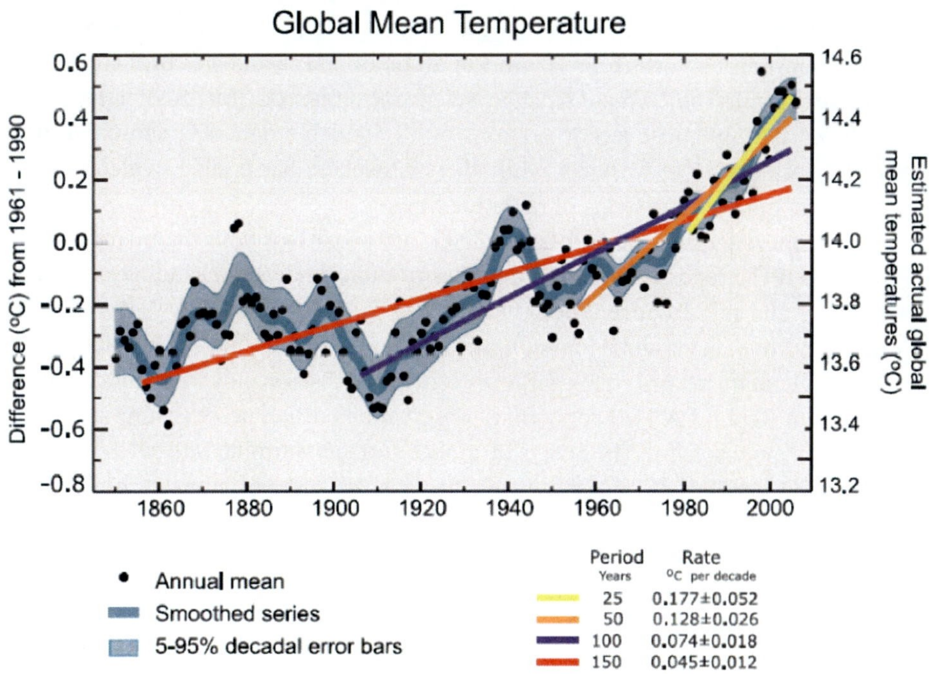

Figure IV.2. Annual global mean temperatures (black dots) with linear fits to the data.
The left-hand axis shows temperature anomalies relative to the 1961 to 1990 average and the right-hand axis shows estimated actual temperatures, both in °C. Linear trends are shown for the last 25 (yellow), 50 (orange), 100 (purple), and 150 years (red). The smooth blue curve shows decadal variations with the decadal 90% error range shown as a pale blue band about that line. The total temperature increase from the period 1850 to 1899 to the period 2001 to 2005 is 0.76 ± 0.19 °C. Source: Solomon et al. (2007).

- The warmest years in the instrumental record of global (ocean and land combined) surface temperatures on record (since the mid-19th century) have mainly occurred in the past 12 years. Including 2007, 7 of the 8 warmest years on record have occurred since 2001 and the 10 warmest years have all occurred since 1995. The warmest two years are 1998 and 2005 (1998 ranks first in one estimate, and 2005 ranks slightly higher in the other two estimates). The years 2002 to 2004 are the third, fourth, and fifth warmest years since 1850. The year 2007 tied for second warmest in the period of instrumental data, behind the record warmth of 2005, according to the National Aeronautics and Space Administration (NASA GISS, 2008a). The global land surface average temperature in 2007 was the warmest on record, according to both NASA and National Oceanic and Atmospheric Administration (NOAA) analyses.
- The global average diurnal temperature range (the average range the temperature spans— from high temperature to low temperature—over the course of a day) has stopped decreasing. Previously (1950s through 1980s), the diurnal temperature range had been decreasing at a rate of approximately 0.1 °C per decade, but the diurnal temperature range has not changed since the 1980s because daytime and nighttime temperatures have increased at approximately the same rate.

Radiosonde (balloon-borne) instruments have been used over the past 50 to 60 years, and satellites have been used over the past 28 years to measure the temperature throughout the troposphere (from Earth's surface to 10–16 km altitude) and in the stratosphere (10–50 km above Earth's surface). The Climate Change Science Program (CCSP) SAP 1.1 (Karl et al., 2006) concluded that the most recent versions of all available data sets show that both the surface and troposphere have warmed, while the stratosphere has cooled, which is in accord with our understanding of the effects of radiative forcing.

The IPCC conclusions (Trenberth et al., 2007) are consistent with the major conclusions of SAP 1.1. The IPCC reported that new analyses of radiosondes and satellite measurements of global lower- and mid-tropospheric temperature show warming rates that are consistent within their respective uncertainties in comparison to one another and are similar to those of the surface temperature record. This largely resolved a discrepancy reported in the IPCC Third Assessment Report (2001a) concerning observations of greater warming at the surface than in the lower atmosphere. The range of global surface warming since 1979 is between 0.16 and 0.18 °C per decade, depending on the data set used. Estimates of tropospheric temperatures measured by satellite show a warming range of 0.12 to 0.19 °C per decade. The agreement between the surface warming rates and those measured by radiosondes for the lower troposphere is not as close as that for satellites; however, the radiosonde record is notably less spatially complete than the surface record, and increasing evidence suggests that it is *very likely* that a number of radiosonde records have a cooling bias, especially in the tropics.

Globally averaged lower-stratospheric temperatures have cooled since 1979, which is one of the expected results from increasing atmospheric concentrations of greenhouse gases. Estimates from adjusted radiosondes, satellites, and reanalyses are in rough agreement, suggesting a lower- stratospheric cooling of between 0.3 and 0.6 °C per decade since 1979.

Like global mean temperatures, U.S. temperatures also warmed during the 20th and into the 21st century. U.S. annual average temperature is now approximately 0.6 °C warmer than at the start of the 20th century, according to NOAA (2008a), with an increased rate of warming over the past 30 years. Temperatures in the period from the 1930s through the 1970s changed little both over the United States and globally. There was a general warming trend from about 1915 to the 1930s.

The NOAA data for the average annual temperature for the contiguous United States shows that 1998 was the warmest year on record, with 2006 and 1934 as the next warmest (within 0.06 °C of the record set in 1998). The average U.S. temperatures in these years fall within the error bars of the analyses (see Figure IV.3). Due to differences in data sets, processing, and analysis, NASA found that 1934 was the warmest on record for the contiguous United States, followed by 1998 and 1921. The fourth warmest year on record in its analysis was 2006. The warmth in 1934 was associated with extreme dryness in the continental interior.

The year 2007 was the 10th warmest year for the contiguous United States since national records began in 1895, according to preliminary data from the NOAA National Climatic Data Center (NOAA, 2008a). Additionally, the past 9 years have all been among the 25 warmest years on record for the contiguous United States, a streak that is unprecedented in the historical record, and the last nine 5-year periods[22] were the warmest 5-year periods in the last 113 years of national records, illustrating the anomalous warmth of the last decade.

U.S. temperature data analyzed by NASA (NASA GISS, 2008b) show similar trends over the past 100 years.

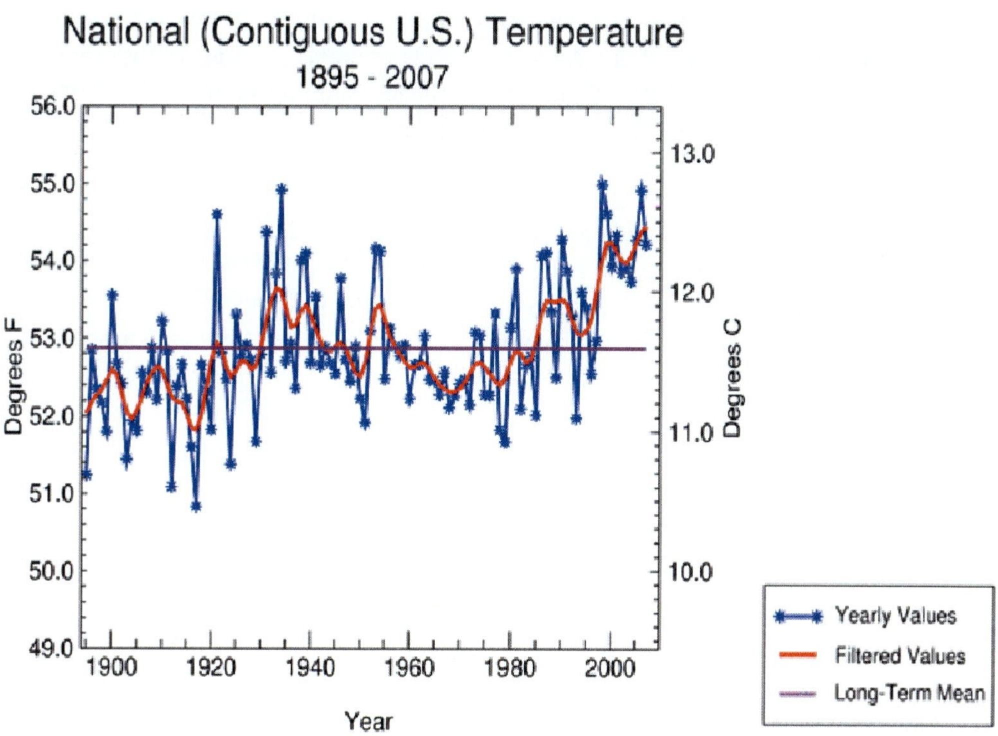

Source: NOAA/NCDC (NOAA, 2008b)

Figure IV.3. Temperature averaged annually over the lower 48 states from 1895 to 2007

Temperature patterns

The rates of temperature change are not the same for different regions around the globe (see Figure IV.4).

- Over the oceans, warming has occurred in both sea surface temperature and nighttime marine air temperature. The average global surface air temperature over the oceans has risen at a rate of 0.13 °C per decade since 1979. Recent warming is strongly evident at all latitudes in sea surface temperatures over each of the oceans (Trenberth et al., 2007).
- Land regions have warmed faster than the oceans—about double the ocean rate after 1979 (more than 0.27 °C per decade), with the greatest warming during winter (December to February) and spring (March to May) in the Northern Hemisphere (Trenberth et al., 2007).
- The warming in the last 30 years is widespread over the globe and is greatest at higher northern latitudes. In the past 100 years, average Arctic temperatures increased at almost twice the global average rate. Arctic temperatures have high decadal variability, and a warm period was also observed from 1925 to 1945 (Trenberth et al., 2007).
- For 1901 to 2005, the rate of warming is statistically significant over most of the world's surface with several exceptions: an area south of Greenland, three smaller regions over the southeastern United States, and parts of Bolivia and the Congo basin. At about 20% of the locations where there is no statistically significant warming, it is *likely* that the lack of significant warming is driven by changes in atmospheric circulation (Karoly and Wu, 2005).
- Enhanced warming relative to the global average has been observed in other locations, notably over the continental interiors of Asia and northwestern North America and over some mid-latitude ocean regions of the Southern Hemisphere as well as southeastern Brazil (see Figure IV.4). Since 1979, warming has been strongest over western North America, northern Europe, and China in winter; over Europe and northern and eastern Asia in spring; over Europe and North Africa in summer; and over northern North America, Greenland, and eastern Asia in autumn (Trenberth et al., 2007).

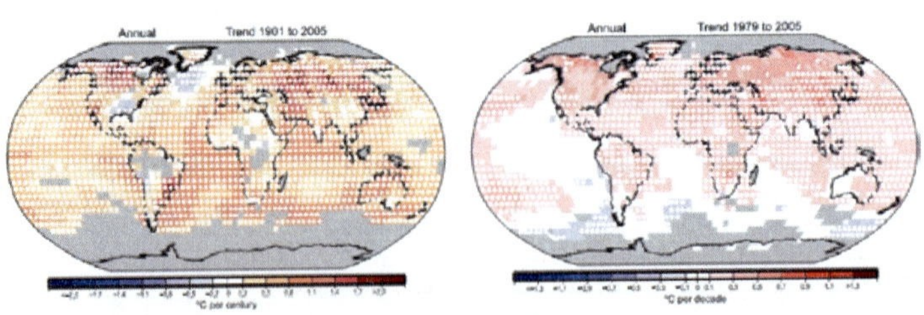

Figure IV.4. Linear trend of annual temperatures for 1901 to 2005 (left; °C per century) and 1979 to 2005 (right; °C per decade). Areas in grey either have insufficient data to produce reliable trends or have slightly negative trends. The minimum number of years needed to calculate a trend value is 66 years for 1901 to 2005 and 18 years for 1979 to 2005. An annual value is available if there are 10 valid monthly temperature anomaly values. The data set used was produced by NCDC from Smith and Reynolds (2005 in Trenberth et al., 2007). Trends significant at the 5% level are indicated by white + marks. Source: Trenberth et al. (2007)

Regional data for North America confirm that warming has occurred throughout most of the United States. The U.S. Historical Climate Network of NOAA's National Climatic Data Center found that for all but 3 of the 11 climate regions, the average temperature increased more than 0.6 °C between 1901 and 2005 (NOAA, 2007b). An area of particularly pronounced warming is Alaska, where the temperature has increased at a rate of 1.8 °C per century. Although the Southeast experienced a very slight cooling trend over the entire period (-.02 °C per century), this region shows warming since 1979. Figure IV.5 shows a map of regional U.S. temperature trends for the period 1901 to 2005, using an analysis of data from the U.S. Historical Climate Network (NOAA Version 1).

The IPCC (Field et al., 2007), describing temperature changes for all of North America in its assessment of regional temperatures, stated that for the period 1955 to 2005, the greatest warming occurred in Alaska and northwestern Canada, with substantial warming in the continental interior and modest warming in the southeastern United States and eastern Canada. Additionally, daily minimum (nighttime) temperatures in North America have warmed more than daily maximum (daytime) temperatures, and spring and winter show the greatest changes in temperature.

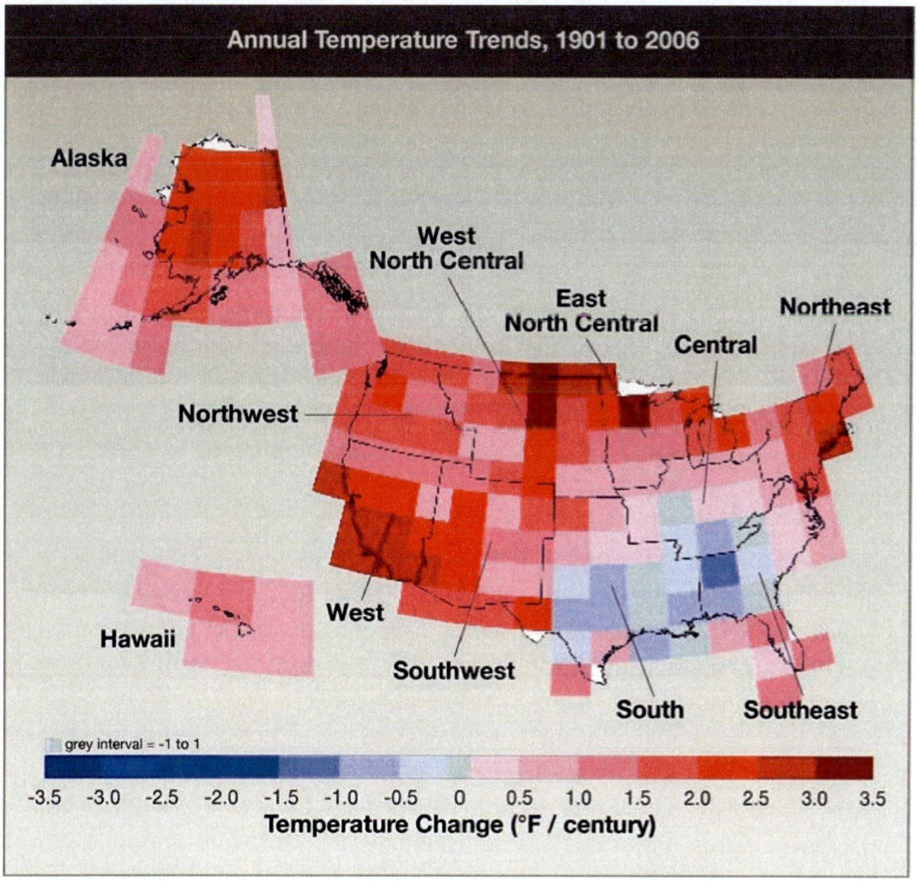

Source: NOAA/NCDC.

Figure IV.5. U.S. trends of annually averaged temperature for the period 1901 to 2006 in °F per century. Data are from the Historical Climate Data Network

Temperature extremes

Climate is defined not only by average temperature and precipitation, but also by the type, frequency, and intensity of extreme events. Observed changes in climate extremes related to temperature, precipitation, tropical cyclones, and sea level apply generally to all parts of the globe, including the United States, although there are some regional and local exceptions due to patterns of natural climate variability (IPCC, 2007d). Substantial changes in the frequency of extreme events are expected from a shift in the mean temperature based on simple statistical reasoning (Solomon et al., 2007). Trends in events that occur at small spatial scales (such as tornadoes, hail, lightning, and dust storms) cannot be determined due to insufficient data (Solomon et al., 2007).

Widespread changes in extreme temperatures have been observed in the last 50 years around the world, and these changes in temperature extremes are consistent with the general warming (Solomon et al., 2007). On a global basis, cold days, cold nights, and frost have generally become less frequent, while hot days, hot nights, and heat waves have become more frequent (IPCC, 2007d). A widespread reduction in the number of frost days in mid-latitude regions, an increase in the number of warm extremes, and a reduction in the number of daily cold extremes are observed in 70 to 75% of the land regions where data are available. The most marked changes are for cold nights (lowest 10%, based on 1961 to 1990), which have become rarer over the 1951 to 2003 period. Warm nights (highest 10%) have become more frequent. Additionally, beginning in the second half of the 20th century, heat waves have increased in duration (Solomon et al., 2007).

North America regional studies consistently show patterns of changes in temperature extremes consistent with a general warming (Trenberth et al., 2007), including intense warming of the lowest daily minimum temperatures over western and central North America (Robeson, 2004 in Trenberth et al., 2007). The IPCC (Trenberth et al., 2007) cautions that although the observed changes of the high- and low-temperature tails of the distributions may be consistent with a simple shift of the entire temperature distribution, the reality is often more complicated.

CCSP SAP 3.3 (Karl et al., 2008) concluded that in the United States there has been a shift towards a warmer climate with an increase in extreme high temperatures and a reduction in extreme low temperatures:

- Since the record hot year of 1998, 6 of the last 10 years (1998–2007) have had annual average temperatures that fall in the hottest 10% of all years on record for the United States. The number of heat waves (extended periods of extremely hot weather) also has been increasing since 1950 (however, the heat waves of the 1 930s remain the most severe in the U.S. historical record).
- There have been fewer unusually cold days in the United States during the last few decades and the last 10 years have had fewer severe cold waves than any other 10-year period in the historical record, which dates back to 1895. There has been a decrease in frost days and a lengthening of the frost-free season over the past century. For the United States as a whole, the average length of the frost-free season over the 1895 to 2000 period increased by almost two weeks.

Surface temperature over the last 2,000 years

The observed warming in the instrumental temperature records (which only began in the late 19th century, when the global network of measurements became large enough for reliable computation of global mean temperatures) can be put in a historical context by examining longer temperature records from sources such as tree rings, corals, ocean and lake sediments, cave deposits, ice cores, boreholes,[23] glaciers, and documentary evidence.

A National Research Council study (NRC, 2006c) assessing the state of scientific efforts to reconstruct surface temperature records for Earth over approximately the last 2,000 years concluded the following:

- Overall, there is a generally consistent picture of temperature trends during the preceding millennium. In particular, relatively warm conditions centered around A.D. 1000 (identified by some as the 'Medieval Warm Period') and a relatively cold period (or 'Little Ice Age') centered around 1700 are evident. (Figure IV.6 shows the aggregate results from several large-scale surface temperature reconstructions.)
- It can be said with a *high level of confidence* (8 out of 10 chance)[24] that global mean surface temperature was higher during the last few decades of the 20th century than during any comparable period during the preceding four centuries. The observed warming in the instrumental record shown in Figure IV.2 supports this conclusion. The IPCC (Solomon et al., 2007) concluded that it is *very likely* that average Northern Hemisphere temperatures during the second half of the 20th century were warmer than any other 50-year period in the last 500 years.
- Less confidence can be placed in large-scale surface temperature reconstructions for the period from A.D. 900 to 1600. Presently available proxy evidence indicates that temperatures at many, but not all, individual locations were higher during the past 25 years than during any period of comparable length since A.D. 900. The uncertainties associated with reconstructing hemispheric mean or global mean temperatures from these data increase substantially backward in time through this period and are not yet fully quantified. On the basis of the NRC's conclusions and other sources of information, the IPCC (Solomon et al., 2007) concluded that average Northern Hemisphere temperatures during the second half of the 20th century were *likely* the warmest in at least the past 1,300 years. It also concluded that a substantial fraction of the decadal temperature variability of the seven centuries preceding 1950 is very likely attributable to natural external forcing (Hegerl et al., 2007).
- Very little confidence can be assigned to statements concerning the hemispheric mean or global mean surface temperature prior to about A.D. 900 because of sparse data coverage and because the uncertainties associated with proxy data and the methods used to analyze and combine them are larger than during more recent time periods.

Reconstructions of Northern Hemisphere surface temperature from several research teams

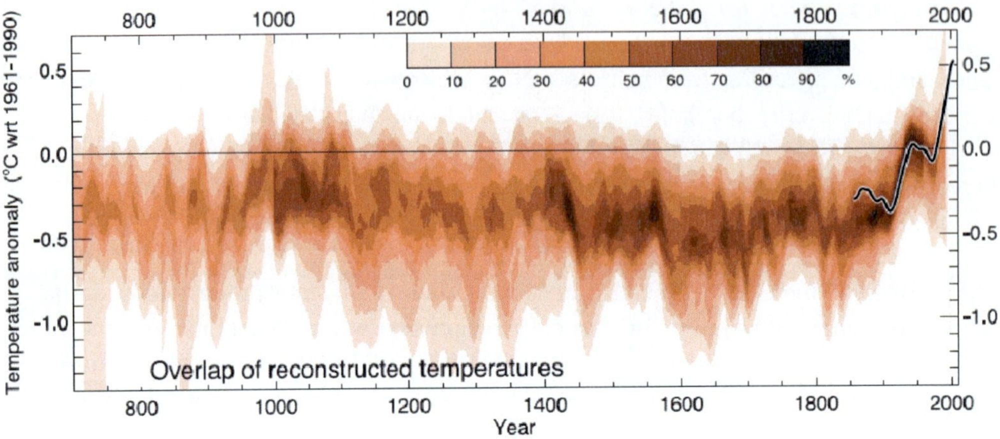

Source: IPCC (2007d).

Figure IV.6. Record of average Northern Hemisphere temperature variation during the last 1,300 years based on reconstructions using multiple climate proxies (e.g., tree ring widths) from 10 different research teams (see IPCC (2007d) for a list of those teams). The darker the shading, the more overlap there is among the various research teams' temperature reconstructions

IV.1.c. Observed precipitation changes

Precipitation trends and patterns

As surface temperatures rise, the evaporation of water vapor increases from oceans and other moist surfaces. Increased evaporation is leading to higher concentrations of water vapor in the atmosphere. Increased atmospheric water vapor tends to produce weather systems that lead to increased precipitation in some areas. At the same time, increased evaporation and evapotranspiration from warming can lead to increased land surface drying and, therefore, increased potential incidence and severity of droughts in other areas.

Observations show that changes are occurring in the amount, intensity, frequency, and type of precipitation. It is noted that precipitation is highly variable spatially and temporally. There is limited data available for large regions of the globe, such that robust long-term trends have not been established for certain regions. A global map of precipitation trends from 1901 to 2005 is provided in Figure IV.7.

Some of the conclusions from the IPCC concerning global precipitation changes are noted below (Trenberth et al., 2007):

- Long-term trends from 1900 to 2005 have shown significant increases in precipitation amount over many large regions, in particular the eastern parts of North and South America, northern Europe, and northern and central Asia.
- The same long-term trends (1900 to 2005) have shown drying in the Sahel, the Mediterranean, southern Africa, and parts of southern Asia.

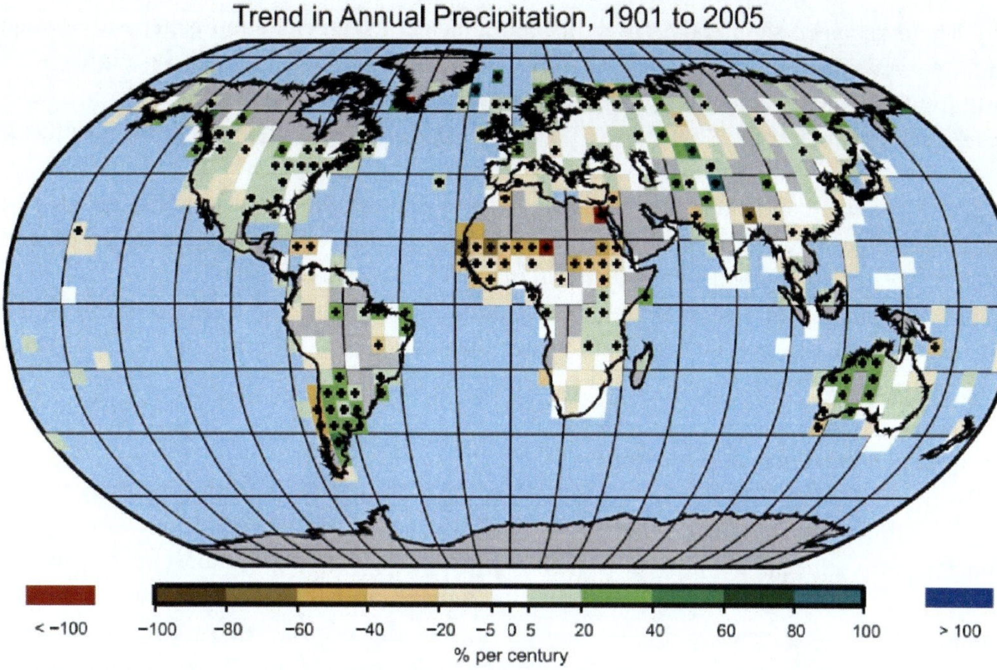

Source: Trenberth et al. (2007; data from NOAA/NCDC).

Figure IV.7. Trends in annual land precipitation amounts for 1901 to 2005 (percent per century). The percentage is based on the means for the 1961 to 1990 period. Areas in grey have insufficient data to produce reliable trends. Trends significant at the 5% level are indicated by black + marks

- More intense and longer droughts have been observed over wider areas since the 1 970s, particularly in the tropics and subtropics. In general terms, drought is a "prolonged absence or marked deficiency of precipitation," a "deficiency of precipitation that results in water shortage for some activity or for some group," or a "period of abnormally dry weather sufficiently prolonged for the lack of precipitation to cause a serious hydrological imbalance" (Heim, 2002 in Trenberth et al., 2007). Increased drying linked with higher temperatures and decreased precipitation has contributed to changes in drought. The regions where droughts have occurred seem to be determined largely by changes in sea surface temperatures, especially in the tropics, through associated changes in the atmospheric circulation and precipitation. Decreased snowpack and snow cover have also been linked to droughts.
- It is *likely* that there have been increases in the number of heavy precipitation events (e.g., 95th percentile events) within many land regions, including regions where there has been a reduction in total precipitation amount. This is consistent with a warming climate and the observations of increasing amounts of water vapor in the atmosphere.
- Rising temperatures have generally resulted in rain rather than snow in locations and seasons where climatological average temperatures for 1961 to 1990 were close to freezing (0 °C), for example the boreal high latitudes.

Over the contiguous United States, annual precipitation totals have increased at an average rate of 6% per century from 1901 to 2005, according to an analysis of data from the NOAA National Climatic Data Center U.S. Historical Climate Network (Version 1; NOAA, n.d.). There has been significant variability in regional U.S. precipitation patterns over time and space. As shown in Figure IV. 8, the greatest increases in precipitation were in the East North Central climate region (12% per century) and the South (11%). The smallest increases in precipitation were in the Southeast (3%), the West North Central (3%) and the Southwest (1%).

Outside of the contiguous United States, Alaska experienced a precipitation increase of about 6% per century (since records began in 1918), and Hawaii experienced a decrease of just over 9% per century (since records begin in 1905).

Despite the overall national trend towards wetter conditions, a severe drought has affected the southwestern United States from 1999 through 2007, and more recently the southeastern United States has experienced severe drought as well (NWS CPC, 2008).

Precipitation extremes and storms

Heavy precipitation events refer to those in the 95th percentile of precipitation events. The IPCC (Solomon et al., 2007) reports that it is *likely* that there have been increases in the number of heavy precipitation events within many land regions, even in those regions where there has been a reduction in total precipitation amount; this is consistent with a warming climate and observed increases in the amount of water vapor in the atmosphere, which have been significant. Increases have also been reported for rarer (1 in 50 year return period) precipitation events, but only a few regions have sufficient data to assess such trends reliably (Trenberth et al., 2007).

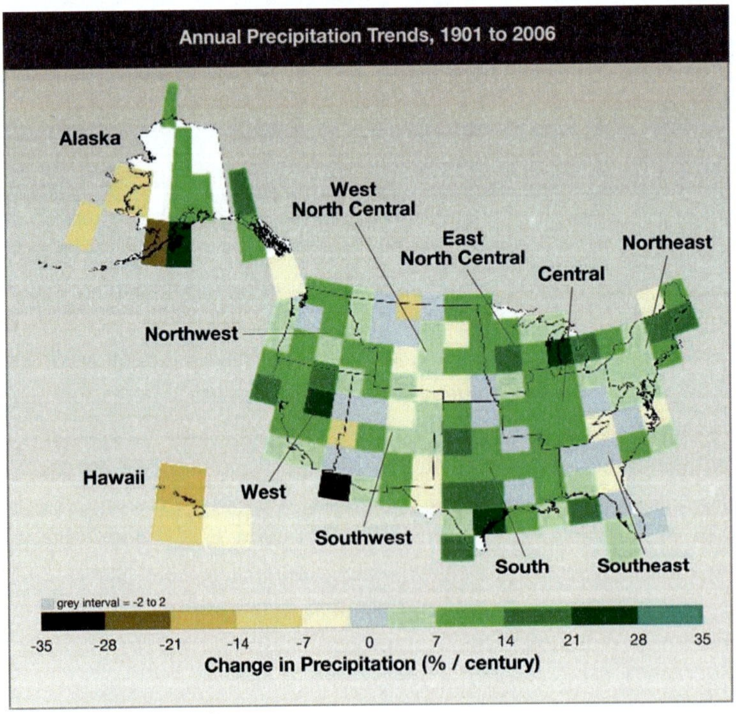

Source: NOAA/NCDC (NOAA, n.d.).

Figure IV.8. U.S. trends in annual precipitation totals for the period 1901 to 2006 in percent per century. Data are from the Historical Climate Data Network

Observations over the contiguous United States show statistically significant increases in heavy precipitation (the heaviest 5%) and very heavy precipitation (the heaviest 1%) of 14% and 20%, respectively, primarily during the last three decades of the 20th century (Kunkel et al., 2003 in Trenberth et al., 2007 and Groisman et al., 2004 in Trenberth et al., 2007). This increase is most apparent over the eastern parts of the country. Some evidence suggests that the relative increase in precipitation extremes is larger than the increase in mean precipitation (Trenberth et al., 2007). CCSP SAP 3.3 (Karl et al., 2008) concluded that very heavy precipitation (the heaviest 1%) in the continental United States increased by 20% over the past century while total precipitation increased by 7%. Additionally, the monsoon season is beginning about 10 days later than usual in Mexico and in general, for the summer monsoon in southwestern North America, there are fewer rain events, but the events are more intense (Karl et al., 2008).

On a global basis, more intense and longer droughts have been observed over wider areas since the 1970s, particularly in the tropics and subtropics (IPCC, 2007d). Increased drying linked with higher temperatures and decreased precipitation has contributed to changes in drought (IPCC, 2007d).

In the western United States, diminishing snow pack and subsequent reductions in soil moisture appear to have contributed to recent drought conditions (Trenberth et al., 2007). This drought has also been attributed to changes in atmospheric circulation associated with warming of the western tropical Pacific and Indian Oceans as well as multi-decadal fluctuations (Hoerling and Kumar, 2003 and McCabe et al., 2004 in Trenberth et al., 2007).

Detection of long-term trends in tropical cyclone activity is complicated by multi-decadal variability and the quality of the tropical cyclone records prior to the start of routine satellite observations in about 1970. Even taking this into account, the IPCC concluded that the observational evidence shows an increase of intense (Category 4 and 5) tropical cyclone activity in the North Atlantic since about 1970, which correlates with increases in tropical sea surface temperatures (Solomon et al., 2007). There is no clear global trend in the total numbers of tropical cyclones since 1970 (IPCC, 2007d). Globally, estimates of the potential destructiveness of hurricanes show a significant upward trend since the mid-1970s, with a trend toward longer lifetimes and greater storm intensity. Such trends are strongly correlated with tropical sea surface temperature (Trenberth et al., 2007). Additionally, there is evidence for an increase in extreme wave height characteristics over the past couple of decades, associated with more frequent and more intense hurricanes (Karl et al., 2008).

A range of models with resolutions ranging from 100 to 20 km predict that future tropical cyclones (typhoons and hurricanes) will become more intense, with larger peak wind speeds and more heavy precipitation associated with ongoing increases in tropical sea surface temperatures. There is less confidence in the projections of tropical cyclone frequency. The apparent increase in the proportion of very intense storms since 1970 in some regions is much larger than simulated by current models for that period (Solomon et al., 2007). According to SAP 3.3 (Karl et al., 2008), rainfall and wind speeds associated with North Atlantic and North Pacific hurricanes are *likely* to increase in response to human-caused warming. Analyses of model simulations suggest that for each 1 °C (1.8 °F) increase in tropical sea surface temperatures, core rainfall rates will increase by 6 to 18% and the surface wind speeds of the strongest hurricanes will increase by about 1 to 8%.

IV.1.d. Observed changes in other physical systems

Of all the observed changes to physical systems assessed by the IPCC (Rosenzweig et al., 2007) for North America (totaling 355), 94% were consistent with changes one would expect with average warming. This section covers changes to the cryosphere (snow and ice), the hydrosphere, and sea level. Changes to the biosphere are covered in Section V.1. The extent to which observed changes discussed here can be attributed to anthropogenic greenhouse gas emissions is discussed in Section IV.2.b.

Cryosphere (snow and ice)

The cryosphere is the 'frozen' component of the climate system, including sea ice, glaciers, snow cover, and permafrost. According to the IPCC (Lemke et al., 2007) and the Arctic Climate Impacts Assessment (ACIA, 2005), the following global-scale physical changes have been observed:

- Overall, snow cover has decreased in most regions, especially in spring and summer. Satellite data from 1966 to the present show decreases in Northern Hemisphere snow cover in every month except November and December, with a stepwise drop of 5% in the annual mean in the late 1980s. In the Southern Hemisphere, the few long records or proxies mostly show either decreases or no changes in the past 40 years or more.
- For North America, snow-covered area increased in November, December, and January from 1915 to the present due to increases in precipitation. Over western North America, however, snow cover decreased during the latter half of the 20th century, especially during the spring (Groisman et al., 2004 in Lemke et al., 2007). In northern Alaska, observations show shifts toward earlier melt by about eight days since the mid-1960s (Stone et al., 2002 in Lemke et al., 2007).
- Freeze-up and breakup dates for river and lake ice have high spatial variability and trends in these dates are generally toward later and earlier occurrences, respectively. However, the average of available data for the Northern Hemisphere spanning the past 150 years shows the freeze-up date for river and lake ice has occurred later at a rate of 5.8 ± 1.6 days per century and the breakup date has occurred earlier at a rate of 6.5 ± 1.2 days per century.
- Annual average Arctic sea ice extent has decreased by 2.7 ± 0.6% per decade, based on satellite data since 1978. Larger decreases have been observed in summer (7.4 ± 2.4% per decade). Figure IV.9 shows the extent of the sea ice loss between 1979 and 2005. Since 2005, Arctic sea ice extent has declined even further. According to the National Snow and Ice Data Center the average sea ice extent for September 2007 was 4.28 million square kilometers, the lowest September on record, shattering the previous record for the month, set in 2005, by 23 percent.[25] In contrast, Antarctic sea ice extent shows no statistically significant average trends (IPCC, 2007d).
- The average sea ice thickness in the central Arctic has *very likely* decreased by up to 1 m from 1987 to 1997, based upon submarine-derived data. Model-based reconstructions support this, suggesting an Arctic-wide reduction of 0.6 to 0.9 m over the same period.

- Mountain glaciers have declined on average in both hemispheres. The strongest mass losses of mountain glaciers (per unit area) have been observed in Patagonia, Alaska, the northwest United States, and southwest Canada. Because of the corresponding large areas, the biggest contributions to sea level rise came from glaciers in Alaska, the Arctic, and the Asian high mountains.
- Since the 1980s, temperatures at the top of the Arctic permafrost layer have generally increased by up to 3 °C. In Alaska, the permafrost base has been thawing at a rate of up to 0.04 m per year since 1992; on the Tibetan Plateau, the rate of permafrost base thawing has been 0.02 m per year since the 1 960s. The maximum area covered by seasonally frozen ground has decreased by 7% in the Northern Hemisphere since 1900, with a decrease in spring of up to 15%.

The Arctic Climate Impact Assessment (ACIA, 2004) found that the area of the Greenland Ice Sheet that experiences some melting increased about 16% from 1979 to 2002, with 2002 experiencing the largest affected area in the record since 1979 (subsequent years since 2004 have seen even more extensive melting).

There are additional effects related to changes in the cryosphere. First, melting of highly reflective snow and ice reveals darker land and ocean surfaces, increasing absorption of the Sun's heat and further warming the planet. Additionally, increases in glacial melt and river runoff add freshwater to the ocean, raising global sea level. There is also the potential positive feedback (in which a result of climate warming in turn causes further climate warming) from the possible release of CH_4 trapped within subsea permafrost. However, available observations do not permit an assessment of changes that might have occurred with respect to this (Lemke et al., 2007).

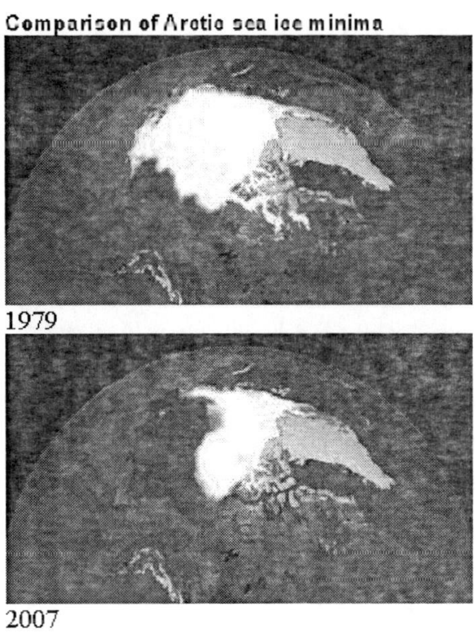

Source: NASA, updated from ACIA (2004).

Figure IV.9. These two images, constructed from satellite data, compare Arctic sea ice concentrations in September of 1979 and 2007

Hydrosphere

The term 'hydrosphere' refers to the component of the climate system consisting of liquid surface and subterranean water, such as rivers, lakes, and underground water. The IPCC (Rosenzweig et al., 2007) summarized several observed changes in these features:

- Documented trends in severe droughts and heavy rains show that hydrologic conditions are becoming more intense in some regions, with indications of intensified droughts in drier regions and evidence for areas of increasing wetness, in particular the Northern Hemisphere high latitudes and equatorial regions. Globally, very dry areas (areas with a Palmer Drought Severity Index (PD SI) rating of less than or equal to -3.0) have more than doubled since the 1970s due to a combination of El Niño–Southern Oscillation events and surface warming (Dai et al., 2004 in Trenberth et al., 2007). Very wet areas (PDSI greater than or equal to +3.0) declined by about 5% since the 1970s. The major contributing factor in this decline was precipitation during the early 1 980s with temperature more important thereafter.
- Over the last century, increasing runoff and streamflow that can be related to climate change has been observed in many regions, particularly in basins fed by glaciers, permafrost, and snowmelt. Evidence shows that average runoff of Arctic rivers in Eurasia has increased, and this increasing runoff has been at least partly correlated with climate warming. There is also evidence of earlier spring snowmelt and increased winter base flow in North America and Eurasia due to enhanced seasonal snowmelt associated with climate warming.
- In the Arctic, lakes forming and then disappearing in permafrost have been observed. Freshwater lakes and rivers are experiencing increased water temperatures and changes in water chemistry, including changes in water acidity and mineral concentrations. Surface and deep lake waters are warming. With this warming, periods of thermal stability are advancing and lengthening. In some cases, this is associated with physical and chemical changes, such as increases in salinity and suspended solids, and a decrease in nutrient content.

Changes in river discharge, as well as in droughts and heavy rains, in some regions indicate that hydrologic conditions have become more intense. However, significant trends in floods and in evaporation and evapotranspiration have not been detected globally. Some local trends in reduced groundwater and lake levels have been reported. However, studies have been unable to separate the effects of variations in temperature and precipitation from the effects of human interventions, such as groundwater management, land use change, and reservoir construction (Rosenzweig et al., 2007).

The IPCC (Rosenzweig et al., 2007) indicates surface water temperatures have warmed by 0.2 to 2 °C in lakes and rivers in North America since the 1960s. Additionally, there is evidence for an earlier occurrence of spring peak river flows and an increase in winter base flow in basins with important seasonal snow cover in North America.

CCSP SAP 4.3 notes that the current hydrologic observing system was not designed for purposes of detecting climate change or its effects on water resources and that many of the data are fragmented, poorly integrated, and in many cases unable to meet the predictive challenges of a rapidly changing climate (Lettenmaier et al., 2008). Despite this, CCSP

(Lettenmaier et al., 2008) does list a number of conclusions related to water resources in the United States:

- During the second half of the 20th century, most of the United States experienced increases in precipitation and streamflow, due to a combination of decadal-scale climate variability and long-term change.
- During the second half of the 20th century, most of the United States experienced decreases in drought severity and duration. However, there is some indication of increased drought severity and duration in the western and southwestern United States, which may have resulted from increased actual evaporation dominating the trend toward increased soil wetness. On a longer time scale, paleoclimatic reconstructions of droughts show that much more severe droughts have occurred over the last 2,000 years than those that have been observed in the instrumental record, notably, the Dust Bowl drought of the 1930s and extensive drought in the 1950s.
- Across the western United States, there is a trend toward reduced mountain snowpack and earlier spring snowmelt runoff peaks, which is *very likely* due to long-term warming, with potential influence from decadal-scale variability (see discussion in Section V.4 about the impacts of these changes).
- As the climate warms, stream temperatures are *likely* to increase, with effects on aquatic ecosystems. There is some evidence that temperatures have increased in some western U.S. streams, although a comprehensive analysis has yet to be conducted. Temperature changes will be most evident during low flow periods, when they are of greatest concern.

Sea level

There is strong evidence that global sea level gradually rose in the 20th century, after a period of little change between A.D. 0 and A.D. 1900, and is currently rising at an increased rate (IPCC, 2007a).

According to the IPCC (Bindoff et al., 2007):

- There is *high confidence* that the rate of sea level rise increased between the mid-19th and mid-20th centuries.
- The average rate of sea level rise measured by tide gauges from 1961 to 2003 was 1.8 ± 0.5 mm per year. For the entire 20th century, the average rate was 1.7 ± 0.5 mm per year.
- There is considerable decadal variability in the rate of global mean sea level rise. The global average rate of sea level rise measured by satellite altimetry during 1993 to 2003 was 3.1 ± 0.7 mm per year, significantly higher than the 20th century average rate. Coastal tide gauge measurements confirm this observation. It is unclear whether the faster rate for 1993 to 2003 is a reflection of short-term variability or an increase in the longer-term trend.

Two major processes lead to changes in global mean sea level on decadal and longer time scales: 1) thermal expansion and 2) the exchange of water between oceans and other reservoirs, including glaciers and ice caps, ice sheets, and other land water reservoirs. The IPCC (Bindoff et al., 2007) concluded:

- Overall, for the period 1961 to 2003, the full magnitude of the observed sea level rise was not satisfactorily explained by the available data sets, where thermal expansion contributed about one-quarter of the observed sea level rise and melting of land ice accounted for less than half of the observed sea level rise. The observing system is much better for recent years (1993 to 2003). In this period, thermal expansion and melting of land ice each account for about half of the observed sea level rise, although there is some uncertainty in the estimates.
- On average, over the period from 1961 to 2003, global ocean temperature from the surface to a depth of 700 m rose by 0.10 °C, contributing an average of 0.4 ± 0.1 mm per year to sea level rise. For the period 1993 to 2003, thermal expansion contributed about 1.6 ± 0.5 mm per year, reflecting a higher rate of warming for this period relative to 1961 to 2003.
- On average, over the period from 1961 to 2003, melting of land ice accounted for approximately 0.7 ± 0.5 mm per year of sea level rise (Lemke et al., 2007). The total contribution from melting ice to sea level change between 1993 and 2003 was 1.2 ± 0.4 mm per year. The rate increased over the 1993 to 2003 period, primarily due to increasing losses from mountain glaciers and ice caps, increasing surface melt on the Greenland Ice Sheet, and faster flow of parts of the Greenland and Antarctic Ice Sheets (Lemke et al., 2007).

Sea level rise is highly non-uniform around the world. In some regions, rates of rise have been as much as several times the global mean, while sea level is falling in other regions. This non- uniformity is driven by thermal expansion and exchanges of water between oceans and other reservoirs, as well as changes in ocean circulation or atmospheric pressure, and geologic processes (Bindoff et al., 2007). According to the IPCC (Bindoff et al., 2007), satellite measurements for the period 1993 to 2003 provide unambiguous evidence of regional variability of sea level change. The largest sea level rise since 1992 has taken place in the western Pacific and eastern Indian Oceans, while nearly all of the Atlantic Ocean shows sea level rise during the past decade with the rate of rise reaching a maximum (over 2 mm per year) in a band running east–northeast from the U.S. East Coast. Sea level in the eastern Pacific and western Indian Oceans has been falling.

U.S. sea level data obtained from the Permanent Service for Mean Sea Level of the Proudman Oceanographic Laboratory[26] reach at least as far back as the early 20th century. These data show that along most of the U.S. Atlantic and Gulf Coasts, sea level has been rising 2.0 to 3.0 mm per year. The rate of sea level change varies from a rise of about 10 mm per year along the Louisiana Coast (due to land sinking) to a drop of tens of millimeters per decade (a few inches per decade) in parts of Alaska (because land is rising). Records from the coast of California indicate that sea levels have risen almost 18 cm during the past century (California Energy Commission, 2006b). Local sea levels can actually be falling in some cases (for example, the Pacific Northwest coast) if the land level is rising more than the sea level (Gamble et al., 2008).

The IPCC (Rosenzweig et al., 2007) concluded that along the U.S. East Coast, 75% of the shoreline that is removed from the influence of spits, tidal inlets, and engineering structures is eroding, which is probably due to sea level rise. It also cites studies reporting losses in coastal wetlands observed in Louisiana, the mid-Atlantic region, and in parts of New England and New York, in spite of recent protective environmental regulations.

CCSP SAP 4.1 (Titus et al., 2008) reported a number of effects of sea level rise relevant to the United States. Approximately one-sixth of the Nation's land close to sea level is in the mid- Atlantic, and as such, the report focuses on the mid-Atlantic states:

- Sea level rise is *virtually certain* (>99% probability) to cause some areas of dry land to become inundated. Approximately 900 to 2,100 km^2 (350 to 800 mi^2) of dry land, half of which is in North Carolina, would be flooded during spring high tides if sea level rises 50 cm (20 in), assuming no shore protection measures are taken.
- Nationally, it is *very likely* that erosion will increase in response to sea level rise, especially in sandy shore environments that comprise all of the mid-Atlantic coast. Within the mid- Atlantic region, it is *virtually certain* (>99% probability) that coastal headlands, spits, and barrier islands will also erode in response to future sea level rise.
- It is *virtually certain* that the Nation's tidal wetlands already experiencing submergence by sea level rise and associated high rates of loss (e.g., Mississippi River Delta in Louisiana, Blackwater River marshes in Maryland) will continue to lose area under the influence of future accelerated rates of sea level rise and changes in other climate and environmental drivers.
- It is *very unlikely* (1 to 10% probability) that there will be a net increase in tidally influenced wetland area on a national scale over the next 100 years, given current wetland loss rates and the few occurrences of new tidal wetland expansion (e.g., Atchafalaya Delta in Louisiana).
- Depending on local conditions, habitat may be lost or migrate inland in response to sea level rise. Loss of tidal marshes would seriously threaten coastal ecosystems, causing fish and birds to move or produce less offspring. Many estuarine beaches may also be lost, threatening species such as the terrapin and horseshoe crab.
- There is evidence for an increase in incidence of extreme high sea level globally and in the United States since 1975 based upon an analysis of 99th percentiles of hourly sea level at 141 stations around the world (Bindoff et al., 2007). A global analysis is not feasible over the entire 20th century because longer records back in time are limited in space and under- sampled in time (Solomon et al., 2007).

IV.2. Attribution of Observed Climate Change to Human Activities at the Global and Continental Scale

The previous section (Section IV. 1) describes observations presented primarily in the recent IPCC Fourth Assessment Report (2007d) of global- and continental-scale temperature increases, changes in other climate variables and physical and biological systems, and the radiative forcing caused by anthropogenic versus natural factors. This section addresses the extent to which observed climate change at the global and continental or national scale can be attributed to *global* anthropogenic emissions of greenhouse gases.

Computer-based climate models are the primary tools used for simulating the probable patterns of climate system response to natural and anthropogenic forcings, such as greenhouse gases, aerosols, and solar intensity (see Section IV. 1 .a). Confidence in these models comes from their foundation in accepted physical principles and from their ability to reproduce observed features of current climate and past climate changes (Meehl et al., 2007) (see Section IV.3.b). Studies to attribute causes of climate change use well-tested models to evaluate whether observed changes are consistent with quantitative responses to different forcings and are not consistent with alternative physically plausible explanations. Attribution studies often use 'fingerprint methods' to rigorously test for the presence of a climatic response to greenhouse gas increases and other forcings, and to evaluate whether similarities between observed patterns and model-simulated fingerprints could have occurred by chance. These methods account for feedbacks enhancing or decreasing the response to individual external influences compared to model-simulated responses.

Difficulties remain in attributing temperature changes at smaller than continental scales and over time scales of less than 50 years. Attribution results at these scales have, with limited exceptions, not been established. Averaging over smaller regions reduces the natural variability less than does averaging over large regions, making it more diffi cult to distinguish between changes expected from external forcing and variability. In addition, temperature changes associated with some modes of variability are poorly simulated by models in some regions and seasons. Furthermore, the small-scale details of external forcing and the response simulated by models are less credible than large-scale features (IPCC, 2007a).

IV.2.a. Attribution of observed climate change to anthropogenic emissions and land use / land cover

There is clear evidence that human-induced warming of the climate system is widespread (Hegerl et al., 2007). Studies to detect climate change and attribute its causes using patterns of observed changes in temperature show clear evidence of human influences on the climate system (Karl et al., 2006). Discernible human influences extend to additional aspects of climate, including ocean warming, continental-average temperatures, temperature extremes, and wind patterns (Hegerl et al., 2007).

Temperature

The IPCC has strengthened its statement on the linkage between greenhouse gases and temperatures with each assessment report over the past two decades. The IPCC's First Assessment Report in 1990 contained little observational evidence of a detectable anthropogenic influence on climate (IPCC, 1990). In its Second Assessment Report in 1995, the IPCC stated the balance of evidence suggests a discernible human influence on the climate of the 20th century (IPCC, 1996). The Third Assessment Report in 2001 concluded that most of the observed warming over the last 50 years is *likely* to have been due to the increase in greenhouse gas concentrations (IPCC, 2001). The conclusion in the IPCC's 2007 Fourth Assessment Report (2007d) is the strongest yet:

> Most of the observed increase in global average temperatures since the mid-20th century is *very likely* due to the observed increase in anthropogenic greenhouse gas concentrations.

The IPCC (Hegerl et al., 2007) cites a number of factors for the increased confidence in the greenhouse gas contribution to the observed warming:

- an expanded and improved range of observations, allowing attribution of warming to be more fully addressed jointly with other changes in the climate system;
- analyses of paleoclimate data that have increased confidence in the role of external influences on climate;
- improvements in the simulation of many aspects of present mean climate and its variability on seasonal to inter-decadal time scales;
- more detailed representations of processes related to aerosol and other forcings in models;
- simulations of 20th-century climate change that use many more models and much more complete inclusion of anthropogenic and natural forcings; and
- multi-model ensembles that increase confidence in attribution results by providing an improved representation of model uncertainty.

The results from climate model simulations evaluated by the IPCC (Figure IV. 10) show that natural forcings alone cannot explain the observed warming for the globe, the global land, and global ocean and that the observed warming can only be reproduced with models that contain both natural and anthropogenic forcings. The IPCC (Hegerl et al., 2007) cites additional new evidence linking warming to increasing concentrations of greenhouse gases:

- Attribution studies have established anthropogenic contributions to changes in surface and atmospheric temperatures (Section IV. 1 .b), changes in temperatures in the upper several hundred meters of the ocean, and changes in sea level pressure (Section IV.1.d).
- Coupled climate models used to predict future climate have been used to reproduce key features of past climates using boundary conditions and radiative forcing for those periods.

It is *very unlikely*, according to the IPCC (Hegerl et al., 2007), that the global pattern of warming observed during the past half century is due to only known natural external causes (solar activity and volcanoes) since the warming occurred in both the atmosphere and ocean and took place when natural external forcing factors would *likely* have produced cooling. The IPCC (Hegerl et al., 2007) also concluded that greenhouse gas forcing alone would *likely* have resulted in warming greater than observed if there had not been an offsetting cooling effect from aerosols and natural forcings during the past half century.

In addition to evidence from surface temperatures, evidence has also accumulated for an anthropogenic influence through the vertical profile of the atmosphere. Fingerprint studies,[27] rather than only linear trend comparisons,[28] have identified greenhouse gas and sulfate aerosol signals in observed surface temperature records, a stratospheric ozone depletion signal in stratospheric temperatures, and the combined effects of these forcing agents in the vertical structure of atmospheric temperature changes (Karl et al., 2006). There is a potentially important inconsistency in the tropics, where most observational data sets show more warming at the surface than in the troposphere. In contrast, almost all model simulations have larger warming aloft than at the surface. The issue is still under investigation and may be due to errors in the observations (Karl et al., 2006).

It is *likely* that there has been a substantial anthropogenic contribution to surface temperature increases since the middle of the 20th century on every continent except Antarctica, which has insufficient observational coverage to make an assessment (Hegerl et al., 2007). Figure IV.11 shows that the observed temperatures over the last century for North America can only be reproduced using model simulations containing both natural and anthropogenic forcings, and not just natural forcings. Attribution studies show that it is *likely* that there has been a substantial human contribution to the surface temperature increase in North America. Fingerprint attribution studies detect a significant anthropogenic contribution to North American temperature change that can be separated from the response to natural forcings and from internal climate variability. As noted above, attribution of temperature changes on scales smaller than continental and for time scales of less than 50 years, with limited exceptions, has not yet been established (Hegerl et al., 2007).

In addition to average temperatures, anthropogenic forcings have also *likely* influenced extremes in temperature (Hegerl et al., 2007). Many indicators of climate extremes, including the annual numbers of frost days, warm and cold days, and warm and cold nights, show changes that are consistent with warming (Hegerl et al., 2007). An anthropogenic influence has been detected in some of these indices, and there is evidence that anthropogenic forcing may have substantially increased the risk of extremely warm summer conditions regionally. The 2003 European heat wave is an example of such temperature extremes (Hegerl et al., 2007).

Additional climate variables

There is evidence of anthropogenic influence in other parts of the climate system. The IPCC noted the following global examples (Hegerl et al., 2007):

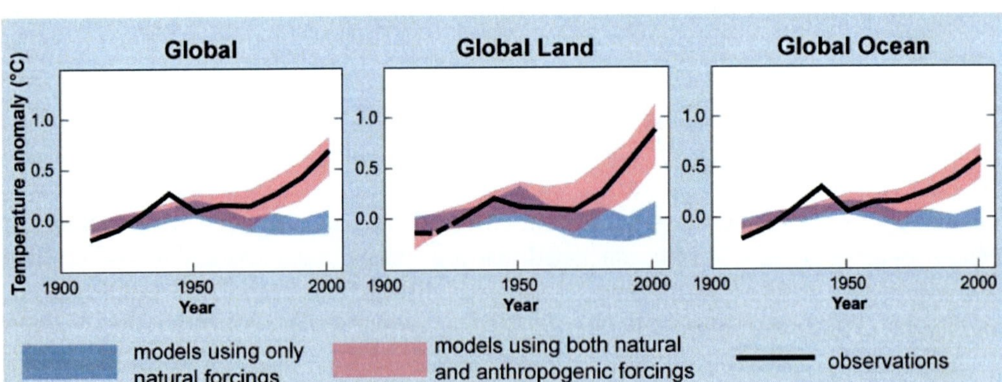

Source: IPCC (2007d).

Figure IV.10. Comparison of observed global-scale changes in surface temperature with results simulated by climate models using natural and anthropogenic forcings. Decadal averages of observations are shown for the period 1906 to 2005 (black line) plotted against the center of the decade and relative to the corresponding average for 1901 to 1950. Lines are dashed where spatial coverage is less than 50%. Blue shaded bands show the 5 to 95% range for 19 simulations from five climate models using only the natural forcings due to solar activity and volcanoes. Red shaded bands show the 5 to 95% range for 58 simulations from 14 climate models using both natural and anthropogenic forcings.

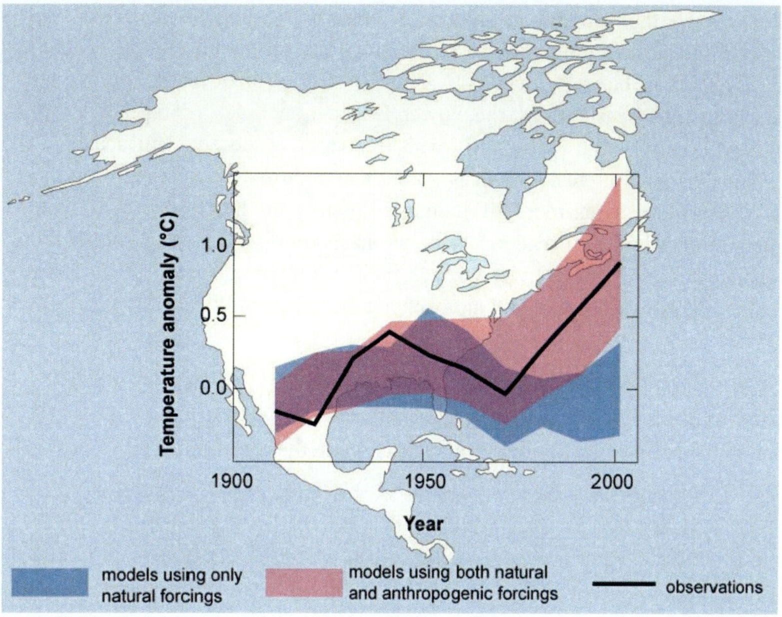

Source: Hegerl et al. (2007).

Figure IV.11. Comparison of observed North American changes in surface temperature with results simulated by climate models using natural and anthropogenic forcings. Decadal averages of observations are shown for the period 1906 to 2005 (black line) plotted against the center of the decade and relative to the corresponding average for 1901 to 1950. Lines are dashed where spatial coverage is less than 50%. Blue shaded bands show the 5 to 95% range for 19 simulations from five climate models using only the natural forcings due to solar activity and volcanoes. Red shaded bands show the 5 to 95% range for 58 simulations from 14 climate models using both natural and anthropogenic forcings

- Recent decreases in Arctic sea ice extent (Section IV. 1 .d) are likely to have been contributed to by anthropogenic forcings.
- Human activity is likely related to trends over recent decades in the Northern and Southern Annular Modes,29 which correspond to sea level pressure reductions over the poles. These modes affect storm tracks, winds, and temperature patterns in both hemispheres. Models reproduce the sign of the Northern Annular Mode trend, but the simulated response is smaller than observed. A realistic trend in the Southern Annular Mode is simulated by models that include both greenhouse gas and stratospheric ozone changes, suggesting this trend is due to human influence on global sea level pressure patterns.
- The latitudinal pattern of change in land precipitation over the 20th century appears to be consistent with the anticipated response to anthropogenic forcing. The same is true for observed increases in heavy precipitation.
- It is more likely than not (greater than 50% probability) that anthropogenic influence has contributed to increases in the frequency of the most intense tropical cyclones (i.e., category 4 and 5).

Regarding this latter issue, the CCSP synthesis and assessment product on climate extremes (Karl et al., 2008), concluded the following regarding hurricane activity specifically in the North Atlantic:

> It is *very likely* that the human-induced increase in greenhouse gases has contributed to the increase in sea surface temperatures in the hurricane formation regions. There is a strong statistical connection between tropical Atlantic sea surface temperatures and Atlantic hurricane activity as measured by the Power Dissipation Index (combines storm intensity, frequency, and duration) and particularly with frequency on decadal timescales over the past 50 years. This evidence suggests a substantial human contribution to recent hurricane activity. However, a confident assessment of human influence on hurricanes will require further studies using models and observations, with emphasis on distinguishing natural from human-induced changes in hurricane activity through their influence on factors such as historical sea surface temperatures, wind shear, and atmospheric vertical stability.

IV.2.b. Attribution of observed changes in physical and biological systems

As detailed in Section IV.2.a, warming in the last 50 years is *very likely* the result of the accumulation of anthropogenic greenhouse gases in the atmosphere. Observed changes in physical systems (e.g., melting glaciers) and biological systems and species (e.g., geographic shift of species) that are shown to change as a result of recent warming can be attributed to anthropogenic greenhouse gas forcing. This section includes the observed changes in physical and biological systems in North America and in other parts of the world, which the IPCC (2007b) concluded are discernibly influenced by recent warming: "Observational evidence from all continents and most oceans shows that many natural systems are being affected by regional climate changes, particularly temperature increases." Furthermore, the IPCC (2007e) concluded that: "Anthropogenic warming over the last three decades has likely had a discernable influence at the global scale on observed changes in many physical and biological systems."

In order to make robust conclusions about the role of anthropogenic climate change in affecting biological and physical systems, climate variability and non-climate drivers (e.g., land use change and habitat fragmentation) all need to be considered. The IPCC (Rosenzweig et al., 2007) concluded from a number of joint attribution studies that the consistency of observed significant changes in physical and biological systems and observed significant warming across the globe *likely* cannot be explained entirely due to natural variability or other confounding non-climate factors.

The physical systems undergoing significant change include the cryosphere (snow and ice systems), hydrologic systems, water resources, coastal zones, and the oceans. A number of effects are reported with *high confidence* by the IPCC (Rosenzweig et al., 2007):

- ground instability in mountain and permafrost regions;
- shorter travel season for vehicles over frozen roads in the Arctic;
- enlargement and increase of glacial lakes in mountain regions and destabilization of moraines damming these lakes;
- changes in Arctic flora and fauna including the sea ice biomes and predators higher in the food chain;
- limitations on mountain sports in lower-elevation alpine areas; and
- changes in indigenous livelihoods in the Arctic.

CCSP SAP 4.3 (Backlund et al., 2008) discusses the role of climate change in determining the levels of snow water equivalent in the Pacific Northwest of North America. It concluded that observed snow water equivalent changes in the western United States for the period 1950 to 2000 can be attributed to temperature rather than precipitation changes (Mote, 2003 and Hamlet et al., 2005 in Backlund et al., 2008).

Regarding biological systems, the IPCC (Rosenzweig et al., 2007) concluded with *very high confidence* that the overwhelming majority of studies of regional climate effects on terrestrial species reveal trends consistent with warming, including:

- poleward and elevational range shifts of flora and fauna;
- earlier onset of spring events, migration, and lengthening of the growing season;
- changes in abundance of certain species, including limited evidence of a few local disappearances; and
- changes in community composition.

See Section V for further description of these changes in the physical and biological systems.

Human system responses to climate change are more difficult to identify and isolate due to the larger role of non-climate factors, such as management practices in agriculture and forestry, and adaptation responses to protect human health against adverse climatic conditions.

CCSP SAP 4.3 (Backlund et al., 2008) concluded that, due to a lack of appropriate observations, it is currently not possible to understand the extent to which climate change is damaging or enhancing the goods and services that ecosystems provide or how additional climate change would affect the future delivery of such goods and services. It concluded:

> [I]n principle it is possible to evaluate both damages and benefits from climate change for any region and/or ecosystem, but such studies will need to be very carefully designed and implemented in order to yield defensible quantification. Until then, we will need to continue to rely on a combination of existing observations made for other purposes and on model output to construct such estimates (Backlund et al., 2008).

IV.3. Projected Future Greenhouse Gas Concentrations and Climate Change

According to the IPCC (2007d), "continued greenhouse gas emissions at or above current rates would cause further warming and induce many changes in the global climate system during the 21st century that would *very likely* be larger than those observed during the 20th century." This section describes future greenhouse gas emission scenarios, the associated changes in atmospheric concentrations and radiative forcing, and the resultant changes in temperature, precipitation, and sea level at global and U.S. scales. An important note is that all future greenhouse gas emission scenarios described in this section assume no new explicit greenhouse gas mitigation policies (neither in the United States nor in other countries) beyond those that were already enacted at the time the scenarios were developed. Future risks and impacts associated with the climate change projections are addressed in Part V.

IV.3.a. Global emission scenarios and associated changes in concentrations and radiative forcing

Greenhouse gas emissions

Section IV. 1.a described a number of different greenhouse gases, aerosols, and other factors that cause radiative forcing changes and thus contribute to climate change. This section discusses the range of published global reference (or baseline) future emission projections, which are primarily drawn from the IPCC *Special Report on Emission Scenarios* (SRES) (IPCC, 2000). The SRES developed a range of long-term global reference scenarios for the major greenhouse gases emitted by human activities and for some aerosols out to the year 2100. The main drivers of emissions are population, economic growth, technological change, and land use activities including deforestation.

The scenarios described in this report were developed and published elsewhere using integrated assessment models. These models integrate socioeconomic and technological determinants of the emissions of greenhouse gases with models of the natural science of Earth system response, including the atmosphere, oceans, and terrestrial biosphere. The detailed underlying assumptions (including final and primary energy by major fuel types) across all scenarios, and across all modeling teams that produced the scenarios, can be found in the SRES (IPCC, 2000). Box IV. 1 provides background information on the different SRES emission scenarios. The IPCC SRES scenarios assume no explicit greenhouse gas mitigation policies beyond those currently enacted and, and having been initiated prior to the inception of the Kyoto Protocal, do not explicitly account for it. None of the scenarios are viewed as more likely than the others, and they were not constructed using formal uncertainty analysis.

Figure IV.12 presents the global SRES projections for the two most significant anthropogenic greenhouse gases: CO_2 and CH_4. The ranges of greenhouse gas emissions in the scenarios widen over time as a result of uncertainties in the underlying drivers, where CO_2 emissions are primarily from the burning of fossil fuels and CH_4 emissions are primarily from biogenic sources such as wetlands, ruminant animals, rice agriculture, and biomass burning. The IPCC (2000) SRES report did not assign probabilities or likelihood to the scenarios, and stated that there is no single most probable, central, or best-guess scenario, either with respect to the scenarios or to the underlying scenario literature. As a result, the IPCC (2000) recommended using the range of scenarios, with their variety of underlying assumptions, in analyses.

Despite the range in future emission scenarios, the majority of all reference-case scenarios project an increase in greenhouse gas emissions across the century and show that CO_2 remains the dominant greenhouse gas over the course of the 21st century. Total *cumulative* (1990 to 2100) CO_2 emissions across the SRES scenarios range from 2,826 gigatons of CO_2 ($GtCO_2$) (or 770 GtC) to approximately 9,322 $GtCO_2$ (or 2,540 GtC).[30] The 90th percentile range in 2100 for *annual* CO_2 emissions is 17 to 135 $GtCO_2$, compared to about 34 $GtCO_2$ emitted in 2000.

Since the IPCC SRES (IPCC, 2000), new scenarios in the literature have emerged, but these generally have a range similar to the emission scenarios from the IPCC SRES. These newer scenarios have used lower values for some drivers of emissions, notably population projections. However, for those studies incorporating these new lower population projections, changes in other drivers, such as economic growth, offset these differences, resulting in little change in overall emission levels (IPCC, 2007c).

CCSP developed several reference-case emission scenarios as part of SAP 2.1 (Clarke et al., 2007). Global projections of CO_2 emissions from the burning of fossil fuels and industrial sources from these three reference cases are shown in Figure IV. 13 for comparison. Box IV.2 provides background information on the reference case scenarios developed by CCSP. The CCSP scenarios, because they were developed more recently than the IPCC SRES scenarios, do account for the implementation of the Kyoto Protocol by participating countries along with the U.S. greenhouse gas intensity goal, but no explicit greenhouse gas mitigation policies beyond the Kyoto Protocol. The emissions in 2100 are approximately 88 $GtCO_2$ (24 GtC). This level of emissions is above the post-SRES IPCC median of 60 $GtCO_2$ (16 GtC), but well within the 90th percentile of the IPCC range. The recent Raupach et al. (2007) study suggests that the actual emissions trajectory since 2000 was close to the highest-emissiof scenario, A1FI. More importantly, the emissions growth rate since 2000 exceeded that for the A1FI scenario.

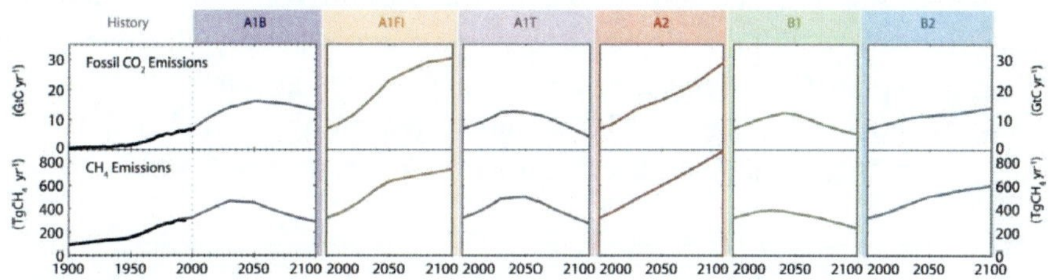

Source: Meehl et al. (2007).

Figure IV.12. Observed and projected global CO_2 and CH_4 emissions for the IPCC Special Report on Emission Scenarios. Projected fossil CO_2 and CH_4 emissions for six illustrative non-mitigation emission scenarios. Historical emissions (black lines) are shown for fossil and industrial CO_2 and for CH_4

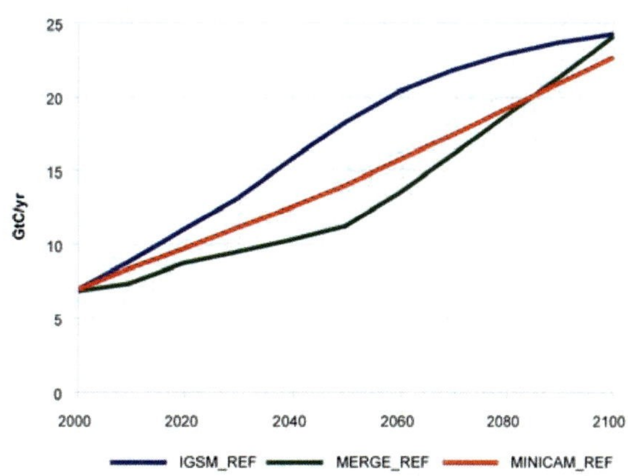

Source: Clarke et al. (2007).

Figure IV.13. Projected global emissions of CO_2 from fossil fuels and industrial sources across CCSP reference scenarios. Global emissions of CO_2 from fossil fuel combustion and other industrial sources, mainly cement production, increase over the century in all three reference scenarios. By 2100, emissions reach 22.5 to 24.0 GtC per year

Box IV.1. IPCC Reference Case Emission Scenarios from the *Special Report on Emission Scenarios*

A1. The A1 storyline and scenario family describes a future world of very rapid economic growth, global population that peaks in mid-century and declines thereafter, and the rapid introduction of new and more efficient technologies. Major underlying themes are convergence among regions, capacity building, and increased cultural and social interactions, with a substantial reduction in regional differences in per capita income. The A1 scenario family develops into three groups that describe alternative directions of technological change in the energy system. The three A1 groups are distinguished by their technological emphasis: fossil intensive (**A1FI**), non-fossil energy sources (**A1T**), or a balance across all sources (**A1B**) (where balanced is defined as not relying too heavily on one particular energy source, on the assumption that similar improvement rates apply to all energy supply and end use technologies).

A2. The A2 storyline and scenario family describes a very heterogeneous world. The underlying theme is self-reliance and preservation of local identities. Fertility patterns across regions converge very slowly, which results in continuously increasing population. Economic development is primarily regionally oriented and per capita economic growth and technological change more fragmented and slower than other storylines.

B1. The B1 storyline and scenario family describes a convergent world with the same global population, which peaks in mid-century and declines thereafter, as in the A1 storyline, but with rapid change in economic structures toward a service and information economy, with reductions in material intensity and the introduction of clean and resource-efficient technologies. The emphasis is on global solutions to economic, social, and environmental sustainability, including improved equity, but without additional climate initiatives.

B2. The B2 storyline and scenario family describes a world in which the emphasis is on local solutions to economic, social, and environmental sustainability. It is a world with continuously increasing global population, at a rate lower than A2, intermediate levels of economic development, and less rapid and more diverse technological change than in the B1 and A1 storylines. While the scenario is also oriented toward environmental protection and social equity, it focuses on local and regional levels.

An illustrative scenario was chosen for each of the six scenario groups: A1B, A1FI, A1T, A2, B1, and B2. All should be considered equally sound.

The SRES scenarios do not include additional climate initiatives, which means that no scenarios are included that explicitly assume implementation of the United Nations Framework Convention on Climate Change or the emissionstargets of the Kyoto Protocol.

BOX IV.2. CCSP REFERENCE CASE EMISSION SCENARIOS FROM SYNTHESIS AND ASSESSMENT PRODUCT 2.1

The three integrated assessment models used in CCSP SAP 2.1 are:

1. The Integrated Global Systems Model (IGSM) of the Massachusetts Institute of Technology's (MIT) Joint Program on the Science and Policy of Global Change.
2. The Model for Evaluating the Regional and Global Effects (MERGE) of greenhouse gas reduction policies developed jointly at Stanford University and the Electric Power Research Institute.
3. The MiniCAM Model of the Joint Global Change Research Institute, a partnership between the Pacific Northwest National Laboratory and the University of Maryland. The MiniCAM model was also used to generate IPCC SRES scenarios.

Each modeling group produced a reference scenario under the assumption that no climate policies are imposed beyond current commitments, namely the 2008–2012 first period of the Kyoto Protocol and the U.S. goal of reducing greenhouse gas emissions per unit of its gross domestic product by 18% by 2012. The resulting reference cases are not predictions or best- judgment forecasts but scenarios designed to provide clearly defined points of departure for studying the implications of alternative stabilization goals. The modeling teams used model input assumptions they considered 'meaningful and plausible'. The resulting scenarios provide insights into how the world might evolve without additional efforts to constrain greenhouse gas emissions, given various assumptions about principal drivers of these emissions such as population increase, economic growth, land and labor productivity growth, technological options, and resource endowments.

In addition to the emission scenarios described above, 18 models were involved in the 21st Study of Stanford University's Energy Modeling Forum on multi-gas mitigation (EMF-21; see de la Chesnaye and Weyant, 2006 in Fisher et al., 2007) to produce a range of emission scenarios that are representative of the literature. Figure IV. 14 illustrates reference-case emission projections for CO_2, CH_4, N_2O, and the fluorinated gases in aggregate (HFCs, PFCs, and SF_6 or 'F-gases') from EMF-21 and from the IPCC SRES (IPCC, 2000). The broad ranges of EMF-21 emissions projections in Figure IV. 14, especially for N_2O and the F-gases, illustrate the uncertainties in projecting these future emissions, which are generally consistent with the range found in the SRES.

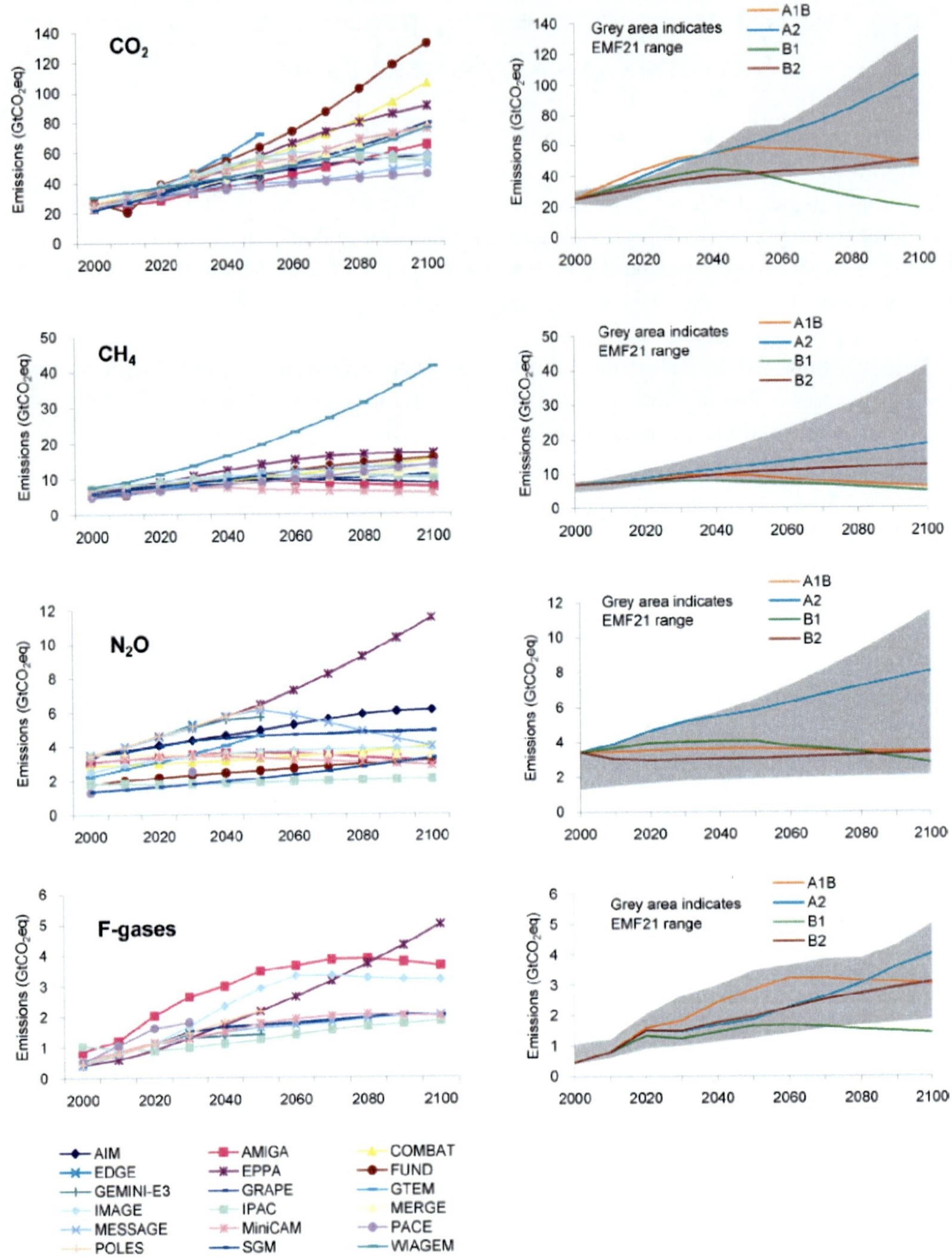

Source: Fisher et al. (2007).

Figure IV.14. EMF-21 and IPCC global emissions projections for CO_2, CH_4, N_2O, and the fluorinated gases. Development of baseline emissions in EMF-21 scenarios developed by a number of different modeling teams (left) and a comparison between EMF-21 and SRES scenarios (right) from de la Chesnaye and Weyant (2006); see also Van Vuuren et al. (2006)

For comparison, Figure IV.15 provides the global CH4 and N2O projections from the three CCSP reference-case scenarios (CCSP, 2007).

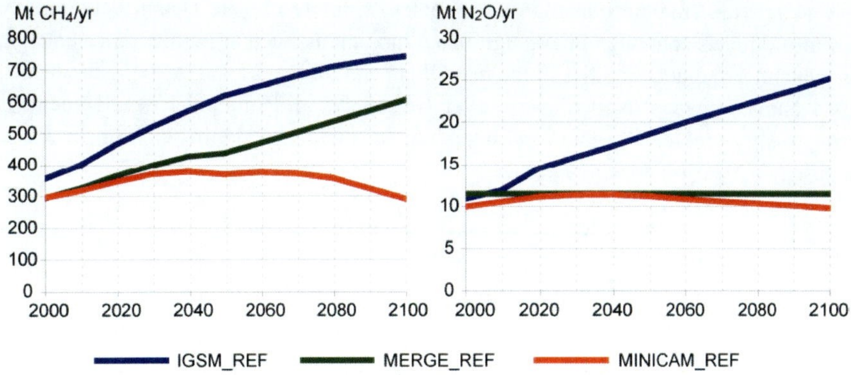

Source: Clarke et al. (2007).

Figure IV.15. Projected global CH_4 and N_2O emissions across three CCSP reference scenarios. Global anthropogenic emissions of CH_4 and N_2O vary widely among the reference scenarios. There is uncertainty in year-2000 CH_4 emissions, with MIT's IGSM reference scenario ascribing more of the emissions to human activity and less to natural sources. Differences in scenarios reflect, to a large extent, different assumptions about whether current emissions rates will be reduced significantly for other reasons, for example, whether higher natural gas prices will stimulate capture of CH_4 for use as a fuel

Future concentration and radiative forcing changes

Figure IV.16 shows the latest IPCC projected increases in atmospheric CO_2, CH_4, and N_2O concentrations for the SRES scenarios, and Figure IV. 17 shows the associated radiative forcing for CO_2 from these scenarios. In general, reference concentrations of CO_2 and other greenhouse gases are projected to increase. This is true even for those scenarios where annual emissions near the end of the century are assumed to be lower than current annual emissions, due to the long atmospheric lifetimes of these gases. The CCSP scenarios show a similar picture of how atmospheric concentrations of the main greenhouse gases and total radiative forcing change over time.

CO_2 is projected to be the largest contributor to total radiative forcing in all periods and the radiative forcing associated with CO_2 is projected to be the fastest growing. The radiative forcing associated with the non-CO_2 greenhouse gases is still significant and growing over time.

IV.3.b. Projected changes in physical components of the climate system

In order to project future changes in the climate system, including temperature, precipitation, and sea level at global and regional scales, academic institutions and government-supported research laboratories in the United States and other countries have developed a number of computer models that simulate the Earth system and that are based on the various emissions scenarios described in Section IV.3.a. The IPCC helps coordinate modeling efforts to facilitate comparisons across models, and synthesizes results published by several modeling teams. According to the IPCC (Meehl et al., 2007):

> [C]onfidence in models comes from their physical basis, and their skill in representing observed climate and past climate changes. Models have proven to be extremely important tools for simulating and understanding climate, and there is considerable confidence that they

are able to provide credible quantitative estimates of future climate change, particularly at larger scales. Models continue to have significant limitations, such as in their representation of clouds, which lead to uncertainties in the magnitude and timing, as well as regional details, of predicted climate change. Nevertheless, over several decades of model development, they have consistently provided a robust and unambiguous picture of significant climate warming in response to increasing greenhouse gases.

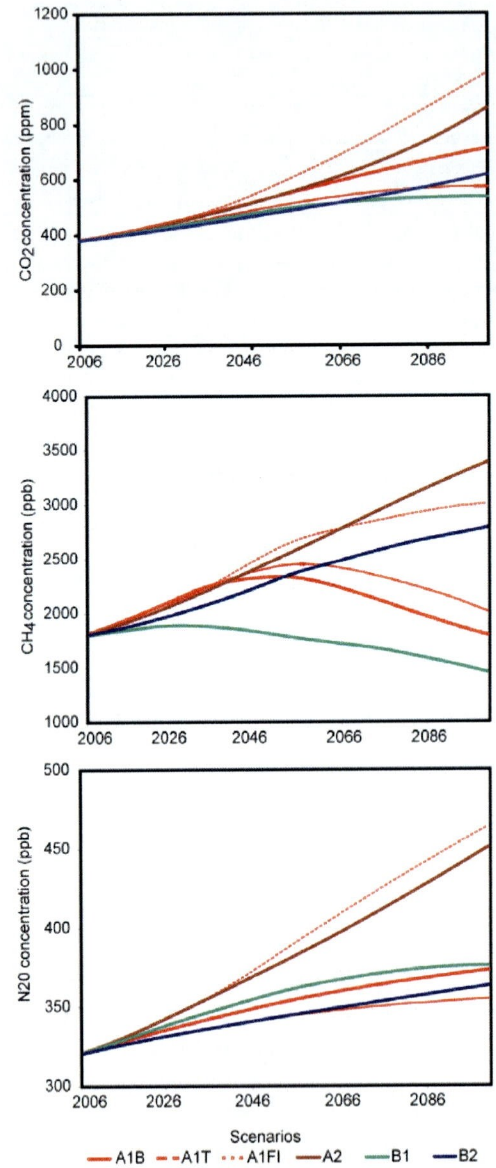

Source: Meehl et al. (2007).

Figure IV.16. Projected global CO_2, CH_4, and N_2O concentrations for the IPCC *Special Report on Emission Scenarios*. Projected fossil CO_2, CH_4, and N_2O concentrations for six illustrative SRES non-mitigation emission scenarios as produced by a simple climate model tuned to 19 atmosphere–ocean general circulation models

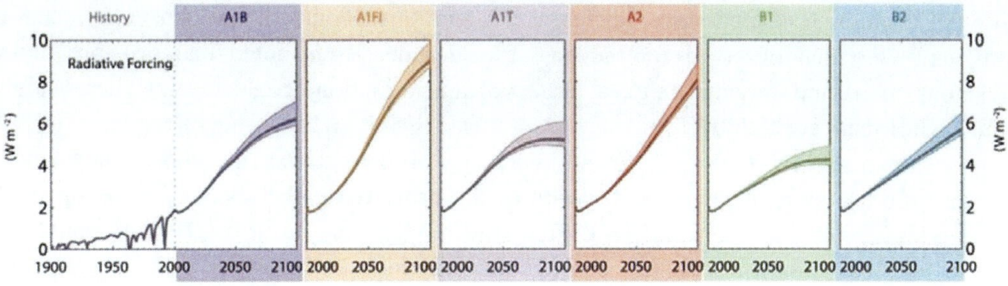

Source: Meehl et al. (2007).

Figure IV.17. Projected radiative forcing from CO_2 for the IPCC *Special Report on Emission Scenarios*. Projected radiative forcing from CO_2 for six illustrative SRES non-mitigation emission scenarios as produced by a simple climate model tuned to 19 atmosphere–ocean general circulation models. The lighter shaded areas depict the change in this uncertainty range if carbon cycle feedbacks are assumed to be lower or higher than in the medium setting

Confidence in changes projected by global models decreases at smaller spatial and temporal scales because many important small-scale processes, in particular clouds, cannot be represented explicitly in models, and so must be included in approximate form as they interact with larger- scale features (Randall et al., 2007). On these scales, natural climate variability is relatively larger. In addition, uncertainties in local forcings and feedbacks also make it difficult to estimate the contribution of greenhouse gas increases to observed small-scale changes (IPCC, 2007d). Some of the most challenging aspects of understanding and projecting regional climate changes (e.g., at the scale of states and counties) relate to possible changes in the circulation of the atmosphere and oceans, and their patterns of variability (Christensen et al., 2007). Nonetheless, the IPCC (2007d) concluded that recent advances in regional-scale modeling have led to higher confidence in projected patterns of warming and other regional-scale features.

Temperature projections

This section describes the temperature projections from a suite of climate models included in the IPCC's Fourth Assessment. Compared to earlier assessments, the latest IPCC assessment uses a larger number of simulations available from a broader range of models to project future climate (IPCC, 2007d). All of the simulations assessed by the IPCC during its Fourth Assessment project that the climate system will warm, for the full range of emissions scenarios.

The IPCC found that increased confidence has been gained in short-term projections by comparing previous IPCC projections of climate change to recent observations. For the next two decades, a global average warming of about 0.2 °C per decade is projected for a range of SRES emission scenarios (IPCC, 2007d; see Figure IV. 18). Committed warming from greenhouse gases already released into the environment accounts for a global average increase of about 0.1 °C per decade during the next two decades because of the time it takes for the climate system, particularly the oceans, to reach equilibrium. This is shown in Figure IV. 18 for the 'Year 2000 Constant Concentrations' scenario that shows projected increases in temperature assuming that the concentrations of all greenhouse gases and aerosols are kept constant at year 2000 levels. Through about 2030, the warming rate is affected little by different scenario assumptions or different assumptions about climate sensitivity, and is consistent with that observed for the past few decades. It is noted that possible future variations in natural forcings (e.g., a large volcanic eruption) could change these values somewhat (Meehl et al., 2007).

As shown in Figure IV. 18, by mid-century (2046 to 2065), the choice of scenario becomes more important for the magnitude of the projected global average warming, with average values of 1.3, 1.8, and 1.7 °C from the models for scenarios B1 (low emissions growth), A1B (medium emissions growth), and A2 (high emissions growth), respectively (Meehl et al., 2007). About one-third of that warming is projected to be due to climate change that is already committed to, again shown in the 'Year 2000 Constant Concentrations' scenario.

By the end of the century (2090 to 2099), projected global average surface warming varies significantly by emissions scenario. The full suite of SRES scenarios provide warming for 2090 to 2099 relative to 1980 to 1999 with a range of 1.8 to 4.0 °C with an uncertainty range of 1.1 to 6.4 °C. The multi-model average warming and associated uncertainty ranges for 2090 to 2099 (relative to 1980 to 1999) for each scenario, as illustrated in Figure IV. 18, are as follows:

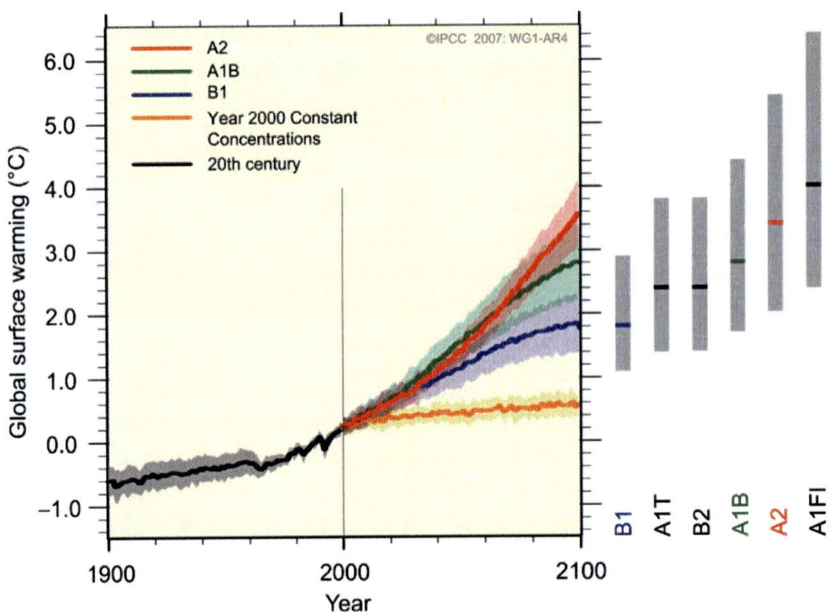

Source: IPCC (2007d).

Figure IV.18. Multi-model averages and assessed ranges for surface warming. Solid lines are multi-model global averages of surface warming (relative to 1980 to 1999) for the scenarios A2, A1B, and B 1, shown as continuations of the 20th-century simulations. Shading denotes the ±1 standard deviation range of individual model annual averages. The orange line is for the experiment where concentrations were held constant at year 2000 values. The grey bars at right indicate the best estimate (solid line within each bar) and the likely range assessed for the six SRES marker scenarios. The assessment of the best estimate and likely ranges in the grey bars includes results from the atmosphere–ocean general circulation models shown in the left part of the figure, as well as results from a hierarchy of independent models and observational constraints

Table IV.3.b.1.

Scenario	Projected Global Average Surface Warming by End of Century Relative to ~1990		Uncertainty Range	
	°C	°F	°C	°F
B1	1.8	3.2	1.1 to 2.9	2.0 to 5.2
A1T	2.4	4.3	1.4 to 3.8	2.5 to 5.7
B2	2.4	4.3	1.4 to 3.8	2.5 to 5.7
A1B	2.8	5.0	1.7 to 4.4	3.1 to 7.9
A2	3.4	6.1	2.0 to 5.4	3.6 to 9.7
A1FI	4.0	7.2	2.4 to 6.4	4.3 to 11.5

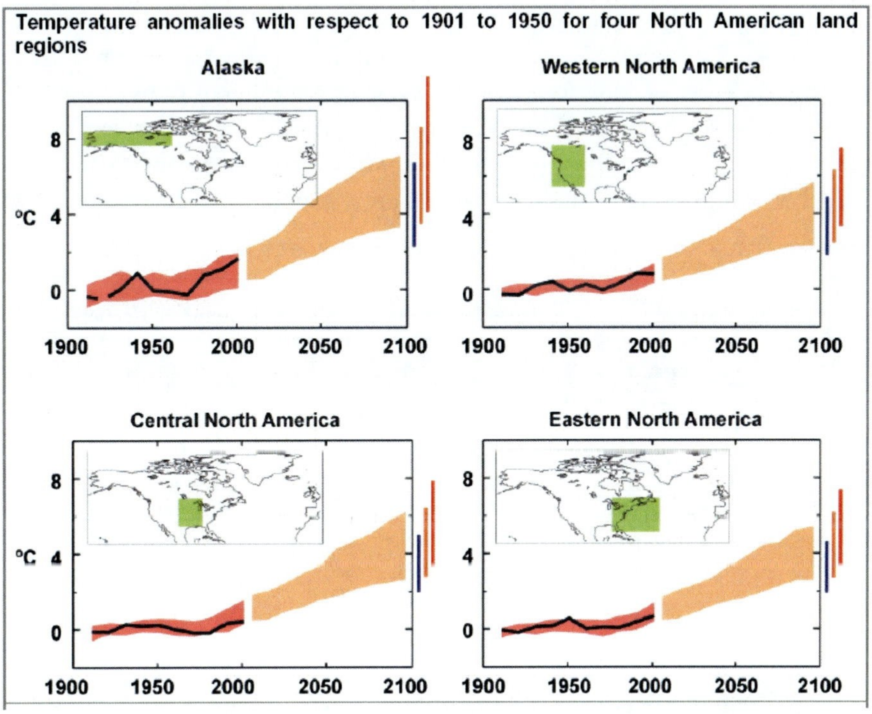

Source: Christensen et al. (2007).

Figure IV.19. Temperature anomalies with respect to 1901 to 1950 for four North American land regions for 1906 to 2005 (black line) and as simulated (red envelope) by multi-model dataset (MMD) models incorporating known forcings, and as projected for 2001 to 2100 by MMD models for the A1B scenario (orange envelope). The bars at the end of the orange envelope represent the range of projected changes for 2091 to 2100 for the B1 scenario (blue), the A1B scenario (orange), and the A2 scenario (red). The black line is dashed where observations are present for less than 50% of the area in the decade concerned

The wide range in these estimates reflects the different assumptions about future anthropogenic emissions of greenhouse gases and aerosols in the various scenarios considered by the IPCC and the differing climate sensitivities of the various climate models used in the simulations (NRC, 2001a).

Geographical patterns of projected warming show greatest temperature increases over land (roughly twice the global average temperature increase) and at high northern latitudes, and less warming over the southern oceans and North Atlantic. These projections are consistent with trends in observations (see Section IV.1.b) during the latter part of the 20th century (Meehl et al., 2007). According to the Arctic Climate Impact Assessment (ACIA, 2004), the Arctic is predicted to warm faster than the global average, with projected Arctic warming of 4 to 7 °C over the next 100 years; this is consistent with the IPCC's conclusions (Christensen et al., 2007).

The IPCC found that all of North America is *very likely* not only to warm during this century (as shown in Figures IV. 19 and IV.20), but to warm more than the global mean warming in most areas (Christensen et al., 2007). The average warming in the United States is projected by nearly all the models used in the IPCC assessment to exceed 2 °C, with 5 out of 21 models projecting average warming in excess of 4 °C. The largest warming is projected to occur in winter over northern parts of Alaska, reaching 10 °C for scenario A1B (moderate emissions scenario relative to other IPCC scenarios) in the northernmost parts (as shown in Figure IV.20), due to the positive feedback from a shorter season of snow cover. In western, central, and eastern regions of North America, the projected warming has less seasonal variation and is not as large, especially near the coast, which is consistent with less warming over the oceans. An increase in surface evaporation is expected to accompany the projected widespread increase in temperature.

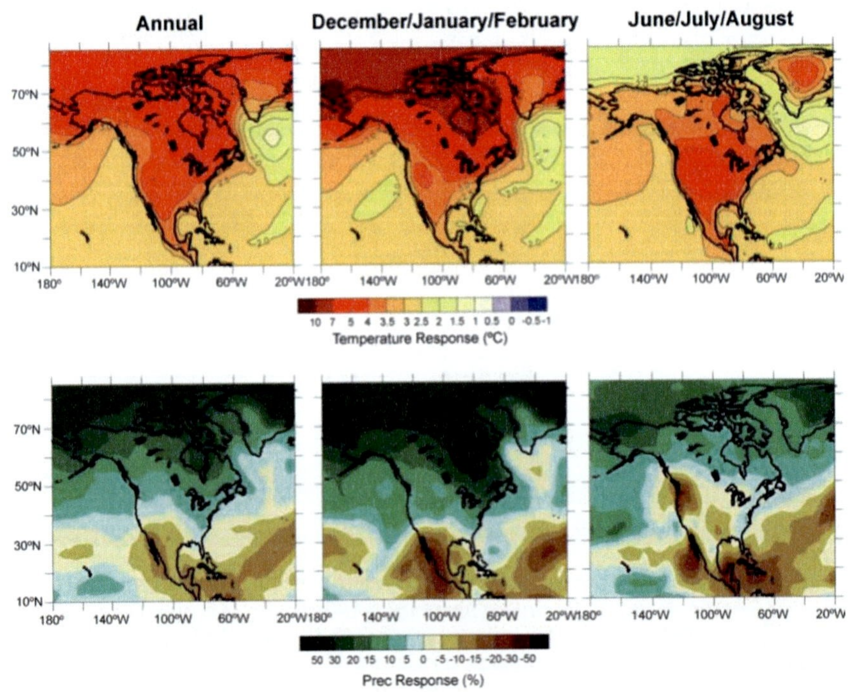

Source: Christensen et al. (2007).

Figure IV.20. Projected temperature and precipitation changes over North America from the MMDA1B simulations. Top row: annual mean, December-January-February, and June-July-August temperature between 1980 to 1999 and 2080 to 2099, averaged over 21 models. Bottom row: same as top, but for fractional change in precipitation

Precipitation projections

Overall, future model projections show that global mean precipitation increases with the warming of the climate (Meehl et al., 2007), but with substantial spatial and seasonal variations. The IPCC concluded that increases in the amount of precipitation are *very likely* at high latitudes, while decreases are *likely* in most subtropical land regions, continuing observed patterns in recent trends in observations.

A widespread increase in annual precipitation is projected by the IPCC over most of the North American continent except the southern and southwestern part of the United States and over Mexico. This is largely consistent with trends in recent decades (as described in Section IV. 1 .c) (Christensen et al., 2007). Figure IV.20 shows that the largest increases are projected over the northern tier of states (including Alaska) associated with a poleward shift in storm tracks, where the magnitude of precipitation increase is projected to be greatest in autumn, whereas winter precipitation is projected to increase by the largest fraction relative to its present amount. In western North America, modest changes in annual mean precipitation are projected, but the majority of models indicate an increase in winter and a decrease in summer. Models show greater consensus on winter increases to the north and on summer decreases to the south. These decreases are consistent with enhanced subsidence and flow of drier air masses in the southwestern United States and northern Mexico. Accordingly, some models project drying in the southwestern United States, and more than 90% of the models project drying in northern and particularly western Mexico. In the northeastern United States, annual mean precipitation is *very likely* to increase. It is uncertain whether summer precipitation will increase or decrease over large portions of the interior United States. On the windward slopes of the mountains in the west, precipitation increases are *likely* to be enhanced due to orographic lifting (ascent of air from a lower elevation to a higher elevation as it moves over rising terrain).

It is *likely* that droughts will continue to be exacerbated by earlier and possibly lower spring snowmelt runoff in the mountainous West, which results in less water available in late summer (Karl et al., 2008). A recent analysis (Milly et al., 2005 in Karl et al., 2008) shows that several atmosphere–ocean general circulation models (AOGCMs) project greatly reduced annual water availability over the southwestern United States and northern Mexico in the future.

Sea level rise projections

Along with increases in global ocean temperatures, the IPCC (2007d) projects sea level rise of between 0.18 and 0.59 m by the end of the century (2090 to 2099) relative to the base period (1980 to 1999). These numbers represent the lowest and highest projections of the 5 to 95% ranges for all SRES scenarios considered collectively and include neither uncertainty in carbon cycle feedbacks nor rapid dynamical changes in ice sheet flow. In all scenarios, the average rate of sea level rise during the 21st century is *very likely* to exceed the 1961 to 2003 average rate (1.8 ± 0.5 mm per year). Again, there are committed effects, where even if greenhouse gas concentrations were to be stabilized, sea level rise would continue for centuries due to the time scales associated with climate processes and feedbacks (IPCC, 2007d). Thermal expansion of ocean water is the primary driver of sea level rise, contributing 70 to 75% of the central estimate for the rise in sea level for all scenarios (Meehl et al., 2007).

Over the coming century, glaciers, ice caps, and the Greenland Ice Sheet are also projected to add to sea level. The Antarctic Ice Sheet, however, is projected by general circulation models to receive increased snowfall without experiencing substantial surface melting, thus gaining mass and reducing sea level rise, according to the IPCC (Meehl et al., 2007). There is potential uncertainty in these projections for the Antarctic Ice Sheet associated with dynamical processes related to ice flow that are not included in current models, but are suggested by recent observations. This could increase the vulnerability of the ice sheets to warming, increasing future sea level rise. Meehl et al. (2007) note that further accelerations in ice flow of the kind recently observed in some Greenland outlet glaciers and West Antarctic ice streams could substantially increase the contribution from the ice sheets, a possibility not reflected in the upper bound of global sea level projections. For example, if ice discharge from these processes were to increase in proportion to global average surface temperature change, it would add at least 0.1 to 0.2 m to the upper bound of sea level rise by 2090 to 2099. Understanding of these processes is limited and there is no consensus on their magnitude (Meehl et al., 2007; IPCC, 2007d).

Sea level rise during the 21st century is projected by the IPCC to have substantial geographic variability due to factors discussed in Section IV. 1 .d that influence changes at the regional scale, including changes in ocean circulation or atmospheric pressure, and geologic processes (Meehl et al., 2007). The patterns simulated by different models are not generally similar in detail, but have some common features, including smaller than global average sea level rise in the Southern Ocean, larger than global average sea level rise in the Arctic, and a narrow band of pronounced sea level rise stretching across the southern Atlantic and Indian Oceans.

For North American coasts, emissions scenario A1B shows sea level rise values close to the global mean, with slightly higher rates in eastern Canada and western Alaska, and stronger positive anomalies in the Arctic. The projected rate of sea level rise off the low-lying U.S. South Atlantic and Gulf Coasts is also higher than the global average. Vertical land motion from geologic processes may decrease (through uplift) or increase (through subsidence) the relative sea level rise at any site (Nicholls et al., 2007).

Cryosphere (snow and ice) projections

The IPCC (Meehl et al., 2007) concluded the following during the Fourth Assessment:

- snow cover and sea ice extent are projected to decrease as the climate warms;
- glaciers and ice caps are projected to lose mass as increases in summer melting outweigh increases in winter precipitation, which will contribute to sea level rise; and
- widespread increases in thaw depth are projected over most permafrost regions globally.

The Arctic Climate Impact Assessment (ACIA, 2004) concluded that snow cover will decline by 10 to 20% by the 2070s and that the projected decreases in snow and ice cover are *very likely* to persist for centuries.

Focusing on North America, snow season length and snow depth are *very likely* to decrease in most of North America, as illustrated in Figure IV.21. The exception is in the northernmost part of Canada, where maximum snow depth is *likely* to increase (Christensen et al., 2007). Sea ice is projected to shrink in both the Arctic and the Antarctic under all SRES scenarios (IPCC, 2007d).

In some projections, Arctic late-summer sea ice disappears almost entirely by the latter part of the 21st century.

Projections of extreme events

Models suggest that human-induced climate change will alter the prevalence and severity of many extreme events such as heat waves, cold waves, storms, floods, and droughts. Projections of global temperature from the IPCC (Meehl et al., 2007) show that it is *very likely* that heat waves will become more intense, more frequent, and longer lasting in a future warm climate, whereas cold episodes are projected to decrease significantly. Meehl and Tebaldi (2004, in Meehl et al., 2007) found that the pattern of future changes in heat waves—which shows the greatest intensity increases over western Europe, the Mediterranean, and the southeastern and western United States—is related in part to circulation changes resulting from an increase in greenhouse gases.

Percent Snow Depth Changes in March

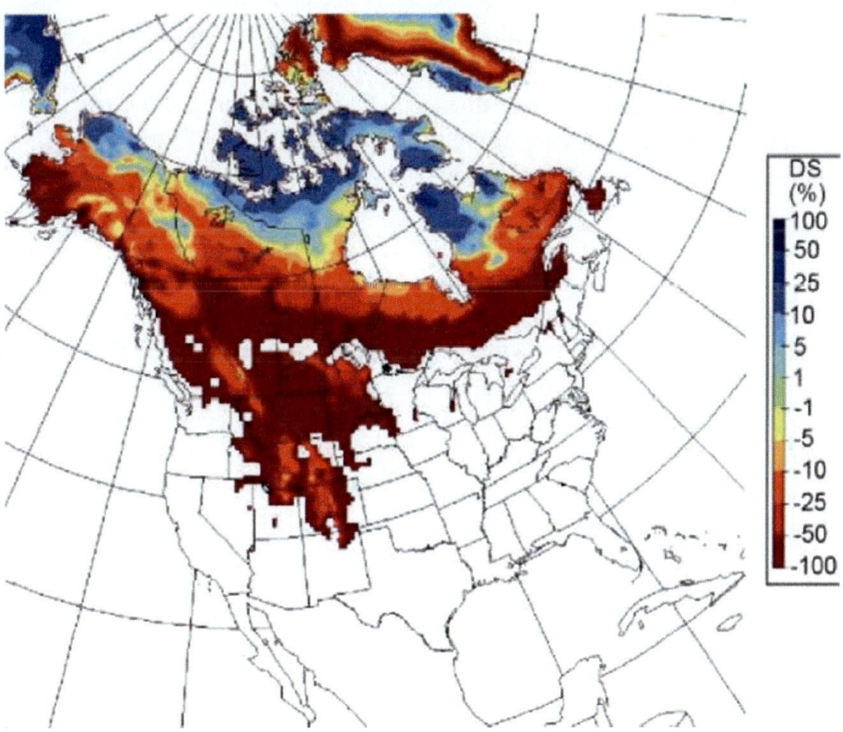

Source: Christensen et al. (2007).

Figure IV.21. Percent snow depth changes in March (only calculated where climatological snow amounts exceed 5 mm of water equivalent), as projected by the Canadian Regional Climate Model (Plummer et al., 2006 in Christensen et al., 2007), driven by the Canadian General Circulation Model, for 2041 to 2070 under the SRES A2 scenario compared to 1961 to 1990

CCSP SAP 3.3 (Karl et al., 2008) concluded for North America that abnormally hot days and nights and heat waves are *very likely* to become more frequent and that cold days and cold nights are *very likely* to become much less frequent[31]. Additionally, the number of days with frost is *very likely* to decrease. Focusing on North America and the United States, the IPCC (Christensen et al., 2007) cites a number of studies that project changes in temperature extremes in the United States. Diffenbaugh et al. (2005 in Christensen et al., 2007) find that the frequency and the magnitude of extreme temperature events change dramatically under a high-end emission scenario (SRES A2), with increases in extreme hot events and decreases in extreme cold events. Bell et al. (2004, in Christensen et al., 2007) examine changes in temperature extremes in their simulations centered on California and find increases in extreme temperature events, prolonged hot spells, and increased diurnal temperature range. Leung et al. (2004, in Christensen et al., 2007) find increases in diurnal temperature range in six sub-regions of the western United States in summer.

Projections of global precipitation extremes and droughts show that the intensity of precipitation events is projected to increase, particularly in tropical and high-latitude areas that experience increases in mean precipitation (Meehl et al., 2007). Even in areas where mean precipitation is projected to decrease (most subtropical regions), precipitation intensity is projected to increase. However, there would be longer periods between rainfall events. The IPCC (Meehl et al., 2007) also found that increases in heavy precipitation events have been linked to increases in flooding. The IPCC (Meehl et al., 2007) projected a tendency for drying in mid-continental areas during summer due to higher temperatures, indicating a greater risk of droughts in those regions. Extreme drought increases from 1% of present-day land area to 30% by the end of the century in the A2 (high emissions growth) scenario (Burke et al., 2006 in Meehl et al., 2007).

Several regional studies in the IPCC project changes in precipitation extremes in parts of the United States. These changes range from a decrease in heavy precipitation in California (Bell et al., 2004 in Christensen et al., 2007) to an increase during winter in the northern Rockies, Cascades, and Sierra Nevada mountain ranges (Leung et al., 2004 in Christensen et al., 2007). For the contiguous United States, Diffenbaugh et al. (2005, in Christensen et al., 2007) found widespread increases in extreme precipitation events under SRES A2 (high emissions growth). CCSP SAP 3.3 (Karl et al., 2008) concluded for North America that, on average, precipitation is *likely* to be less frequent but more intense, and precipitation extremes are *very likely* to increase. For example, for a mid-range emission scenario, daily precipitation so heavy that it now occurs only once every 20 years is projected to occur approximately every 8 years by the end of this century over much of eastern North America (Karl et al., 2008).

In its Fourth Assessment, the IPCC concluded that it is *likely* that future tropical cyclones (typhoons and hurricanes) will become more intense, with larger peak wind speeds and more heavy precipitation associated with ongoing increases in tropical sea surface temperatures (IPCC, 2007d; Meehl et al., 2007). Some modeling studies have projected a decrease in the number of tropical cyclones globally due to increased stability of the tropical atmosphere in a warmer climate, characterized by fewer weak storms and greater numbers of intense storms. A number of modeling studies have also projected a general tendency for more intense but fewer storms outside the tropics, with a tendency towards more extreme wind events and higher ocean waves in several regions associated with these deepened cyclones (Meehl et al., 2007).

Some of the sectoral impacts of extreme events for the United States are summarized throughout section V of this report.

Abrupt climate change projections

Abrupt climate change refers to sudden (of the order of decades) large changes in some major component of the climate system, with rapid, widespread effects. Abrupt climate changes are an important consideration because, if triggered, they could occur so quickly and unexpectedly that human or natural systems would have difficulty adapting to them (NRC, 2002). This section focuses on the general risks of abrupt climate change globally, with some discussion of potential regional implications where information is available. Potential abrupt climate change implications for the United States are not discussed in the Section V because they cannot be predicted with confidence, particularly for specific regions.

An abrupt climate change occurs when the climate system is forced to cross some threshold, which triggers a transition into a new state (NRC, 2002). Crossing systemic thresholds may lead to large and widespread consequences (Schneider et al., 2007). As detailed by the NRC (2002), the triggers for abrupt climate change can be forces that are external and/or internal to the climate system, including

- changes in Earth's orbit,[32]
- a brightening or dimming of the Sun,
- melting or surging ice sheets,
- strengthening or weakening of ocean currents, and
- emissions of climate-altering gases and particles into the atmosphere.

More than one of these triggers can operate simultaneously, since all components of the climate system are linked.

Scientific data show that abrupt changes in the climate at the regional scale have occurred throughout history and are characteristic of the Earth's climate system (NRC, 2002). During the last glacial period, abrupt regional warming events (*likely* up to 16 °C within decades over Greenland) and cooling events occurred repeatedly over the North Atlantic region (Jansen et al., 2007). These warming events *likely* had some large-scale effects such as major shifts in tropical rainfall patterns and redistribution of heat within the climate system, but it is improbable that they were associated with large changes in global mean surface temperature.

The NRC (2002) concluded that anthropogenic forcing could increase the risk of abrupt climate change, making the following statements:

- Greenhouse warming and other human alterations of the Earth system may increase the possibility of large, abrupt, and unwelcome regional or global climatic events.
- Abrupt changes of the past are not fully explained yet, and climate models typically underestimate the size, speed, and extent of those changes.
- Therefore, future abrupt changes cannot be predicted with confidence, and climate surprises are to be expected.

Changes in weather patterns (sometimes referred to as weather regimes or natural modes) can result from abrupt changes that might occur spontaneously due to dynamic interactions in the atmosphere/ice–ocean system, or from crossing a threshold from slow external forcing (as described above) (Meehl et al., 2007). In a warming climate, changes in the frequency and amplitudes of these patterns might not only evolve rapidly, but also trigger other processes that lead to abrupt climate change (NRC, 2002). Examples of these patterns include the El Niño– Southern Oscillation (ENSO)[33] and the North Atlantic Oscillation/Arctic Oscillation (NAO/AO).[34]

Scientists have investigated the possibility of an abrupt slowdown or shutdown of the Atlantic meridional overturning circulation (MOC) triggered by greenhouse gas forcing. The MOC transfers large quantities of heat to the North Atlantic and Europe, so an abrupt change in the MOC could have important implications for the climate of this region (Meehl et al., 2007). However, according to the IPCC (Meehl et al., 2007), the probability of an abrupt change in (or shutdown of) the MOC during the 21st century is *very unlikely*. However, longer-term changes in the MOC cannot be assessed with confidence. The slowdown in the MOC projected by most models is gradual, so the resulting decrease in heat transport to the North Atlantic and Europe would not be large enough to reverse the warming that results from the increase in greenhouse gases.

The rapid disintegration of the Greenland Ice Sheet, which would raise sea levels 7 m, is another commonly discussed abrupt change. Although models suggest the complete melting of the Greenland Ice Sheet would only require sustained warming in the range of 1.9 to 4.6 °C (relative to pre-industrial temperatures), it is expected to be a slow process that would take many hundreds of years to complete (Meehl et al., 2007).

A collapse of the West Antarctic Ice Sheet, which would raise sea levels 5 to 6 m, has been discussed as a low-probability, high-impact response to global warming (NRC, 2002; Meehl et al., 2007). The weakening or collapse of ice shelves, caused by melting on the surface or by melting at the bottom by a warmer ocean, might contribute to a potential destabilization of the West Antarctic Ice Sheet. Recent satellite and *in situ* observations of ice streams behind disintegrating ice shelves highlight some rapid reactions of ice sheet systems (Lemke et al., 2007). Ice sheet models are only beginning to capture the small-scale dynamic processes that involve complicated interactions with the glacier bed and the ocean at the perimeter of the ice sheet (Meehl et al., 2007). These processes are not represented in the models used by the IPCC to project sea level rise, which suggest Antarctica will gain mass due to the likelihood of increasing snowfall, reducing sea level rise. It is noted that recent studies find no significant continent-wide trends in accumulation over the past several decades (Lemke et al., 2007). But it is possible that acceleration of ice discharge could become dominant, causing a net positive contribution. Given these competing factors, there is presently no consensus on the long-term future of the West Antarctic Ice Sheet or its contribution to sea level rise (Meehl et al., 2007).

IV.3.c. Land cover change as an environmental driver

The SAP 4.3 report from CCSP (Backlund et al., 2008) concluded that global climate change effects will be superimposed on and modify those resulting from land use and land cover patterns in ways that are as of yet uncertain.

A literature review of the relationship between climate change and land use indicates that land use change has had a much greater effect on ecosystems than has climate change and that the vast majority of land use changes have little to do with climate or climate change (Dale, 1997 in Backlund et al., 2008). In the near term, climate fluctuation and change will influence ecosystems primarily through the impact of land use on ecosystems, and the response of ecosystems to land use (Backlund et al., 2008). Changes in land cover can make restoration prohibitively costly, and in some cases, impossible. This is particularly the case for native rangeland that is disturbed, whether intentionally through intensive agriculture or unintentionally through climate change.

As discussed in SAP 4.3 (Backlund et al., 2008), land use changes are an important driver of changes in arid lands. Today's arid lands reflect a legacy of historic land uses, and future land use practices will arguably have the greatest impact on arid land ecosystems in the next two to five decades. Grazing has traditionally been the most extensive land use in arid regions. However, land use has significantly shifted to exurban development and recreation in recent decades. Thus, in addition to grazing, the response of arid lands to climate change will be influenced by environmental pressures related to air pollution, nitrogen deposition, energy development, motorized off-road vehicles, feral pets, and horticultural invasive species. Many plants and animals in arid ecosystems are near their physiological limits for tolerating temperature and water stress, such that even slight changes in stress will have significant consequences. Projected climate changes will increase the sensitivity of arid lands to disturbances such as grazing and fire. Invasion of non-native grasses will increase fire frequency. In the near-term, fire effects will exceed climate effects on ecosystem structure and function.

The current land use system in the United States requires high resource inputs, from the use of synthetic fertilizer on crops to the transport of crops to animal feeding operations. In addition to being inefficient with regard to fuel use, this system creates environmental problems ranging from erosion to degradation of water supplies due to high nutrient levels. SAP 4.3 (Backlund et al., 2008) discussed specific details related to the effects of land use changes:

- Land cover and land use patterns are changing. For example, as exurban development spreads to previously undeveloped areas, U.S. forests are becoming increasingly fragmented, further raising fire risk and compounding the effects of summer drought, pests, and warmer winters.
- It is *unlikely* that the hydrologic trends detected in the various studies reviewed above can be attributed, at least in large part, to land cover and land use change, but sufficient questions remain that it cannot be definitively ruled out.
- There has been long-standing controversy in determining the relative contribution of climatic and anthropogenic factors as drivers of desertification. Local fence line contrasts argue for the importance of land use (e.g., changes in grazing and fire regimes). Vegetation change in areas with no known change in land use argues for climatic drivers.

The Arctic Climate Impact Assessment (ACIA, 2004) also discusses projected changes in land cover, stating that as the climate warms, the tree line is expected to move northward and to higher elevations. In these projections, forests will replace significant amounts of existing tundra and tundra vegetation will move into polar deserts. These changes are *likely* to increase carbon uptake, which would partially offset warming effects. However, the reduced reflectivity associated with the vegetation land cover is *likely* to outweigh this, causing further warming.

Ultimately, the ability to attribute effects to either climate change or land use changes is hampered by the lack of observations of land use changes. As noted by SAP 4.3 (Backlund et al., 2008), there is no coordinated national network for monitoring changes in land resources associated with climate change, most disturbances (storms, insects, and diseases), and changes in land cover/land use. Thus, separating climate effects from other environmental stresses is difficult but in some cases feasible. For example, when detailed water budgets exist, the effects of land use, climate change, and consumptive use on water levels can be calculated. While climate effects can be difficult to quantify at small scales, sometimes regional effects can be separated. For example, regional trends in productivity, estimated using satellite methods, can often be assigned to regional trends in climate versus land use, although on any individual small- scale plot, climate effects may be primary or secondary. In other cases, scientific understanding is sufficiently robust that models in conjunction with observations can be used to estimate climate effects. Overall, SAP 4.3 (Backlund et al., 2008) recommended that more refined analysis and/or monitoring systems designed specifically for detecting climate change effects would provide more detailed and complete information and probably capture a range of more subtle impacts, which might, in turn, lead to early warning systems and more accurate forecasts of potential future changes.

IV.3.d. Effects on and from stratospheric ozone

Stratospheric ozone protects Earth's surface from much of the Sun's biologically harmful ultraviolet radiation. According to the WMO (2007), climate change that results from changing greenhouse gas concentrations will affect the evolution of the ozone layer through changes in chemical transport, atmospheric composition, and temperature. In turn, changes in stratospheric ozone can affect the weather and climate of the troposphere. The coupled interactions between the changing climate and the ozone layer are complex, and scientific understanding is incomplete (WMO, 2007). Specific information on climate change effects on and from stratospheric ozone in the United States has not been assessed. Except where indicated, the findings in this section apply generally to the globe with a focus on polar regions.

The *2006 Scientific Assessment of Ozone Depletion* (WMO, 2007) concluded that it is *unlikely* that total ozone averaged over the region 60° S to 60° N will decrease significantly below the low values of the 1 990s, because the abundances of ozone-depleting substances have peaked and are in decline. The current best estimate is that ozone between 60° S and 60° N will return to pre- 1980 levels around the middle of the 21st century.

Effects of elevated greenhouse gas concentrations on stratospheric ozone

WMO (2007) concluded that future concentrations of stratospheric ozone are sensitive to future levels of the well-mixed greenhouse gases. WMO (2007) found that there are several competing effects:

- Future increases in greenhouse gas concentrations, primarily CO_2, will contribute to the average cooling in the stratosphere. Stratospheric cooling is expected to slow gas-phase ozone depletion reactions and increase ozone.
- Enhanced CH_4 emissions (from warmer and wetter soils) are expected to enhance ozone production in the lower stratosphere.

- An increase in N_2O emissions is expected to reduce ozone in the middle and upper stratosphere.

Two-dimensional models that include coupling between all of these well-mixed greenhouse gases and temperature project that ozone levels between 60° S and 60° N will return to 1980 values up to 15 years earlier than projected by models that are uncoupled (Bodeker et al., 2006). The impact of stratospheric cooling on ozone might be the opposite in polar regions, where cooling could cause increases in polar stratospheric clouds, which, given enough halogens, would increase ozone loss (Bodeker et al., 2006).

Concentrations of stratospheric ozone are also sensitive to stratospheric water vapor concentrations, which may remain relatively constant or increase (Baldwin et al., 2007). Increases in water vapor would cause increases in hydrogen oxide (HO_x) radicals, affecting ozone loss processes (Baldwin et al., 2007). Several studies (Dvortsov and Solomon, 2001 and Shindell, 2001 in Baldwin et al., 2007) suggest that increasing stratospheric water vapor would delay ozone layer recovery. Increases in stratospheric water vapor could also increase spring ozone depletion in the polar regions by raising the temperature threshold for the formation of polar stratospheric clouds (WMO, 2007).

The possible effects of climate change on stratospheric ozone are further complicated by possible changes in climate dynamics. Climate change can affect temperatures, upper-level winds, and storm patterns, which, in turn, impact planetary waves that affect the stratosphere (Baldwin et al., 2007). Changes in the forcing and propagation of planetary waves[35] in the polar winter are a major source of uncertainty for predicting future levels of Arctic ozone loss (Austin et al., 2003 in Baldwin et al., 2007).

Effects of stratospheric ozone changes on climate

The WMO (2007) found that changes in the temperature and circulation of the stratosphere affect climate and weather in the troposphere. The dominant tropospheric response, simulated in models and identified in analyses of observations, comprises changes in the strength of mid-latitude westerly winds. The mechanism for this response is not well understood.

Modeling experiments, which simulate observed changes in stratospheric ozone and combined stratospheric ozone depletion and greenhouse gas increases, also suggest that Antarctic ozone depletion, through its effects on the lower-stratospheric vortex, has contributed to the observed surface cooling over interior Antarctica and warming of the Antarctic Peninsula, particularly in summer (Baldwin et al., 2007). While the physics of these effects are not well understood, the simulated pattern of warming and cooling is a robust result seen in many different models, and is well supported by observational studies.

As the ozone layer recovers, tropospheric changes that have occurred as a result of ozone depletion are expected to reverse. However, temperature changes due to increasing greenhouse gas concentrations may offset this reversal (Baldwin et al., 2007).

SECTION V. ANALYSIS OF THE EFFECTS OF GLOBAL CHANGE ON THE NATURAL ENVIRONMENT AND HUMAN SYSTEMS

In the Intergovernmental Panel on Climate Change (IPCC) Fourth Assessment Report, Field et al. (2007)[1] concluded:

> [The] United States and Canada will experience climate changes through direct effects of local changes (e.g., temperature, precipitation, and extreme weather events), as well as through indirect effects, transmitted among regions by interconnected economies and migrations of humans and other species. Variations in wealth and geography, however, lead to an uneven distribution of likely impacts, vulnerabilities and capacities to adapt [North America chapter of Working Group II].

Observed and potential effects of these global changes are discussed in the following sections:

V.1 Biological diversity, ecosystems, and the natural environment;
V.2 Agriculture;
V.3 Forestry;
V.4 Water resources;
V.5 Social systems and settlements;
V.6 Human health;
V.7 Energy production, use, and distribution; and
V.8 Transportation.

This discussion draws heavily on the Climate Change Science Program (CCSP) Synthesis and Assessment Products (SAPs) 4.3, 4.4, 4.5, 4.6, and 4.7. These products included an extensive examination of the relevant scientific literature to examine the effects of climate change on U.S. agriculture, land and water resources, biodiversity, social systems, human health, and transportation. Other important source documents include the following chapters in the IPCC Working Group II volume of the Fourth Assessment Report: "Assessment of observed changes and responses in natural and managed systems" (Rosenzweig et al., 2007); "North America" (Field et al., 2007); "Ecosystems, their properties, goods and services" (Fischlin et al., 2007); "Food, fiber, and forest products" (Easterling et al., 2007); "Freshwater resources and their management" (Kundzewicz et al., 2007); "Coastal systems and low-lying areas" (Nicholls et al., 2007); "Industry, settlement, and society" (Wilbanks et al., 2007a); "Human health" (Confalonieri et al., 2007); "Polar regions" (Anisimov et al., 2007); and "Small islands" (Mimura et al., 2007).

Recent years have seen a substantial increase in scientific research on global change, yielding an improved understanding of emerging effects and areas of vulnerability (Field et al., 2007). Multiple stressors are contributing to these emerging changes. As discussed in the previous section, there is a robust scientific consensus that human-induced climate change is occurring. The Fourth Assessment Report concluded that "the understanding of anthropogenic warming and cooling influences on climate has improved since the Third Assessment Report, leading to *very high confidence* (at least a 9 out of 10 chance)[36] that the global average net effect of human activities since 1750 has been one of warming" (IPCC,

2007a). According to the IPCC, terrestrial ecosystems and marine and freshwater systems show that recent warming is strongly affecting natural biological systems (*very high confidence*) (Rosenzweig et al., 2007). Other stressors that are also affecting ecosystems include changing patterns of land use and management, invasive species, air and water pollution, and disturbances such as wildland fires, floods, and outbreaks of plant pathogens and vegetation defoliators. These stressors may cause effects on their own, or they may interact with other stressors, resulting in nonlinear impacts (Field et al., 2007).

V.1. Biological Diversity, Ecosystem Composition, and the Natural Environment

Ecosystems provide a number of goods (e.g., food, fiber, fuel, pharmaceutical products) and services (e.g., cycling of water and nutrients, regulation of weather and climate, removal of waste products, sustaining biological diversity, providing recreational and spiritual opportunities) to society. Changing global conditions have implications for the health of terrestrial and aquatic ecosystems and the goods and services required by our growing population. The IPCC (Field et al., 2007) found that North American animals are responding to climate change, with effects observed on phenology, migration, reproduction, dormancy, and geographic range (Parmesan and Yohe, 2003; Root et al., 2003, 2005; Parmesan and Galbraith, 2004).

The potential impacts of climate change on ecosystems have long have been a concern to the scientific community (Peters and Lovejoy, 1992; IPCC, 1990). Substantial research has examined the effects of climate change on vegetation and wildlife, leading to the conclusion that the changing climate is already having real and demonstrable impacts on a variety of ecosystem types (Janetos et al., 2008). In their summary of findings, the IPCC (Working Group II) recently concluded, "Recent studies have allowed a broader and more confident assessment of the relationship between observed warming and impacts than was made in the Third Assessment." That assessment concluded that "there is *high confidence* [about an 8 out of 10 chance[37]] that recent regional changes in temperature have had discernible impacts on many physical and biological systems." In the Fourth Assessment Report, the IPCC concluded, "Observational evidence from all continents and most oceans shows that many natural systems are being affected by regional climate changes, particularly temperature increases" (IPCC, 2007b). CCSP SAP 4.3 (Backlund et al., 2007) concluded that it is *very likely* that climate changes—including temperature increases, increasing CO_2 levels, and altered patterns of precipitation—are already affecting U.S. water resources, agriculture, land resources, and biodiversity. This report also concluded that it is *very likely* that climate change will increase in importance as a driver for changes in biodiversity over the next several decades, although for most ecosystems it is not currently the largest driver of change.

Of the 455 biological observations assessed by the IPCC, 92% were consistent with the changes expected due to average warming (Rosenzweig et al., 2007). Indeed, an analysis of 866 peer- reviewed papers exploring the ecological consequences of climate change found that nearly 60% of the 1,598 species studied exhibited shifts in their distributions and/or phenologies over the 20- and 140-year timeframes (Parmesan and Yohe, 2003). In their review of the research on effects on North American ecosystems, Field et al. (2007) wrote,

"Over the 21st century, changes in climate will cause species to shift north and to higher elevations and fundamentally rearrange North American ecosystems. Differential capacities for range shifts and constraints from development, habitat fragmentation, invasive species, and broken ecological connections will alter ecosystem structure, function, and services."

The following sections discuss specific conclusions regarding current findings and trends. Specifically, they address: 1) ecosystem distribution and phenology, 2) ecosystem services, 3) ecosystem effects from extreme events, 4) ecosystem effects from sea level rise, and 5) vulnerable ecosystems.

V.1.a. Ecosystem distribution and phenology

Ecosystem distribution

As described in SAP 4.3 (Janetos et al., 2008), evidence from two meta-analyses (Parmesan and Yohe, 2003; Root et al., 2003) and a synthesis (Parmesan, 2006) suggests that recent climatic warming has significantly contributed to a long-term, large-scale alteration of animal and plant populations across a broad array of taxa (Root and Schneider, 2006; Root et al., 2003; Parmesan, 2006). In North America, climate warming is expected to result in shifts of species ranges poleward and upward along elevational gradients (Parmesan, 2006). A key obstacle to detecting change is the inadequacy of the current set of observations and inventories.

Many animals, including most mammals, have evolved powerful mechanisms to regulate their physiology. Therefore, they will primarily experience climate change effects through pathways involving their food source, habitat, and predators, rather than through direct effects of climate change on body temperature (Schneider and Root, 1996). As warming occurs, species may migrate to higher elevations where more suitable temperatures exist. This is possible where habitat connectivity (the degree to which a habitat is physically linked with other suitable areas for a particular species) exists and other biotic and abiotic conditions are appropriate. However, species that require higher-elevation habitat, such as alpine ecosystems, often have nowhere to migrate.

To date, relatively few studies have been conducted at a scale that encompasses an entire species' range. These studies have evaluated distributional shifts of amphibians (Pounds et al., 1999, 2006), pikas (Beever et al., 2003), birds (Dunn and Winkler, 1999), and butterflies (Parmesan, 1996, 2006). The majority of studies infer range shifts from observations made at a smaller scale, either from a small portion of the species range or from changes in species abundance within local communities (Parmesan, 2006).

The IPCC (Field et al., 2007) found:

- "Many North American species have shifted their ranges, typically to the north or to higher elevations (Parmesan and Yohe, 2003).
- Edith's checkerspot butterfly has become locally extinct in the southern, low-elevation portion of its western North American range but has extended its range 90 km north and 120 m higher in elevation (Parmesan, 1996; Croier, 2003; Parmesan and Galbraith, 2004).
- Red foxes have expanded northward in northern Canada, leading to retreat of competitively subordinate arctic foxes (Hersteinsson and Macdonald, 1992)."

Janetos et al. (2008) review substantial research examining warming-induced shifts in distribution among butterflies. As climate changes, the distribution of butterflies in North America is shifting northward, including a contraction at the southern end of their historical range, and to higher elevations. In a synthesis of prior research, Parmesan (2006) found that 30 to 75% of studied butterfly species had expanded northward, less than 20% had contracted southward, while the remainder was stable.

The review (Janetos et al., 2008) also identified similar results that have been identified in European species. Sixty-three percent of a sample of 35 non-migratory butterflies had shifted their ranges to the north by 35 to 240 km during the 20th century while 3% had shifted to the south (Parmesan et al., 1999). In central Spain, Wilson et al. (2007) documented that the richness and composition of butterfly species between 1967 to 1973 and 2004 to 2005 shifted uphill by 293 m. Ultimately, these shifts resulted in a net decline in species richness over approximately 90% of the study region (Wilson et al., 2007). Over a 19-year study period, Franco et al. (2006) found climate change was a driver of local extinction for three species of butterflies in Britain. In addition, these results indicate that range boundaries shifted 70 to 100 km northward for *Aricia artaxerxes* and *Erebia aethiops* and 130 to 150 m uphill for *E. epiphron* in a region with estimated latitudinal and elevational temperature shifts of 88 km northward and 98 m uphill in the same time period. Additional research on 51 British butterfly species found those with northern and/or montane distributions have disappeared from low elevations while colonizing higher-elevation sites, but did not find evidence for a systematic shift northward across all species (Hill et al., 2002). Model results project 65 and 24% declines in range sizes for northern and southern species, respectively, for the period 2070 to 2099.

Effects on phenology

Phenology is the study of the times of recurring natural phenomena, for example, the date of emergence of leaves and flowers, the first flight of butterflies, and the first appearance of migratory birds.

Substantial evidence indicates natural systems are being demonstrably affected by climate change. Impacts include changes in the timing of the onset, completion, and length of the growing season; phenology; primary production, and as discussed above, species distributions and diversity (Parmesan and Yohe, 2003). Field-based analyses of multiple species indicate average shifts in the arrival of key spring indicators of 2.3 days per decade across all species (Parmesan and Yohe, 2003) with shifts as great as 5.1 days per decade (Root et al., 2003).

Growing season length and primary production shifts

Research reviewed by Janetos et al. (2008) suggests a significant lengthening of the growing season and higher net primary productivity in the higher latitudes of North America where temperature increases are relatively high. Global satellite data since 1981 indicate an earlier onset of spring across the temperate latitudes by 10 to 14 days (Zhou at al., 2001; Lucht et al., 2002) as well as an increase in summer photosynthetic activity (Zhou et al., 2001). Similarly, an analysis of climate variables in the higher latitudes in Europe indicates a lengthening of the growing season of 1.1 to 4.9 days per decade since 1951 (Menzel et al., 2003). Numerous field studies have documented consistent earlier leaf expansion (Beaubien and Freeland, 2000; Wolfe et al., 2005) and earlier flowering (Schwartz and Reiter, 2000; Cayan et al., 2001) across different species and ecosystem types. Some of the specific examples contained in the IPCC (Field et al., 2007) are:

- earlier flowering in lilac by 1.8 days per decade from 1959 to 1993 at 800 North American study sites (Schwart and Reiter, 2000);
- earlier flowering in honeysuckle by 3.8 days per decade in the western United States (Cayan et al., 2001);
- earlier leaf expansion in apple and grape by 2 days per decade in 72 northeastern United States sites (Wolfe et al., 2005); and
- earlier leaf expansion by 2.6 days per decade since 1900 in trembling aspen growing in the region around Edmonton, Alberta, Canada (Beaubien and Freeland, 2000).

Research reviewed in Janetos et al. (2008) found that between 1982 and 1998, net primary production increased nearly 10% in the continental United States (Boisvenue and Running, 2006). The largest documented increases were in croplands and grasslands of the central United States, as a result of improved water balance (Lobell et al., 2002; Nemani et al., 2002; Hicke and Lobell, 2004). In contrast, forest productivity increased by less than 1% per decade and is generally limited by low temperature and short growing seasons in the higher latitudes and elevations (Caspersen et al., 2000; McKenzie et al., 2001; Joos et al., 2002; Boisvenue and Running, 2006). However, productivity has decreased in forested regions subject to drought from climate warming since 1895 (McKenzie et al., 2001) and subalpine regions (e.g., Monson et al., 2005; Sacks et al., 2007). Recently, widespread mortality over 12,000 km^2 of lower-elevation forest in the southwest United States demonstrated the impacts of increased temperature and the associated multiyear drought (Breshears et al., 2005) even as productivity at tree line had increased previously (Swetnam and Betancourt, 1998). In addition, in summarizing research on North American forests, the IPCC indicated forest productivity "can be influenced indirectly by climate through effects on disturbance, especially from wildfire, storms, insects and diseases" (Field et al., 2007). The IPCC also noted the large increases in area burned by wildfires over the last three decades.

Shifts in phenology

Regarding the effects of climate change on wildlife in North America, Field et al. (2007) made the following conclusions:

> Warmer springs have led to earlier nesting for 28 migrating bird species on the east coast of the U.S. (Butler, 2003) and to earlier egg laying for Mexican jays (Brown et al., 1999) and tree swallows (Dunn and Winkler, 1999). In northern Canada, red squirrels are breeding 18 days earlier than 10 years ago (Reale et al., 2003). Several frog species now initiate breeding calls 10 to 13 days earlier than a century ago (Gibbs and Breisch, 2001). In lowland California, 70% of 23 butterfly species advanced the date of first spring flights by an average 24 days over 31 years (Forister and Shapiro, 2003).

Specific effects are reviewed in greater detail in CCSP SAP 4.3, *The Effects of Climate Change on Agriculture, Land Resources, Water Resources, and Biodiversity* (Janetos et al., 2008). Key findings are presented below.

Migratory birds (adapted from Janetos et al., 2008) For migratory birds, the timing of arrival at breeding territories and overwintering grounds is an important determinant of reproductive success, survivorship, and fitness. Climate variability on interannual and longer time scales can influence migration behavior by altering the timing of arrival and/or departure. Disruptions in the phenology of migration to summer and wintering areas have been noted for long-distance, continental migrations as well regional, local, or elevational migrations (Dunn and Winkler, 1999; Winkler et al., 2002; Butler, 2003; Cotton, 2003). Reviewing arrival dates of short- and long-distance migrants, Jonzen et al. (2006) found that, while both exhibited changes in migration timing, long-distance migrants have advanced their spring arrival more than have short-distance migrants, based on data from 1980 to 2004. In a review of multiple data sets from the United Kingdom, Germany, Switzerland, and Denmark, Thorup et al. (2007) found advances in spring arrival across all insectivorous songbirds included in the study (with similar trends among studies and locations), while the timing of autumn departure was variable.

While the advancement of spring migration phenology is one specific effect of climate change (Root et al., 2003), ultimately, the indirect effects may be more important in determining the long-term impacts on species persistence and diversity. Of particular importance is the potential for mismatches in the timing of migration, breeding, and peak food availability. There is no *a priori* reason to expect migrants and their respective food sources to shift their phenologies at the same rate. A differential shift can lead to mistimed reproduction and reduced population success (Stenseth and Mysterud, 2002; Visser et al., 2004, 2006; Visser and Both, 2005).

Reviewing the relevant literature, Janetos et al. (2008) concluded the responses are likely to be highly complex depending on species-specific traits, characteristics of local microhabitats, and aspects of local microclimates. These points are illustrated by a long-term study of the migratory pied flycatcher (*Ficedula hypoleuca*), which exhibited an advance in laying date, hatching date, and clutch size in response to the advancing peak abundance of their food supply. However, the timing of food availability and subsequent impacts were not uniform across their range. Populations of the flycatcher have declined by about 90% over the past two decades in areas where the food for provisioning nestlings peaks early in the season but not in areas with a late food peak (Both, 2006).

Butterflies (adapted from Janetos et al., 2008) The migration of butterflies in the spring is highly correlated with spring temperatures. Research has documented earlier arrivals in 16 of 23 species in central California (Forister and Shapiro, 2003), 26 of 35 species in the United Kingdom (Roy and Sparks, 2000), and all studied species (17) in Spain (Stefanescu et al., 2004).

Changes in timing of migrations and distributions are expected to present resource mismatches that will influence population success and the probability of extinction. To date, a few studies have linked population extinctions directly to climate change (McLaughlin et al., 2002, Franco et al., 2006). Substantial research found that phenological asynchrony in butterfly–host interactions in California led to population extinctions of the checkerspot butterfly (*E. editha*) during extreme drought and low-snowpack years (Singer and Ehrlich, 1979; Ehrlich et al., 1980; Singer and Harter, 1996; Thomas et al., 1996). A modeling experiment for two populations of checkerspot butterfly suggested the decline was hastened by increasing variability in precipitation associated with climate change in a region that allowed no distributional shifts because of persistent habitat fragmentation (McLaughlin et al., 2002). These results are supported by additional model projections that suggest butterfly species could adapt to changing conditions by shifting their ranges in response to warming. However, limited habitat availability and connectivity may limit migration ability resulting in significant population declines (Hill et al., 2002).

Amphibians (adapted from Janetos et al., 2008) There is evidence that amphibian breeding is occurring earlier in some regions, driven by increasing temperatures (Blaustein et al., 2001; Gibbs and Breisch, 2001; Beebee, 2002). However, responses are complex. Statistical tests (Blaustein et al., 2002) indicated that half of the 20 species examined by Beebee (1995), Reading (1998), Gibbs and Breisch (2001), and Blaustein et al. (2001) are breeding earlier. Even in the half not exhibiting statistically significant earlier breeding, they showed biologically important trends toward breeding earlier that, if continued, are expected to become statistically significant (Blaustein et al., 2002). These findings suggest the influence of climate on amphibian breeding patterns in many species. However, while some temperate-zone frog and toad populations show a trend toward breeding earlier, others do not. For example, Fowler's toad (*Bufo fowleri*) a late breeder, has bred progressively later in spring over the past 15 years on the north shore of Lake Erie (Blaustein et al., 2001).

Projected impacts

Specific conclusions from the IPCC (Fischlin et al., 2007) regarding global-scale changes included the following:

- The resilience of many ecosystems is *likely* to be exceeded this century by an unprecedented combination of climate change, associated disturbances (e.g., flooding, drought, wildfire, insects, and ocean acidification), and other global change drivers (e.g., land use change, pollution, and over-exploitation of resources) (*high confidence*).
- Over the course of this century, net carbon uptake by terrestrial ecosystems is *likely* to peak before mid-century and then weaken or even reverse, thus amplifying climate change (*high confidence*).
- Approximately 20 to 30% of plant and animal species assessed so far are *likely* to be at increased risk of extinction if increases in global average temperature exceed 1.5 to 2.5 °C (*medium confidence*, about a 5 out of 10 chance[38]).
- For increases in global average temperature exceeding 1.5 to 2.5 °C and accompanying atmospheric carbon dioxide (CO_2) concentrations, major changes are projected in ecosystem structure and function, species' ecological interactions, and species' geographical ranges— with predominantly negative consequences for biodiversity, and ecosystem goods and services, such as water and food supply (*high confidence*).
- The progressive acidification of oceans due to increasing atmospheric CO_2 is expected to have negative impacts on marine shell-forming organisms (e.g., corals) and their dependent species (*medium confidence*).

Specific to North American ecosystems, Field et al. (2007) drew the following conclusions:

- Net primary productivity is projected to increase at high latitudes with mixed results at mid-latitudes driven by whether precipitation increases are great enough to offset increased evapotranspiration demands (Bachelet et al., 2001; Berthelot et al., 2002; Gerber et al., 2004; Woodward and Lomas, 2004b).
- The areal extent of ecosystems limited by drought is projected to increase by 11% per degree Celsius of warming in the continental United States (Bachelet et al., 2001).
- Ecosystems in the northeastern and southeastern United States are *likely* to become carbon sources, while those in the western United States are *likely* to remain carbon sinks (Bachelet et al., 2004).
- Several simulations (Cox et al., 2000; Berthelot et al., 2002; Fung et al., 2005) indicate that, over the 21st century, warming will lengthen growing seasons. Despite some decreased sink strength, resulting from greater water limitations in western forests and higher respiration in the tropics, this projected lengthening of growing seasons would sustain forest carbon sinks in North America (*medium confidence*).
- Overall forest growth in North America will *likely* increase modestly (10 to 20%) as a result of extended growing seasons and elevated CO_2 over the next century (M. Morgan et al., 2001), but with important spatial and temporal variations (*medium confidence*).
- A 2 °C temperature increase in the Olympic Mountains (United States) would cause dominant tree species to shift upward in elevation by 300 to 600 m, causing temperate species to replace subalpine species over 300 to 500 years (Zolbrod and Peterson, 1999). As for widespread species, temperature increases would have variable effects throughout their range. For example, for lodgepole pine an increase of 3 °C would increase growth in the northern part of its range, decrease growth in the middle, and result in substantial mortality in its southern extent (Rehfeldt et al., 2001).
- Bioclimatic modeling based on output from five general circulation models suggests that, over the next century, vertebrate and tree species richness will decrease in most parts of the conterminous United States, even though long-term trends (over millennia) ultimately favor increased richness in some taxa and locations (Currie, 2001).

Population and community dynamics

Ecosystems and species are influenced by the interaction of multiple factors including climate variables, land use and management, disturbance, and invasive species. Increases in plant productivity resulting from the effects of rising concentrations of atmospheric CO_2 may partially offset these adverse effects. In California, temperature increases greater than 2 °C may lead to the conversion of shrubland into desert and grassland ecosystems and evergreen conifer forests into mixed deciduous forests (Fischlin et al., 2007). In SAP 4.3, Janetos et al. (2008) also conclude that projected increases in CO_2 are expected to stimulate the growth of most plant species. Some invasive plants are expected to respond with greater growth rates than noninvasive plants. While some invasive plants may have higher growth rates and greater maximal photosynthetic rates relative to native plants under increased CO_2, definitive evidence of a general benefit of CO_2 enrichment to invasive plants over natives has not emerged.

The sea ice biome accounts for a large proportion of primary production in polar waters and supports a substantial food web. In the Northern Hemisphere, projections of ocean biological response to climate warming by 2050 show contraction of the highly productive marginal sea ice biome by 42% (Fischlin et al., 2007). Recent decreases in Arctic sea ice minima have been more rapid than in the past few decades.[39] In the Bering Sea, primary productivity in surface waters is projected to increase, the ranges of some cold-water species will shift north, and ice-dwelling species will experience habitat loss (ACIA, 2004).

Species-level projections

After reviewing studies on the projected impacts of climate change on species, the IPCC made the following global-scale conclusions (Fischlin et al., 2007 and references therein):

- Projected impacts on biodiversity are significant and of key relevance, since global losses in biodiversity are irreversible (*very high confidence*).
- Richness of endemic species (those that are unique to their location or region and are not found anywhere else on Earth) is highest where regional paleoclimatic changes have been subtle, providing circumstantial evidence of their vulnerability to projected climate change (*medium confidence*). With global average temperature changes of 2 °C above pre-industrial levels, many terrestrial, freshwater, and marine species (particularly endemics across the globe) are at a far greater risk of extinction than in the geological past (*medium confidence*).
- Approximately 20 to 30% of species (with a global uncertainty range from 10 to 40%, but varying among regional biota from as low as 1% to as high as 80%) will be at increasingly high risk of extinction by 2100.
- Based on relationships between habitat area and biodiversity, on average 15 to 37% (ranging from 9% under the most optimistic assumptions to 52% under the most pessimistic) of plant and animal species in a global sample may be "committed to extinction" by 2050, although actual extinctions will be strongly influenced by human forces and could take centuries (Thomas et al., 2004)

In North America, climate change impacts on inland aquatic ecosystems will range from the effects of increased temperature and CO_2 concentration on vegetation growth to the effects associated with alterations in hydrologic systems resulting from changes in precipitation regimes and melting glaciers and snowpack (Fischlin et al., 2007). For many freshwater animals, such as amphibians, migration to breeding ponds and the production of eggs is intimately tied to temperature and moisture availability. Asynchronous timing of breeding cycles and pond drying due to the lack of precipitation can lead to reproductive failure. Differential responses among species in arrival or persistence in ponds are expected to lead to changes in community composition and nutrient flow in ponds (Fischlin et al., 2007).

Bioclimatic modeling based on output from five general circulation models suggests that, over the next century, vertebrate and tree species richness will decrease in most parts of the conterminous United States, even though long-term trends (millennial scale) ultimately favor increased richness in some taxa and locations (Field et al., 2007). Changes in the composition of plant species in response to climate change can increase ecosystem vulnerability to other disturbances, including fire and biological invasion.

On small oceanic islands with cloud forests or high-elevation ecosystems, such as the Hawaiian Islands, extreme elevation gradients exist, ranging from nearly tropical to alpine environments. In these ecosystems, anthropogenic climate change, land use changes, and biological invasions will work synergistically to drive several species (e.g., endemic birds) to extinction (Mimura et al., 2007).

According to the IPCC, climate change (*very high confidence*) and ocean acidification due to the effects of elevated CO_2 concentrations (*medium confidence*) will impair a wide range of planktonic and other marine organisms that use aragonite to make their shells or skeletons (Fischlin et al., 2007). These impacts could result in potentially severe ecological changes in tropical and coldwater marine ecosystems where carbonate-based phytoplankton and corals are the foundation for the trophic system (Schneider et al., 2007). Moreover, Nicholls et al. (2007) wrote that corals have low adaptive capacity to thermal stress. Sea surface temperature increases of 1 to 3 °C are projected to result in more frequent bleaching events and widespread mortality, if there is not thermal adaptation or acclimatization by corals and their algal symbionts. However, the ability of coral reef ecosystems to withstand the impacts of climate change will depend to a large degree on the extent of degradation from other anthropogenic pressures and the frequency of future bleaching events (Nicholls et al., 2007).

For the Arctic, the IPCC (Anisimov et al., 2007 and references therein) made the following conclusions:

- Decreases in the abundance of keystone species[40] are expected to be the primary factor in causing ecological cascades[41] and other changes in ecological dynamics.
- Arctic animals are *likely* to be most vulnerable to warming-induced drying of small water bodies; changes in snow cover and freeze–thaw cycles that affect access to food (e.g., polar bear dependence on sea ice for seal hunting) and protection from predators (e.g., snow hare camouflage in snow); changes that affect the timing of behavior (e.g., migration and reproduction); and influx of new competitors, predators, parasites, and diseases.
- In the past, subarctic species have been unable to live at higher latitudes because of harsh conditions. Warming induced by climate change will increase the rate at which subarctic species are able to establish. Some non-native species, such as the North American mink, will become invasive, while other species that have already colonized some Arctic areas are *likely* to expand into other regions. The spread of non-native, invasive plants will *likely* have adverse impacts on native plant species. For example, experimental warming and nutrient addition has showed that native mosses and lichens become less abundant when non-native plant biomass increases.
- Bird migration routes and timing are *likely* to change as the availability of suitable habitat in the Arctic decreases.
- Loss of sea ice will affect species, such as harp seals, that depend on it for survival.
- Climate warming is *likely* to increase the incidence of pests, parasites, and diseases such as musk ox lungworm and abomasal nematodes of reindeer.

V.1.b. Ecosystem services

The IPCC (Fischlin et al., 2007) concluded that ecosystems provide many goods and services that are of vital importance for the functioning of the biosphere, and provide the basis for the delivery of tangible benefits to human society. The Millennium Ecosystem Assessment (MEA, 2005b) defines these to include supporting, provisioning, regulating, and cultural services. The IPCC (Fischlin et al., 2007) described these as follows:

> i. *Supporting services*, such as primary and secondary production, and biodiversity, a resource that is increasingly recognized to sustain many of the goods and services that humans enjoy from ecosystems. These provide a basis for three higher-level categories of services.
>
> ii. *Provisioning services*, such as products (cf. Gitay et al., 2001), i.e., food (including game, roots, seeds, nuts and other fruit, spices, fodder), fibre (including wood, textiles) and medicinal and cosmetic products (including aromatic plants, pigments).
>
> iii. *Regulating services*, which are of paramount importance for human society such as (a) carbon sequestration, (b) climate and water regulation, (c) protection from natural hazards such as floods, avalanches or rock-fall, (d) water and air purification, and (e) disease and pest regulation.
>
> iv. *Cultural services*, which satisfy human appreciation of ecosystems and their components.

Projected impacts

Increasing temperatures and shifting precipitation patterns, along with the effects of elevated CO_2 concentrations, sea level rise, and changes in climatic variability, will affect the quantity and quality of these services. By the end of the 21st century, climate change and its impacts may be the dominant driver of biodiversity loss and changes in ecosystem services globally (MEA, 2005a).

As noted in SAP 4.3, biological diversity provides a fundamental underpinning for these services (Janetos et al., 2008). While not specific to the United States, a major finding of the MEA from a global perspective was that 16 of 24 analyzed ecosystem services were being used in unsustainable ways. The MEA evaluated the relative magnitudes and importance of a number of different drivers of changes in ecosystems, and whether the importance of those drivers was likely to increase, decrease, or remain the same over the next several decades. The conclusion was that although climate change was not currently the most important driver of change in many ecosystems, it was one of the only drivers whose importance was expected to continue to increase in all ecosystems over the next several decades.

As an example of an ecosystem service experiencing current changes, in SAP 4.3 Janetos et al. (2008) note the increasing recognition of, and recent declines in abundance, observed for some pollinators, particularly the introduced honey bee (*Apis mellifera*) (NRC, 2006b). The economic significance of pollination is underscored by the fact that about three-quarters of the world's flowering plants depend on pollinators, and that almost one-third of the food that we consume results from their activity. The majority of pollinators are insects, whose distributions, phenology, and resources are all being affected by climate change (Inouye, 2007). Unfortunately, with the exception of honey bees and butterflies, few data are available on the abundance and distribution of pollinators, so it has been difficult to assess their status and the changes that they may be undergoing (NRC, 2006b).

V.1.c. Extreme events

Many effects of climate change on U.S. ecosystems and wildlife may emerge most strongly through changes in the intensity and the frequency of extreme events (Fischlin et al., 2007). In their review of the literature, Fischlin et al. (2007) identify multiple impacts of extreme events, such as hurricanes, tropical storms, and wildfires. These events can have substantial immediate effects, such as mass mortality in vegetation and wildlife populations, while also contributing significantly to alterations in species distribution, abundance, and fitness following the disturbance (Parmesan et al., 2000 in Fischlin et al., 2007). For example, the aftermath of a hurricane can cause coastal forest to die from storm surge-induced salt deposition, or wildlife may find it difficult to find food, thus lowering the chance of survival (Wiley and Wunderle, 1994 in Fischlin et al., 2007). Droughts play an important role in forest dynamics as well, causing pulses of tree mortality (Breshears et al., 2005 in Fischlin et al., 2007). Heat waves can contribute to both short-term (e.g., vegetation desiccation, mortality, reduced primary production, and increased wildfire extent) and long-term effects (e.g., depletion of oxygen in aquatic systems and conversion of vegetation type), as evidenced by the 2003 European heat wave (e.g., Beniston, 2004; Schär et al., 2004; Gobron et al., 2005; Ciais et al., 2005 in Fischlin et al., 2007). Greater magnitude and frequency of extreme events will alter disturbance regimes in coastal ecosystems, leading to changes in diversity and ecosystem function (e.g., Bertness and Ewanchuk, 2002 in Fischlin et al., 2007). Species inhabiting salt marshes, mangroves, and coral reefs are *likely* to be particularly vulnerable to these effects (Fischlin et al., 2007). To date, more than half of the original salt marsh habitat in the United States has been lost (Kennish, 2001).

Research reviewed by the IPCC (Field et al., 2007) indicated:

- Extreme events can add to other stresses on ecological integrity (Scavia et al., 2002; Burkett et al., 2005), including shoreline development and nitrogen eutrophication (Bertness et al., 2002).
- Many coastal ecosystems in North America are potentially exposed to storm-surge flooding (Titus and Richman, 2001; Titus, 2005); one area in which these extremes are evident is the San Francisco area, where 140 years of tide gauge data suggest an increase in severe winter storms since 1950 (Bromirski et al., 2003).
- Impacts on coastal communities and ecosystems can be more severe when major storms occur in short succession, limiting the opportunity to rebuild natural resilience (Forbes et al., 2004).
- Recent winters with less ice in the Great Lakes and Gulf of St. Lawrence have increased coastal exposure to damage from winter storms (Forbes et al., 2002).

Other potential extreme events influenced by a warming climate include wildfires, floods, drought, insects, and pathogens. Research summarized by the IPCC (Field et al., 2007) indicated:

- Since 1980, an average of 22,000 km^2 per year has burned in U.S. wildfires, almost twice the 1920 to 1980 average of 13,000 km^2 per year (Schoennagel et al., 2004).
- The forested area burned in the western United States from 1987 to 2003 is 6.7 times the area burned from 1970 to 1986 (Westerling et al., 2006).

- In Canada, burned area has exceeded 60,000 km^2 per year three times since 1990, twice the long-term average (Stocks et al., 2002).
- Wildfire-burned area in the North American boreal region increased from 6,500 km^2 per year in the 1 960s to 29,700 km^2 per year in the 1990s (Kasischke and Turetsky, 2006).
- Human vulnerability to wildfires has also increased, with a rising population in the wildland– urban interface.
- A warming climate encourages wildfires through a longer summer period that dries fuels, promoting easier ignition and faster spread (Running, 2006). Westerling et al. (2006) found that in the last three decades, the wildfire season in the western United States has increased by 78 days, and burn durations of fires greater than 1,000 hectares in area have increased from 7.5 to 37.1 days, in response to a spring/summer warming of 0. 87 °C.

The IPCC (Field et al., 2007) noted that insects and diseases are a natural part of ecosystems; however, changing conditions are increasing the severity of their effects. Forests often experience insect epidemics that kill trees over large regions. The lifecycles of these insects are climate-sensitive (Williams and Liebhold, 2002). As found by the IPCC (Field et al., 2007), warming temperatures are contributing to changes in insect lifecycles and may be contributing to more widespread epidemics. Specifically, many northern insects have a two-year lifecycle and warmer winter temperatures have allowed a greater proportion of overwintering larvae to survive. For example:

- "Recently, spruce budworm in Alaska has completed its lifecycle in one year, rather than the previous two (Volney and Fleming, 2000).
- Mountain pine beetle has expanded its range in British Columbia into areas previously too cold (Carroll et al., 2004).
- Insect outbreaks often have complex causes. Susceptibility of the trees to insects is increased when multiyear droughts degrade the trees' ability to generate defensive chemicals (Logan et al., 2003). Recent dieback of aspen stands in Alberta was caused by light snowpacks and drought in the 1980s, triggering defoliation by tent caterpillars, followed by woodboring insects and fungal pathogens (Hogg et al., 2002)."

Other key findings from SAP 4.3 (Janetos et al., 2008) included the following:

- "Evidence is beginning to accumulate that links the spread of pathogens to a warming climate. For example, the chytrid fungus (Batrachochytrium dendrobatidis) is a pathogen that is rapidly spreading world-wide and decimating amphibian populations. A recent study by Pounds and colleagues (2006) showed that widespread amphibian extinction in the mountains of Costa Rica is positively linked to global climate change."
- To date, geographic range expansion of pathogens related to warming temperatures have been the most easily detected (Harvell et al., 2002), perhaps most readily for arthropod-borne infectious disease (Daszak et al., 2000). However, a recent literature

review found additional evidence gathered through field and laboratory studies that support hypotheses that latitudinal shifts of vectors and diseases are occurring under warming temperatures.

- Nonetheless, invasive plants in general may better tolerate a wider range of environmental conditions and may be more successful in a warming world because they can migrate and establish in new sites more rapidly than native plants and they are not usually limited by pollinators or seed dispersers (Vila et al., 2007).
- Finally, it is critical to recognize that other elements of climate change (e.g., nitrogen deposition and land conversion) will play a significant role in the success of invasive plants in the future, either alone or under elevated CO2 (Vila et. al., 2007).

Projected impacts

The IPCC highlights the following projections that are expected to have significant implications for ecosystems:

- Some studies project widespread increases in extreme precipitation (Christensen et al., 2007), with greater risks of not only flooding from intense precipitation, but also droughts from greater temporal variability in precipitation.
- In general, projected changes in precipitation extremes are larger than changes in mean precipitation (Meehl et al., 2007).
- Future trends in hurricane frequency and intensity remain very uncertain. Climate models with sufficient resolution to depict some aspects of individual hurricanes tend to project some increases in both peak wind speeds and precipitation intensities (Meehl et al., 2007). The pattern is clearer for extra-tropical storms, which are likely to become more intense, but perhaps less frequent, leading to increased extreme wave heights in the mid-latitudes (Meehl et al., 2007).

Impacts on ecosystem structure and function may be amplified by changes in extreme meteorological events and increased disturbance frequencies (Field et al., 2007). Ecosystem disturbances, caused either by humans or by natural events, accelerate both loss of native species and invasion of exotics (Sala et al., 2000). Changes in plant species composition in response to climate change can facilitate other disturbances, including fire (Smith et al., 2000) and biological invasion (Zavaleta and Hulvey, 2004). In North America, disturbances like wildfire and insect outbreaks are increasing and are *likely* to intensify in a warmer future with drier soils and longer growing seasons (*very high confidence*) (Field et al., 2007).

V.1.d. Sea level rise

The IPCC (Rosenzweig et al., 2007) concluded that many "coastal regions are already experiencing the effects of relative (local) sea-level rise, from a combination of climate-induced sea-level rise, geological and anthropogenic-induced land subsidence, and other local factors." Overall, sea levels have been rising at a rate of 1.7 to 1.8 mm per year over the last century, with an increased rate of 3 mm per year over the last decade (e.g., Church et al., 2004; Bindoff et al., 2007 in Rosenzweig et al., 2007). They also conclude that sea level rise may be contributing to coastal erosion across the eastern United States (e.g., Zhang et al., 2004 in Rosenzweig et al., 2007). In addition, regional sea level rise has contributed to increased storm surge impacts along the North American eastern coast, although there has not been an increase in storm events (Zhang et al., 2000 in Rosenzweig et al., 2007). Due to low elevation and relative sea level rise, the U.S. Gulf Coast is particularly vulnerable to storm surges (Rosenzweig et al., 2007). These findings illustrate the interactive effects of climate impacts through rising sea levels and extreme storm events. Drawing on results like this, the IPCC concluded with *high confidence* that "coasts are experiencing the adverse consequences of hazards related to climate and sea level" (Parry et al., 2007).

For small islands, the coastline is long relative to island area. As a result, many resources and ecosystem services are threatened by a combination of human pressures and climate change effects including sea level rise, increases in sea surface temperature, and possible increases in extreme weather events (Mimura et al., 2007).

Climate change and sea level rise affect sediment transport in complex ways, as described by the IPCC (Nicholls et al., 2007). Erosion and ecosystem loss is affecting many parts of the U.S. coastline, but it remains unclear to what extent these losses result from climate change as opposed to land loss associated with relative sea level rise due to subsidence and other human drivers (Nicholls et al., 2007).

Coastal wetland loss is also occurring where these ecosystems are squeezed between natural and artificial landward boundaries and rising sea levels, a process known as 'coastal squeeze' (Field et al., 2007). The degradation of coastal ecosystems, especially wetlands and coral reefs, can have serious implications for the well-being of societies dependent on them for goods and services (Nicholls et al., 2007).

Engineering structures, such as bulkheads, dams, levees, and water diversions, limit sediment supply to coastal areas. Wetlands are especially threatened by sea level rise when insufficient amounts of sediment from upland watersheds are deposited on them. If sea level rises slowly, the balance between sediment supply and morphological adjustment can be maintained if a salt marsh vertically accretes (builds up through accumulated sediments and other materials), or a lagoon infills, at the same rate. However, an acceleration in the rate of sea level rise may mean that coastal marshes and wetlands cannot keep up, particularly where the supply of sediment is limited (e.g., where coastal floodplains are inundated after natural levees or artificial embankments are overtopped) (Nicholls et al., 2007).

Although open coasts have been the focus of research on erosion and shore stabilization technology, sheltered coastal areas in the United States are also vulnerable and suffer secondary effects from rising seas (NRC, 2006a). For example, barrier island erosion in Louisiana has increased the height of waves reaching the shorelines of coastal bays. This has enhanced erosion rates of beaches, tidal creeks, and adjacent wetlands. The impacts of accelerated sea level rise on gravel beaches have received less attention than sandy beaches. However, these systems are threatened by sea level rise, even under high wetland accretion rates. The persistence of gravel and cobble-boulder beaches will also be influenced by storms, tectonic events, and other factors that build and reshape these highly dynamic shorelines (Nicholls et al., 2007).

According to the IPCC, most of the world's sandy shorelines retreated during the past century and climate change-induced sea level rise is one underlying cause (Nicholls et al., 2007). To date in the United States, more than 50% of the original salt marsh habitat has already been lost (Kennish, 2001 in Field et al., 2007). In Mississippi and Texas, over half of the shorelines have eroded at average rates of 2.6 to 3.1 m per year since the 1970s, while

90% of the Louisiana shoreline has eroded at a rate of 12.0 m per year (Nicholls et al., 2007 and references therein).

In the Arctic, coastal stability is affected by factors common to all areas (i.e., shoreline exposure, relative sea level change, climate, and local geology), and by factors specific to the high latitudes (i.e., low temperatures, ground ice, and sea ice) (Anisimov et al., 2007). Adverse impacts have already been observed along Alaskan coasts and traditional knowledge points to widespread coastal change in Alaska. Rising temperatures in Alaska are reducing the thickness and spatial extent of sea ice. This creates more open water and allows for winds to generate stronger waves, which increase shoreline erosion. Sea level rise and thawing of coastal permafrost exacerbate this problem. Higher waves will create even greater potential for this kind of erosion damage (ACIA, 2004).

Projected impacts

The U.S. coastline is long and diverse with a wide range of coastal characteristics (Shaw et al., 1998; Dyke and Peltier, 2000; Zervas, 2001 in Field et al., 2007). Yet, relative sea level is rising in many areas (O'Reilly et al., 2005 in Field et al., 2007). Sea level rise changes the shape and location of coastlines by moving them landward along low-lying contours and exposing new areas to erosion (NRC, 2006a). Coasts subsiding due to natural or human-induced causes will experience larger relative rises in sea level. In some locations, such as deltas and coastal cities (e.g., the Mississippi Delta and surrounding cities), this effect can be significant (Nicholls et al., 2007). Rapid development, including an additional 25 million people in the coastal United States over the next 25 years, will further reduce the resilience of coastal areas to rising sea levels (Field et al., 2007). Superimposed on the impacts of erosion and subsidence, the effects of rising sea level will exacerbate the loss of waterfront property and increase vulnerability to inundation hazards (Nicholls et al., 2007).

Up to 21% of the remaining coastal wetlands in the U.S. mid-Atlantic region are potentially at risk of inundation between 2000 and 2100 (Field et al., 2007 and references therein). The IPCC (Field et al., 2007) concluded with *high confidence* that the rates of coastal wetland loss, in the Chesapeake Bay and elsewhere, will increase with accelerated sea level rise, in part due to 'coastal squeeze.' Salt marsh biodiversity is *likely* to decrease in northeastern marshes through expansion of non-native species such as *Spartina alterniflora*. The IPCC (Field et al., 2007) projects that many U.S. salt marshes in less-developed areas can potentially keep pace with sea level rise through vertical accretion.

Other conclusions drawn by the IPCC (Parry et al., 2007) included:

- Coasts are *very likely* to be exposed to increasing risks in future decades due to many compounding climate change factors (e.g., sea level rise, increase in sea surface temperatures, increased tropical storm intensity, and ocean acidification) (*very high confidence*).
- The impact of climate change on coasts is exacerbated by increasing human-induced pressures (*very high confidence*).
- Adaptation costs for vulnerable coasts are much less than the costs of inaction (*high confidence*).
- The unavoidability of sea level rise, even in the longer term, frequently conflicts with present-day human development patterns and trends (*high confidence*).

Specifically regarding North America coasts, the IPCC (Field et al., 2007) found:

- Coastal communities and habitats will be increasingly stressed by climate change impacts interacting with development and pollution (*very high confidence*). Sea level is rising along much of the coast, and the rate of change will increase, exacerbating the impacts of progressive inundation, storm surge flooding, and shoreline erosion.
- Storm impacts are *likely* to be more severe, especially along the Gulf and Atlantic Coasts. Salt marshes, other coastal habitats, and dependent species are threatened by sea level rise, fixed structures blocking landward migration, and changes in vegetation. Population growth and rising value of infrastructure in coastal areas increases vulnerability to climate variability and future climate change.

V.1.e. Vulnerable ecosystems

Coastal and marine ecosystems

Effects of elevated CO_2 concentrations and climate change on marine ecosystems include ocean warming, increased thermal stratification, reduced upwelling, sea level rise, increased wave height and frequency, loss of sea ice, increased risk of diseases in marine biota, and decreases in the pH and carbonate ion concentration of the surface oceans. Warmer ocean temperatures will contribute to changes in upwelling dynamics and decreased primary production along the California Current. In 2005, the upwelling season was delayed by three months (it began in August rather than a normal start in April to May), leading to a significant reduction in plankton production. The recruitment success among fish, birds, and mammals dependent upon the plankton was reduced significantly.

The United States has extensive coral reef ecosystems in both the Caribbean Sea and the Pacific Ocean. Coral reefs are very diverse ecosystems, home to a complex of species that support both local and global biodiversity and human societies. The sensitivity of coral reefs is increasingly recognized. With lower pH, aragonite (calcium carbonate) that is used by many organisms to make their shells or skeletons will decline or become under-saturated, affecting coral reefs and other marine calcifiers (e.g., pteropods—marine snails). Reef systems are increasingly vulnerable due to increasing thermal stress, interacting with the effects of various other stressors including increasing storm intensity, overfishing, pollution, and the introduction of invasive species leading to replacement of corals by other organisms in some locations (Fischlin et al., 2007; Nicholls et al., 2007).

In SAP 4.3, Janetos et al. (2008) indicate that it has been estimated that coral reefs provide $30 billion in annual ecosystem service value (Cesar et al., 2003). The effects of climate change in marine systems is highlighted by the 2006 listing of two species of corals in the Caribbean as Threatened under the Endangered Species Act (Federal Register, 2006). The major threats that motivated the listings of elkhorn (*Acropora palmata*) and staghorn (*A. cervicornis*) corals were disease, elevated sea surface temperatures, and hurricanes—all of which relate to climate change and its effects (Mann and Emanuel, 2006; Muller et al., 2007).

The northern Bering Sea along the Alaskan coast supports some of the highest benthic faunal biomass densities in the world's oceans (Janetos et al., 2008). Rising air and seawater temperatures have caused reductions in sea ice cover and primary productivity in benthic (the deepest environment of a water body, usually the seabed or lake floor) ecosystems. These rising temperatures are leading to a northward shift in the boundary between Arctic (northern Bering Sea) and subarctic (southern Bering Sea) waters resulting in a pelagic-dominated marine ecosystem that was previously confined to the southeastern Bering Sea (Anisimov et al., 2007). Changing conditions have led to significant reductions in seabird and marine mammal populations, increases in pelagic fish, occurrences of previously rare algal blooms, abnormally high water temperatures, and smaller salmon runs in coastal rivers (ACIA, 2004; Grebmeier et al., 2006 in Anisimov et al., 2007). Plants and animals in polar regions are also vulnerable to attacks from pests and parasites that develop faster and are more prolific in warmer and moister conditions (Anisimov et al., 2007).

Arctic sea ice ecosystem

(adapted from Janetos et al., 2008)

Changes in the Arctic are resulting in substantial shifts in habitat, especially for sea ice-dependent species. The sea ice, which provides habitat both below and above the ocean, has been in retreat for at least 30 years (Rothrock et al., 2003; Stroeve et al., 2005). Models project an Arctic Ocean free of summer sea ice by the end of this century (Overpeck et al., 2005), with increasing evidence suggesting this could happen by 2050 (Holland et al., 2006) and some models suggesting that it could happen as soon as 2040 (Holland et al., 2006). Ice loss to date is already causing measurable changes in polar bear and ringed seal populations and fitness (Stirling et al., 1999; Derocher et al., 2004; Ferguson et al., 2005).

Sea ice seasonally covered as much as 15 million square kilometers of the Arctic Ocean before it began declining in the 1 970s. For millennia, that ice has been integral to an ecosystem that provides for polar bears and the indigenous people. The ice also strongly influences the climate, oceanography, and biology of the Arctic Ocean and surrounding lands. Further, sea ice influences global climate in several ways, including via its high albedo and its role in atmospheric and oceanic circulation. Organisms that depend on sea ice ranging from ice algae to seals and polar bears will diminish in number or become extinct. Many changes have already been observed and are predicted to accelerate along with the rates of climate change.

At the base of the sea ice ecosystem are epontic algae adapted to very low light levels. Blooms of those algae on the undersurface of the ice are the basis of a food web leading through zooplankton and fish to seals, whales, polar bears, and people. Sea ice also strongly influences winds and water temperature, both of which influence upwelling and other oceanographic phenomena whereby nutrient-rich water is brought up to depths at which there is sufficient sunlight for phytoplankton to make use of those nutrients.

Among the more southerly and seasonally ice-covered seas, the Bering Sea produces the Nation's largest commercial fish harvests as well as supporting subsistence economies of Alaskan Natives. Ultimately, the fish populations depend on plankton blooms regulated by the extent and location of the ice edge in spring. Naturally, many other organisms, such as seabirds, seals, walruses, and whales, depend on primary production, mainly in the form of those plankton blooms. As Arctic sea ice continues to diminish, the location, timing, and species makeup of the blooms is changing in ways that appear to favor marked changes in community composition.

There are an estimated 20,000 to 25,000 polar bears (*Ursus maritimus*) worldwide, mostly inhabiting the annual sea ice over the continental shelves and inter-island archipelagos of the circumpolar Arctic (IUCN Polar Bear Specialist Group, 2006). Polar bears are specialized predators that hunt ice-breeding seals and are therefore dependent on sea ice for survival. After emerging in spring from a five- to seven-month fast in nursing dens, females require immediate nourishment and thus depend on close proximity between land and sea ice before the sea ice breaks up.

The IPCC noted that continuous access to sea ice allows bears to hunt throughout the year, but in areas where the sea ice melts completely each summer, they are forced to spend several months in tundra fasting on stored fat reserves until freeze-up (Fischlin et al., 2007). The two Alaskan populations (Chukchi Sea: ~2,000 individuals in 1993, and southern Beaufort Sea: ~1,500 individuals in 2006) are vulnerable to large-scale, dramatic seasonal fluctuations in ice movements because of the associated decreases in abundance of and access to prey and increases in the amount of energy needed for hunting (FWS, 2007). The IPCC projects that with a warming of 2.8 °C above pre-industrial temperatures and associated declines in sea ice, polar bears will face a high risk of extinction. Other ice-dependent species (e.g., walruses for rest and small whales for protection from predators) face similar consequences, not only in the Arctic but also in the Antarctic (Fischlin et al., 2007).

Recent modeling of reductions in sea ice cover and polar bear population dynamics predicted declines within the coming century that (though varied by population) totaled 66% of all polar bears (Amstrup et al., 2007 in Janetos et al., 2008). In 2005, the World Conservation Union (IUCN) Polar Bear Specialist Group concluded that the IUCN Red List classification for polar bears should be upgraded from 'Least Concern' to 'Vulnerable' based on the likelihood of an overall decline in the size of the total population of more than 30% within the next 35 to 50 years (Fischlin et al., 2007). In January 2007, the U.S. Fish and Wildlife Service determined that sufficient scientific evidence exists of a global threat to polar bears to warrant proposing it for listing as a threatened species under the Endangered Species Act (1973) (FWS, 2007). This decision was based in part on future risks to the species from climate change (Fischlin et al., 2007).

Temperate montane ecosystems

(adapted from Janetos et al., 2008)

In SAP 4.3, Janetos et al. (2008) wrote about the sensitivity of temperate montane ecosystems to global change. These ecosystems are characterized by cooler temperatures and often increased precipitation compared to surrounding lowlands. Consequently, much of that precipitation falls in the form of snow, which serves to insulate the ground from freezing air temperatures, stores water that is released as the snow melts during the following growing season, and triggers vertical migration by animal species that cannot survive in deep snow. With increasing temperatures, more precipitation is falling as rain (Johnson, 1998). Diaz et al. (2003) found that, over the past three to five decades, all the major continental mountain chains exhibited upward shifts in the height of the freezing level.

These environmental changes are also resulting in the disappearance of glaciers in most montane areas around the world. The changes in patterns and abundance of meltwater from these glaciers have significant implications for the one-sixth of the world's population that is dependent upon glaciers and melting snowpack for water supplies (Barnett et al., 2005b). Plant and animal communities are also affected as glaciers recede, exposing new terrain for colonization in an ongoing process of succession (e.g., for spider communities, see Gobbi et al., 2006). One group of organisms whose reproductive phenology is closely tied to snowmelt is amphibians, for which this environmental cue is apparently more important than temperature (Corn, 2003). Hibernating and migratory species that reproduce at high altitudes during the summer are also being affected by the ongoing environmental changes. For example, marmots are emerging a few weeks earlier than they used to in the Colorado Rocky Mountains, and robins are arriving from wintering grounds weeks earlier in the same habitats (Inouye et al., 2000). Species such as deer, bighorn sheep, and elk, which move to lower altitudes for the winter, may also be affected by changing temporal patterns of snowpack formation and disappearance.

Flowering phenology has been advancing in these habitats (Inouye and Wielgolaski, 2003) as well as in others at lower altitudes, mirroring what is occurring at higher latitudes (Wielgolaski and Inouye, 2003). There is a very strong correlation between the timing of snowmelt, which integrates snowpack depth and spring air temperatures, and the beginning of flowering by wildflowers in the Colorado Rocky Mountains (e.g., Inouye et al., 2002, 2003). The abundance of flowers can have effects on a variety of consumers, including pollinators (Inouye et al., 1991), herbivores, seed predators, and parasitoids, all of which are dependent on flowers, fruits, or seeds. An unexpected consequence of earlier snowmelt in the Rocky Mountains has been the increased frequency of frost damage to montane plants, including the loss of new growth on conifer trees, of fruits on some plants such as *Erythronium grandiflorum* (glacier lilies), and of flower buds of other wildflowers (e.g., *Delphinium* spp., *Helianthella quinquenervis*, etc.) (Inouye, 2008). Over time, this damage may lead to significant demographic consequences.

V.1.f. Adaptation options

SAP 4.4 (Julius et al., 2008) concludes that "while there will always be uncertainties associated with the future path of climate change, the response of ecosystems to climate impacts, and the effects of management, it is both possible and essential for adaptation to proceed using the best available science." The term 'adaptation' in this document refers to adjustments in human social systems (e.g., management) in response to climate stimuli and their effects. Ecosystem management always occurs in the context of desired ecosystem conditions or natural resource management goals. SAP 4.4 presents a series of analyses of adaptation options for several different types of federally managed ecosystems. On the basis of these analyses, the report concludes that:

- Many existing best management practices for 'traditional' stressors of concern have the added benefit of ameliorating climate change exacerbations of those stressors.
- Seven 'adaptation approaches' can be used for strategic adjustment of best management practices to maximize ecosystem resilience to climate change. Here, resilience refers to the amount of change or disturbance that a system can absorb without undergoing a fundamental shift to a different set of processes and structures. The seven approaches are:

  - **Protecting key ecosystem features**, which involves focusing management protections on structural characteristics, organisms, or areas that represent important 'underpinnings' or 'keystones' of the overall system.

- **Reducing anthropogenic stresses**, which is the approach of minimizing localized human stressors (e.g., pollution, fragmentation) that hinder the ability of species or ecosystems to withstand climatic events.

- **Representation**, which refers to protecting a portfolio of variant forms of a species or ecosystem so that, regardless of the climatic changes that occur, there will be areas that survive and provide a source for recovery.

- **Replication**, which centers on maintaining more than one example of each ecosystem or population within a management system, such that if one area is affected by a disturbance, replicates in another area provide insurance against extinction and a source for recolonization of affected areas.

- **Restoration**, which is the practice of rehabilitating ecosystems that have been lost or compromised.

- **Refugia**, which are areas that are less affected by climate change than other areas and can be used as sources of 'seed' for recovery or as destinations for climate- sensitive migrants.

- **Relocation**, which refers to human-facilitated transplantation of organisms from one location to another in order to bypass a barrier (e.g., urban area).

- Levels of confidence in these adaptation approaches vary and are difficult to assess, yet are essential to consider in adaptation planning.
- The success of adaptation strategies may depend on recognition of potential barriers to implementation and creation of opportunities for partnerships and leveraging.
- The Nation's adaptive capacity can be increased through expanded collaborations among ecosystem managers.
- The Nation's adaptive capacity can be increased through creative re-examination of program goals and authorities.
- Establishing current baselines, identifying thresholds, and monitoring for changes will be essential elements of any adaptation approach.
- Beyond 'managing for resilience,' the Nation's capability to adapt will ultimately depend on our ability to be flexible in setting priorities and 'managing for change.'

V.2. Agriculture and Food Production

In SAP 4.3, Hatfield et al. (2008) write that the food and agriculture sector includes a wide range of plant and animal production systems. The United States Department of Agriculture (USDA) classifies 116 plant commodity groups as agricultural products, as well as four livestock groupings (beef cattle, dairy, poultry, and swine) and products derived from animal production (e.g., cheese and eggs). These diverse commodities are produced in a variety of climates, regions, and soils. However, regardless of where they are grown, crops and livestock are affected by temperature, precipitation, CO_2, and water availability. Indeed, variability in yield from year to year is mostly (and strongly) related to weather effects during the growing season (Hatfield et al., 2008). These variations also affect crops and livestock through their effects on insects, disease, and weeds. Not only does U.S. agriculture produce necessary products, it has substantial economic value ($200 billion in 2002). Just over half (52%) derives from livestock, with the rest generated by crops (21% from fruit and nuts, 20% from grain and oilseed, 2% from cotton, and 5% from other commodity production).

The agricultural sector within the United States is sensitive to both short-term climate variability and long-term climate change. Productivity is driven by the interaction of a variety of variables including temperature, radiation, precipitation, humidity, and wind speed (Easterling et al., 2007). The latitudinal distribution of crops around the world is a function of climatic conditions including the total season precipitation and patterns of variability (Olesen and Bindi, 2002), as well as photoperiod (e.g., Leff et al., 2004).

Vulnerability of the U.S. agricultural sector to climate change is a function of many interacting factors including pre-existing climatic and soil conditions, changes in pest competition, water availability, and the sector's capacity to cope and adapt through management practices, seed and cultivar technology, and changes in economic competition among regions. The IPCC (Easterling et al., 2007) found that the growth, development, and yield of crops are dependent upon their responses to their climatic environment (Porter and Semenov, 2005). Particular crops are suited to a particular range of conditions, thus production is reduced and damage can occur when thresholds are exceeded, even for short periods in some cases (Wheeler et al., 2000; Wollenweber et al., 2003 in Easterling et al., 2007).

This section addresses how observed and projected climate change may affect U.S. food production and agriculture, including crop yields, irrigation requirements, effects from extreme events, pests and weeds, livestock production (e.g., milk and meat), and fisheries.

V.2.a. Crop yields and productivity

The productivity of most agricultural enterprises has increased dramatically over recent decades due to cumulative effects from technology, fertilizers, innovations in seed stocks and management techniques, and changing climate influences. Given the interaction of these various factors, it is difficult to identify the specific impact from any one factor on specific yield changes. The largest changes are probably due to technological innovations (Hatfield et al., 2008). However, weather events are a major factor in annual crop yield variation.

Reviewing crop yields in North America, the IPCC (Field et al., 2007) found the following:

- Yields of major commodity crops in the United States have increased consistently over the last century, typically at rates of 1 to 2% per year (Troyer, 2004), with significant variations across regions and between years.
- In the midwestern United States from 1970 to 2000, corn yield increased 58% and soybean yield increased 20%, with annual weather fluctuations resulting in year-to-year variability (Hicke and Lobell, 2004).

- For twelve major crops in California, climate fluctuations over the last 20 years have not had large effects on yield, though they have been a positive factor for oranges and walnuts and a negative factor for avocados and cotton (Lobell et al., 2006).

Increasing temperatures

(adapted from Hatfield et al., 2008)

In SAP 4.3, Hatfield et al. (2008) found that crop species differ in their cardinal temperatures (critical temperature range) for lifecycle development. There is a base temperature at which growth commences and an optimum temperature when the plant develops as fast as possible. Generally, increasing temperatures accelerate progression of a crop through its lifecycle (phenological) phases up to a species-dependent optimum temperature beyond which development slows.

Yield responses to temperature vary among species based on the specific temperature requirements. As temperatures increase, plants with a lower optimum temperature will exhibit significant decreases in yield before those with a higher optimum. Moreover, high temperatures will interact with other factors, such as low water availability, to further reduce productivity.

Research reported in Hatfield et al. (2008) found variable reductions in maize yield. One study found a 17% reduction per 1 °C increase across the United States (although this study did not include effects of water availability) (Lobell and Asner, 2003). Another study found that the response of global maize production to both temperature and rainfall over the period 1961 to 2002 was reduced 8.3% per 1°C warming (Lobell and Field, 2007).

Soybean has cardinal temperatures that are somewhat lower than those of maize. Responses to increasing temperatures are regionally dependent. Yield may actually increase 2.5% with a 1.2 °C rise in the upper Midwest, but would decrease 3.5% for 1.2 °C increase in the South (Boote et al., 1996, 1997). Lobell and Field (2007) reported a 1.3% decline in soybean yield per 1 °C increase in temperature, taken from global production against global average temperature during July to August, weighted by production area.

For wheat, Lawlor and Mitchell (2000) found that a 1 °C rise would shorten the reproductive phase by 6%, grain filling duration by 5%, and would reduce grain yield and harvest index proportionately. Bender et al. (1999) analyzed spring wheat grown at nine sites in Europe and found a 6% decrease in yield per 1 °C temperature rise. Lobell and Field (2007) reported a 5.4% decrease in global mean wheat yield per 1 °C increase in temperature. In addition, wheat illustrates the nonlinear effects that may occur as temperatures increase. In the wheat-growing regions of the Great Plains, yield is estimated to decline 7% per 1 °C increase in air temperature between 18 and 21 °C, and about 4% per 1 °C increase in air temperature above 21 °C, not considering any reduction in photosynthesis or grain-set.

The response of rice to temperature has been well studied (Baker and Allen, 1993a,b; Baker et al., 1995; Horie et al., 2000). Baker et al. (1995) summarized many of their experiments from sunlit controlled-environment chambers and concluded that the optimum mean temperature for grain formation and grain yield of rice is 25 °C. They found that grain yield is reduced about 10% per 1 °C temperature increase above 25 °C, until reaching zero yield at 35 to 36 °C mean temperature (Baker and Allen, 1993a; Peng et al., 2004). Mean air temperature during the rice grain-filling phase in summer in the southern United States and many tropical regions is about 26 to 27 °C. These are above the 25 °C optimum, which illustrates that temperatures elevated above current ones are projected to reduce rice yields in the United States and tropical regions by about 10% per 1 °C rise, or about 12% for a 1.2 °C rise.

Reviewing the literature for North America, the IPCC (Field et al., 2007) found the following regarding increasing temperatures:

- In the Corn and Wheat Belts of the United States, yields of corn and soybeans from 1982 to 1998 were negatively affected by warm temperatures, decreasing 17% for each 1 °C of warm-temperature anomaly (Lobell and Asner, 2003).
- In California, warmer nights have enhanced the production of high-quality wine grapes (Nemani et al., 2001), but additional warming may not result in similar increases.

Increasing CO_2

As with their responses to temperature, crops respond differently to increasing CO_2 concentrations. The evidence for maize response to CO_2 is sparse and questionable (Hatfield et al., 2008). On its own, the expected increment of CO_2 increase over the next 30 years is anticipated to have a negligible effect (i.e., 1%) on maize production (Leakey et al., 2006). In contrast, based on the metadata summarized by Ainsworth et al. (2002), a doubling of atmospheric CO_2 concentrations is expected to yield a 38% increase in soybean yield. In the midwestern United States, an atmospheric CO_2 increase from 380 to 440 ppm is projected to increase yield by 7.4%. For wheat, a cool-season cereal, doubling atmospheric CO_2 concentrations (350 to 700 ppm) increased grain yield by about 31%, averaged over many data sets (Amthor, 2001). For rice, doubling atmospheric CO_2 concentrations (330 to 660 ppm) increased grain yield by about 30% (Horie et al., 2000).

The certainty level of biomass and yield response of these C3 crops[42] to CO_2 is *likely* to *very likely,* given the large number of experiments and the general agreement in response across the different C3 crops.

V.2.b. Water use and irrigation requirements

(adapted from Hatfield et al., 2008)

Projected trends have conflicting effects on likely water needs. Increasing temperatures and a lengthening of the growing season will contribute to increased water demand. However, increasing CO_2 concentrations will contribute to reduced stomatal conductance (speed of water vapor evaporation from plant pores) and decreased demand (Ainsworth and Long, 2005; Ainsworth and Rogers, 2007).

The net irrigation requirement is the difference between seasonal evapotranspiration for a well- watered crop and the amounts of precipitation and soil water storage that are available during a growing season. Some researchers have attempted to estimate future changes in irrigation water requirements based on projected climate changes (including rainfall changes) from general circulation models and estimates of decreased stomatal conductance due to elevated atmospheric CO_2 (e.g., Allen et al., 1991; Izaurralde et al., 2003). For corn, Izaurralde et al. (2003) calculated that by 2030, irrigation requirements will vary from a reduction of 1% in the lower Colorado Basin to an increase of 451% in the lower Mississippi Basin, because of rainfall variation. Given the variation in the sizes and baseline irrigation requirements of U.S. river basins, a representative figure for the overall U.S. increase in irrigation requirements is 64% if stomatal effects are ignored or 35% if they are included.

Research suggests the impacts of climate change on irrigation water requirements may be large (Easterling et al., 2007). The IPCC considered this to be a new, robust finding since the Third Assessment Report in 2001. There is an expected increase in irrigation demand due to climate change in the majority of world regions including the United States due to decreased rainfall in certain regions and/or increased evapotranspiration arising from increased temperatures. In modeling studies of future climate change, additional irrigation is often assumed in order to counterbalance the potential adverse yield effects of significant temperature increases (Easterling et al., 2007).

V.2.c. Climate variability and extreme events

The IPCC (Easterling et al., 2007) found that "short-term natural extremes, such as storms and floods, interannual and decadal climate variations, as well as large-scale circulation changes, such as the El Niño–Southern Oscillation (ENSO), all have important effects on crop, pasture and forest production (Tubiello, 2005)." Indeed, the authors cite recent research that indicates El Niño-like conditions result in higher probabilities of reduced farm incomes across most of Australia (O'Meagher, 2005 in Easterling et al., 2007). They also mention a recently recognized correlation between the winter North Atlantic Oscillation (NAO) and conditions that favor improved wheat quality in the United Kingdom (Atkinson et al., 2005 in Easterling et al., 2007) but result in drought conditions and reduced summer growth in grasslands (Kettlewell et al., 2006 in Easterling et al., 2007).

In addition to changes in average climatic variables, such as temperature and precipitation, it is important to examine the potential for altered variability in extreme events such as extended heat waves, droughts, and floods. The potential for these events to change in frequency and magnitude introduces a key uncertainty regarding the yield of U.S. agriculture even under modest climate change. On this issue, the IPCC (Easterling et al., 2007) drew the following conclusion:

> Recent studies indicate that climate change scenarios that include increased frequency of heat stress, droughts and flooding events reduce crop yields and livestock productivity beyond the impacts due to changes in mean variables alone, creating the possibility for surprises. Climate variability and change also modify the risks of fires, and pest and pathogen outbreaks, with negative consequences for food, fiber and forestry (*high confidence*).

The adverse effects on crop yields due to droughts and other extreme events may offset the beneficial effects of elevated atmospheric CO_2, moderate temperature increases over the near term, and longer growing seasons.

On this topic, the IPCC (Field et al., 2007) reported that during the past decade, agriculture in North America has been exposed to many severe weather events. They concluded: "More variable weather, coupled with out-migration from rural areas and economic stresses, has increased the vulnerability of the agricultural sector overall, raising concerns about its future capacity to cope with a more variable climate (Senate of Canada, 2003; Wheaton et al., 2005)."

Drought events are already a frequent occurrence, especially in the western United States. Vulnerability to extended drought is, according to the IPCC (Field et al., 2007), increasing across North America as population growth and economic development increase demands from agricultural, municipal, and industrial uses, resulting in frequent over-allocation of water resources. While often associated with the western United States, the eastern region has also experienced droughts and attendant reductions in water supply, changes in water quality and ecosystem function, and challenges in allocation (Field et al., 2007).

Regarding future precipitation, projections suggest decreasing average annual precipitation in the southwestern United States but increases over the rest of North America (Christensen et al., 2007). Increasing precipitation can also cause challenges, depending on its intensity, amount, and timing. Some studies project widespread increases in extreme precipitation (Christensen et al., 2007), with greater risks of not only flooding from intense precipitation, but also droughts from greater temporal variability in precipitation.

In SAP 4.3, Hatfield et al. (2008) examine specific impacts and make the following conclusions:

- Historical data for many parts of the United States indicate an increase in the frequency of high-precipitation events (e.g., >5 cm in 48 hours), and this trend is projected to continue for many regions.
- Excessive rainfall results in delayed spring planting, leading to a reduction in profits through a premium available for early season production of high-value horticultural crops such as melon, sweet corn, and tomatoes.
- Flooding during the growing season can result in crop losses associated with anoxia and increased susceptibility to root diseases, while also resulting in additional impacts including soil compaction (due to use of heavy farm equipment on wet soils), and can lead to additional runoff and leaching of nutrients and chemicals into surface and groundwater.
- Concentrating a greater proportion of rainfall in high precipitation events will increase the likelihood of water deficiencies at other times because of reductions in rainfall frequency (Hatfield and Prueger, 2004).
- Storm events that deliver heavy rainfall are also often accompanied by strong wind gusts, increasing the potential for lodging of crops and reduced productivity.
- Wetter conditions at harvest time may result in reduced quality of many crops.
- Regarding temperature variation, short-term, episodic temperature increases can also lead to substantial effects when occurring just prior to, or during, critical crop pollination phases. Crop sensitivity and ability to compensate during later, improved weather varies across crops.

Easterling et al. (2007) include a brief description of agricultural losses sustained during the 2003 European heat wave (temperatures up to 6 °C above means accompanied by 300 mm precipitation deficits). There were substantial reductions in crop yield: 36% for maize in Italy's Po valley, and 30% for maize, 25% for fruit, and 21% for wheat in France (Ciais et al., 2005). Uninsured economic losses were estimated at € 13 billion (Sénat, 2004).

V.2.d. Pests and weeds

Pests and weeds can reduce crop yields, cause economic losses to farmers, and require management control options. How climate change (elevated atmospheric CO_2, increased temperatures, altered precipitation patterns, and changes in the frequency and intensity of extreme events) may affect the prevalence of pests and weeds is an issue of concern for food production and the agricultural sector. Recent warming trends in the United States have led to earlier spring activity by insects and proliferation of some species (Easterling et al., 2007). Additionally, research suggests that increased climate extremes may promote plant disease and pest outbreaks (Alig et al., 2004; Gan, 2004).

In particular, the IPCC review indicated that interactions between CO_2, temperature, and precipitation will play an important role in determining plant damage from pests in future decades (Stacey and Fellows, 2002; Chen et al., 2004; Salinari et al., 2006; Zvereva and Kozlov, 2006 in Easterling et al., 2007). However, to date, most studies continue to investigate pest damage as a separate function of either elevated ambient CO_2 concentrations or temperature. Pests and weeds are additional factors that, for example, are often omitted when projecting the stimulatory effect of elevated CO_2 on crop yields. Research on the combined effects of elevated atmospheric CO_2 and climate change on pests, weeds, and disease is still insufficient for U.S. and world agriculture (Easterling et al., 2007).

V.2.e. Projections for agriculture production

For North American agriculture, the IPCC (2007b) concluded with *high confidence*: "Moderate climate change [1–3 °C] in the early decades of the century is projected to increase aggregate yields of rain-fed agriculture by 5–20%, but with important variability among regions. Major challenges are projected for crops that are near the warm end of their suitable range or which depend on highly utilized water resources." Field et al. (2007) further explained that crops currently near their climate thresholds, such as wine grapes in California, are *likely* to suffer decreases in yields, quality, or both, with even modest warming (*medium confidence*) (Hayhoe et al., 2004; White et al., 2006). As for regional variations, increased climate sensitivity is anticipated in the southeastern United States as well as in the U.S. Corn Belt (Carbone et al., 2003), but not in the Great Plains (Mearns et al., 2003 in Field et al., 2007). Increased average warming leads to an extended growing season, especially for northern regions of the United States.

However, for global agriculture, the IPCC concluded: "At lower latitudes, especially seasonally dry and tropical regions, crop productivity is projected to decrease for even small local temperature increases (1–2 °C), which would increase the risk of hunger. (*medium confidence*)." Further warming is projected to have increasingly negative impacts in all regions (Easterling et al., 2007).

Reviewing recent integrated assessment models that explore interacting impacts of climate and economic factors on agriculture, water resources, and biome boundaries in the conterminous United States (e.g., Edmonds and Rosenberg, 2005; Izaurralde et al., 2005; Smith et al., 2005) led Field et al. (2007) to conclude that scenarios with decreased precipitation create important challenges, restricting the availability of water for irrigation and at the same time increasing water demand for irrigated agriculture and urban and ecological uses. There are still considerable uncertainties regarding precipitation changes. In addition, Field et al. (2007) found that the critical importance of specific agro-climatic events (e.g., last frost) introduces uncertainty in future projections (Mearns et al., 2003).

The IPCC (Easterling et al., 2007) concluded with *high confidence*: "Projected changes in the frequency and severity of extreme climate events will have more serious consequences for food and forestry production, and food insecurity, than will changes in projected means of temperature and precipitation." The authors cited modeling studies that suggest increasing frequency of crop loss from extreme events may counteract increased crop yields from rising temperatures.

There is still debate and uncertainty about the sensitivity of crop yields in the United States and other world regions to the direct effects of elevated atmospheric CO_2 levels. However, the IPCC (Easterling et al., 2007) concluded that elevated CO_2 levels are expected to contribute to small beneficial impacts on crop yields. The IPCC confirmed the general conclusions from its previous Third Assessment Report in 2001. Experimental research on crop responses to elevated atmospheric CO_2 through the FACE (Free Air CO_2 Enrichment)[43] experiments indicate that, at ambient CO_2 concentrations of 550 ppm (approximately double the concentration of preindustrial times), crop yields increase under unstressed conditions by 10 to 25% for C_3 crops, and by 0 to 10% for C_4 crops (*medium confidence*). Crop model simulations under elevated CO_2 are consistent with these ranges (*high confidence*) (Easterling et al., 2007). High temperatures, water and nutrient availability, and ozone exposure, however, can significantly limit the direct stimulatory effects of CO_2.

Analysis in SAP 4.3 suggests the complexity of these results. Hatfield et al. (2008) examined the interaction between temperature, precipitation, and atmospheric CO_2. Conclusions vary across crops and regions, from decreased yield in some cases to increases in other scenarios. Specifically, research reviewed in Hatfield et al. (2008) found that in many cases weeds respond more positively to increasing CO_2 than cash crops, particularly C_3 invasive weeds that reproduce by vegetative means (roots, stolons, etc.) (Ziska, 2003; Ziska and George, 2004). Indeed, in all weed–crop competition studies where the photosynthetic pathway is the same, weed growth is favored as CO_2 increases (Ziska and Runion, 2006a). Moreover, increasing atmospheric CO_2 levels may reduce the efficacy of glyphosate, the most widely used herbicide in the United States (Ziska et al., 1999). Increasing temperatures also increase the potential for northward expansion of weed species' ranges (Patterson et al., 1999).

SAP 4.3 also considered the effects of climate changes on the range of beneficial and harmful insects, microbes, and other organisms associated with agro-ecosystems. Temperature is the single most important factor affecting insect ecology, epidemiology, and distribution, while plant pathogens will be highly responsive to humidity and rainfall, as well as temperature (Coakley et al., 1999). To control the same insects, current trends demonstrate substantially greater applications of insecticide in warmer, more southern regions of the United States, compared to cooler, higher-latitude regions. For example, the frequency of pesticide sprays for control of lepidopteran insect pests in sweet corn currently ranges from 15 to 32 applications per year in Florida (Aerts et al., 1999), to 4 to 8 applications in Delaware (Whitney et al., 2000), and to 0 to 5 applications per year in New York (Stivers, 1999). Warmer winters are projected to increase populations of insect species that are currently marginally overwintering in high latitude regions, such as flea beetles (*Chaetocnema pulicaria*), which act as a vector for bacterial Stewart's Wilt (*Erwinia sterwartii*), an economically important corn pathogen (Harrington et al., 2001).

Moreover, an overall increase in humidity and frequency of heavy rainfall events projected for many parts of the United States will tend to favor some leaf and root pathogens (Coakley et al., 1999). However, these effects may be counterbalanced in some regions by an increase in short- to medium-term drought, which will decrease the duration of leaf wetness and reduce some forms of pathogen attack on leaves.

Considering this range of factors, Easterling et al. (2007) concluded that the vulnerability of North American agriculture to climatic change is multi-dimensional and is determined by interactions among pre-existing conditions, stresses stemming from climate change (e.g., changes in pest competition, water availability), and the sector's capacity to cope with multiple, interacting factors, including economic competition from other regions as well as advances in crop cultivars and farm management (Parson et al., 2003 in Field et al., 2007). Water access is the major factor limiting agriculture in southeast Arizona, but farmers in the region perceive that technologies and adaptations such as crop insurance have recently decreased vulnerability (Vasquez-Leon et al., 2002 in Field et al., 2007). Areas with marginal financial and resource endowments (e.g., the U.S. northern plains) are especially vulnerable to climate change (Antle et al., 2004 in Field et al., 2007). Unsustainable land use practices will tend to increase the vulnerability of agriculture in the U.S. Great Plains to climate change (Polsky and Easterling, 2001 in Field et al., 2007).

V.2.f. Livestock

Climate change has the potential to influence livestock productivity in a number of ways. Elevated atmospheric CO_2 concentrations can affect forage quality. Thermal stress can directly affect the health of livestock animals. An increase in the frequency or magnitude of extreme events can lead to livestock loss. Furthermore, climate change may affect the spread of animal diseases. In the 2007 Fourth Assessment Report, the IPCC generated a number of new conclusions in this area compared to the Third Assessment Report in 2001. The following conclusions can be applied to United States and other livestock-producing regions (Easterling et al., 2007):

- Elevated atmospheric CO_2 can increase the carbon to nitrogen ratio in forages and thus reduce the nutritional value of those grasses, which in turn affects animal weight and performance. Under elevated CO_2, a decrease of C_4 grasses and an increase of C_3 grasses (depending upon the plant species that remain) may occur, which could potentially reduce or alter the nutritional quality of the forage grasses available to grazing livestock. However, the exact effects on both types of grasses and their nutritional quality still need to be determined.
- Increased climate variability (including extremes in both heat and cold) and droughts may lead to livestock loss. The impact on animal productivity due to increased variability in weather patterns will *likely* be far greater than effects associated with the average change in climatic conditions.
- Thermal stress reduces productivity and conception rates and is potentially life threatening to livestock. According to one study reviewed by the IPCC (Frank et al., 2001), the U.S. percentage decrease in swine, beef, and dairy milk production in 2050 averaged 1.2, 2.0, and 2.2%, respectively, using one climate model, and 0.9, 0.7, and 2.1%, respectively, using a different climate model.
- Increasing temperatures may enable the spread of animal diseases from low to mid-latitudes, resulting in new threats and potentially reduced health for livestock (White et al., 2003; Anon, 2006; van Wuijckhuise et al., 2006).

In SAP 4.3, CCSP (Backlund et al., 2008) concluded that higher temperatures will *very likely* reduce livestock production during the summer season, but these losses will *very likely* be partially offset by warmer temperatures during the winter season. The further examination of the potential for thermal stress in SAP 4.3 (Hatfield et al., 2008) included the following:

- As environmental conditions result in core body temperature beyond normal diurnal boundaries, the animal must begin to conserve or dissipate heat to maintain homeostasis through behavioral, physiological, and metabolic thermoregulatory processes (Mader et al., 1997; M. Davis et al., 2003).
- These activities often result in declines in physical activity and an associated decline in eating and grazing activity as well as shifts in cardiac output, blood flow to extremities, and passage rate of digesta.
- Production losses in domestic animals are largely attributed to altered feed intake and increases in maintenance requirements to sustain a constant body temperature. (Mader et al., 2002; M. Davis et al., 2003; Mader and Davis, 2004).
- Voluntary feed intake increases (after a one- to two-day decline) by as much as 30% under cold stress and decreases by as much as 50% almost immediately under heat stress (NRC, 1987).
- Temperatures beyond the ability of the animal to dissipate result in reduced performance (i.e., production and reproduction), health, and well-being if adverse conditions persist (Hahn et al., 1992; Mader, 2003). Such reductions can result in substantial economic losses.
- Nighttime recovery is an essential element of survival when severe heat challenges occur (Hahn and Mader, 1997; Amundson et al., 2006).
- Heat wave events have resulted in documented livestock losses (Hahn and Mader, 1997; Hahn et al., 1999, 2001). Animal losses are particularly high for rapid changes in temperature due to extreme events when the animals are unable to acclimate[44] (High Plains Regional Climate Center, 2000).

V.2.g. Freshwater and marine fisheries

Fisheries are sensitive to changes in temperature and water supply, which affect flows of rivers and streams, as well as lake levels. Climate change can interact with other factors that affect the health of fish and productivity of fisheries (e.g., habitat loss and land use change). The IPCC (Easterling et al., 2007) found the following:

- Increased temperatures can lead to seasonal increases in fish growth but may also increase risks at the upper end of their thermal tolerance zone. The specific effects of interactions between increasing temperature and other global changes including declining pH are unclear.
- Direct effects of increasing temperature on marine and freshwater systems are already occurring, with rapid poleward shifts in some regions. Future shifts are expected to continue. Local extinctions are occurring at the edges of current ranges, particularly for freshwater and anadromous species such as salmon and sturgeon (Friedland et al., 2003; Reynolds et al., 2005).

- Observed changes in primary production will affect fisheries as their effects are transferred through the food chain. While results vary regionally, evidence indicates reduced nutrient supply to the upper productive layer of the Pacific and Atlantic Oceans (McPhaden and Zhang, 2002; Curry and Mauritzen, 2005).
- Climate change has been implicated in mass mortalities of a wide variety of aquatic species, but limited data exists to confirm the correlation generally (Harvell et al., 1999). However, research does confirm the poleward spread of two protozoan parasites from the Gulf of Mexico to Delaware Bay and north, resulting in mass mortalities of oysters (Hofmann et al., 2001).
- Climate change is expected to result in the largest economic impacts in the fisheries sector in central and northern Asia, the western Sahel, and coastal, tropical regions of South America (Allison et al., 2005) as well as some small and medium-sized island states (Aaheim and Sygna, 2000).
- Regional climate variability can severely affect the productivity and species composition of major marine fisheries such as tuna (skipjack, yellowfin, and albacore) and Peruvian anchovy (Barber, 2001; Lehodey et al., 2003). Models need to be improved to account for decadal trends in addition to ENSO events as they will have a greater impact on the food web.

In SAP 4.3, Janetos et al. (2008) further examined the impacts of the NAO, the Pacific Decadal Oscillation (PDO), and ENSO on marine ecosystems.

- The NAO has been strongly positive since the 1980s, resulting in dramatic impacts on northeast Atlantic ecosystems. These impacts include increased flow of oceanic water into the English Channel and North Sea, a northward shift in the distribution of warm water zooplankton species (Beaugrand, 2004), and concomitant changes in dominance in fish communities from whiting (hake) to sprat (similar to a herring). Similar ecosystem shifts were observed in the Baltic Sea, where drastic changes in both zooplankton and fish communities have been observed (Kenny and Mollman, 2006).
- In the North Pacific, the PDO refers to the east–west shifts in location and intensity of the Aleutian Low in winter (Mantua et al., 1997). Widespread ecological changes have been observed, including increased productivity of the Gulf of Alaska when the PDO is in positive phase, resulting in dramatic increases in salmon production (Mantua et al., 1997) and a reversal of demersal fish community dominance from a community dominated by shrimps to one dominated by pollock (Anderson and Piatt, 1999). Associated changes in the California Current ecosystem include dramatic decreases in zooplankton (McGowan et al., 1998) and salmon (Pearcy, 1991) when the PDO changed to a positive phase in 1977. There is also evidence that the large oscillations in sardine and anchovy populations are associated with PDO shifts, such that during positive (warm) phases, sardine stocks are favored but during negative (cool) phases, anchovy stocks dominate (e.g., Chavez et al., 2003).
- ENSO events negatively affect zooplankton and fish stocks, resulting in a collapse of anchovy stocks in offshore ecosystems of Peru. Losses of anchovies, which are harvested for fishmeal, affect global economies because fishmeal is an important

component of chicken feed as well as high-protein supplements in aquaculture feed. In waters off the West Coast of the United States, plankton and fish stocks may collapse due to sudden warming of the waters (by 4 to 10 °C) as well as through poleward advection of tropical species into temperate zones. Many of the countries most affected by ENSO events are developing countries in South America and Africa, with economies that are largely dependent upon agricultural and fishery sectors as a major source of food supply, employment, and foreign exchange.

The IPCC (Field et al., 2007 and references therein) reviewed a number of North American studies showing how freshwater fish are sensitive to, or are being affected by, observed changes in climate:

- Cold- and cool-water fisheries, especially salmonids, have been declining as warmer/drier conditions reduce their habitat. The sea-run salmon stocks are in steep decline throughout much of North America.
- Pacific salmon have been appearing in Arctic rivers.
- Salmonid species have been affected by warming in U.S. streams.
- Success of adult spawning and survival of fry brook trout are closely linked to cold groundwater seeps, which provide preferred temperature refuges for lake-dwelling populations. Rates of fish egg development and mortality increase with temperature increases within species-specific tolerance ranges.

In addition, Janetos et al. (2008) identified impacts in the Bering Sea, which is among the more southerly seasonally ice-covered seas. The Bering Sea produces the Nation's largest commercial fish harvests and supports subsistence economies of Alaskan Natives. Ultimately, the fish populations depend on plankton blooms regulated by the extent and location of the ice edge in spring. As Arctic sea ice continues to diminish, the location, timing, and species makeup of the blooms is changing in ways that appear to promote marked changes in community composition. Specifically, the spring melt of sea ice in the Bering Sea has long favored the delivery of organic material to a benthic community of bivalve mollusks, crustaceans, and other organisms. Those benthic organisms, in turn, are important food for walruses, gray whales, bearded seals, eider ducks, and many fish species. The earlier ice melt resulting from a warming climate, however, leads to later phytoplankton blooms that are largely consumed by zooplankton near the sea surface, vastly decreasing the amount of organic material reaching the benthos. This is expected to result in a radically altered community of organisms favoring a different suite of upper-level consumers and altering the subsistence and commercial harvests of fish and other marine organisms.

Regarding the impacts of future climate change, the IPCC concluded, with *high confidence* for North America, that coldwater fisheries will *likely* be negatively affected, warm-water fisheries will generally benefit, and the results for cool-water fisheries will be mixed, with gains in the northern and losses in the southern portions of ranges (Field et al., 2007). A number of specific impacts by fish species and region in North America are projected (Field et al., 2007 and references therein):

- Salmonids, which prefer cold water, are *likely* to experience the most negative impacts.

- Arctic freshwaters are *likely* to be most affected, as they will experience the greatest warming.
- Many warm-water and cool-water species will shift their ranges northward or to higher altitudes.
- In the continental United States, coldwater species will *likely* disappear from all but the deeper lakes; cool-water species will be lost mainly from shallow lakes; and warm-water species will thrive, except in the far south, where temperatures in shallow lakes will exceed survival thresholds.

Climate variability and change can also affect fisheries in coastal and estuarine waters, although non-climatic factors, such as overfishing and habitat loss and degradation, are already responsible for reducing fish stocks (Nicholls et al., 2007). Coral reefs, for example, are vulnerable to a range of stresses and for many reefs, thermal stress thresholds will be crossed, resulting in bleaching, with severe adverse consequences for reef-based fisheries (Nicholls et al., 2007). Increased storm intensity, temperature, and saltwater intrusion in coastal water bodies can also adversely affect coastal fisheries production.

V.3. Land Resources

The Nation has a large and diverse land base that can be classified based on the type of vegetation supported. In the following sections, we discuss two major land types in the United States: forests, and grasslands and shrublands.

In SAP 4.3, Ryan et al. (2008) found that climate strongly influences both forests and arid lands by shaping broad patterns of ecological communities, vegetation, and wildlife species, their productivity, and the ecosystem goods and services they provide. The interaction of vegetation and climate is a fundamental tenet of ecology. Substantial research has documented how vegetation has changed with climate over the past several thousand years, providing not only a historic description of vegetation but also a predictive ability to anticipate upcoming conditions by drawing on these historic trends. However, because current changes are occurring at an unprecedented rate, in many cases these previously identified relationships and trends may no longer adequately describe future conditions. For example, emerging changes have been identified in the physical climate, levels of CO_2 in the atmosphere, nitrogen deposition, local areas of ozone pollution, species invasions, and disturbance patterns. These factors cause important changes themselves, but their interactions are difficult to predict. This is particularly so because these interactions represent novel combinations beyond society's experience base.

Terrestrial ecosystems are dynamic. Disturbances (such as drought, storms, insect outbreaks, grazing, and fire) have a strong influence on ecological communities and landscapes. In many cases, the timing, magnitude, and frequency of disturbances are influenced by climate. Accordingly, a changing climate will lead to changes in disturbances and ultimately the composition, structure, and function of ecosystems (including productivity, water yield, erosion, and carbon storage) as well as susceptibility to future disturbance across U.S. land resources (Dale et al., 2001).

Disturbance may reset and rejuvenate some ecosystems in some cases and cause enduring change in others. In SAP 4.3, Ryan et al. (2008) provided two examples. First, climate may favor the spread of invasive exotic grasses into arid lands where the native vegetation is too sparse to carry a fire. When these areas burn, they typically convert to non-native monocultures and the native vegetation is lost. In another example, drought may weaken trees and make them susceptible to insect attack and death—a pattern that recently occurred in the Southwest. In these forests, drought and insects converted large areas of mixed pinyon–juniper forests into juniper forests. However, fire is an integral component of many forest ecosystems, and many tree species (such as the lodgepole pine forests that burned in the Yellowstone fires of 1988) depend on fire for regeneration. Ryan et al. (2008) conclude that climate effects on disturbance are expected to shape terrestrial ecosystems as much as the direct effects of climate.

V.3.a. Forests

Forests occur globally and, although forest types are quite diverse, they are all characterized by extensive tree cover. Regarding global forests, research reviewed in Easterling et al. (2007) indicates:

- Globally, forest ecosystems are found on 41.6 million km^2, making up approximately 30% of all land.
- Of forestland, 42% occurs in the tropics, 25% in temperate zones, and 33% in the boreal zone (e.g., Sabine et al., 2004).
- Forests are among the most productive terrestrial ecosystems.
- Currently about one-quarter of anthropogenic CO_2 emitted globally comes from deforestation activities, primarily in tropical and subtropical regions (Houghton, 2003).
- Notwithstanding, forests still sequester the largest fraction of terrestrial ecosystem carbon stocks, recently estimated at 1,640 petagrams of carbon (PgC),[45] equivalent to about 220% of atmospheric carbon (Sabine et al., 2004).
- Forests provide a number of goods (in addition to timber products) that are important for subsistence livelihoods (Gitay et al., 2001; Shvidenko et al., 2005). They also provide key ecosystem services including providing habitat for an increasing fraction of biodiversity (particularly in areas subject to land use pressures: Hassan et al., 2005; MEA, 2005a); carbon sequestration; climate regulation; soil and water protection or purification (>75% of the world's usable freshwater supplies come from forested catchments: Shvidenko et al., 2005); and recreational, cultural, and spiritual benefits (Reid et al., 2005).

In SAP 4.3, Ryan et al. (2008) described the U.S. forest resource in the following manner. Forests cover about 3 million square kilometers (740 million acres) of the United States, making up approximately one-third of the land cover. While occurring in every state, forests are most prevalent in the humid eastern United States, the West Coast, at higher elevations in the interior West and Southwest, and along riparian corridors in the plains states (Zhu and Evans, 1994). The U.S. forest resource is quite diverse. In the eastern United States, there are 1.54 million square kilometers (380 million acres) of forestland—most of which

(83%) is privately owned and 74% of which is broadleaf forest. The 1.46 million square kilometers (360 million acres) of forestland in the western United States are split between public (5 7%) and private ownership and largely consist of coniferous forest types (78%) (USDA Forest Service and U.S. Geological Survey, 2002).

As with global forests, U.S. forests provide a range of goods and ecosystem services important to the well-being of the people of the United States, including raw material for wood and paper products as well as many non-consumptive values and uses. Many Americans are strongly attached to their forests (Ryan et al., 2008). While all forests have considerable economic value, many values are not easily quantified (Costanza et al., 1997; Daily et al., 2000; Krieger, 2001; MEA, 2005b). A changing climate will alter forests and their ability to continue to provide these goods and services at current levels. In addition, impacts increase as human population, recreation, and tourism increase in forested regions across the nation.

V.3.b. Grasslands and shrublands

As defined in the *State of the Nation's Ecosystems* report these lands include diverse ecosystem types including the sagebrush steppes of the northern Rockies, the prairies of the Midwest and the Great Plains, the deserts of the Southwest and the intermountain West, the Alaskan tundra and shrublands, and the scrublands of Florida (Heinz Center, 2002). These ecosystems cover more than 3.1 million square kilometers (770 million acres), making up more than a third of the U.S. land base. A majority are privately owned, particularly east of the Rocky Mountains. These ecosystems are very diverse; example vegetation types include annual grasslands and chaparral in California, sagebrush/bunchgrass in the Great Basin, hot-desert shrublands of New Mexico, plains grasslands of mid-America, the oak savanna of Texas, wet grasslands of Florida, and tundra of Alaska.

Fischlin et al. (2007) found these systems are generally rich in grazing, browsing, and other fauna (especially but not only in Africa), and are strongly controlled by fire (Bond et al., 2005 and/or grazing regimes (Scholes and Archer, 1997; Fuhlendorf et al., 2001). In many cases, disturbance regimes are human-managed (e.g., Sankaran, 2005), although fire regimes depend also on seasonality of ignition events and rainfall-dependent accumulation of flammable material (Brown et al., 2005).

The IPCC (Fischlin et al., 2007) drew the following conclusions regarding these systems:

- These systems appear more sensitive than previously thought to variability of, and changes in, major climate change drivers.
- The CO_2-fertilization impact and warming effect of rising atmospheric CO_2 have contrasting effects on their dominant functional types (trees and C_3 grasses may benefit from rising CO_2 but not from warming; C_4 grasses may benefit from warming, but not from CO_2 fertilization). Uncertain, nonlinear, and rapid changes in ecosystem structure and carbon stocks are *likely.*
- Carbon stocks are *very likely* to be strongly reduced under more frequent disturbance, especially by fire, and disturbance and drought impacts on vegetation cover may exert regional feedback effects.
- On average, grasslands are *likely* to show reduced carbon sequestration (due to enhanced soil respiratory losses through warming, fire regime changes, and increased

rainfall variability). However, possible regional gains in woody cover (through CO_2 fertilization and increased plant carbon stocks) cannot be excluded.

- Scientific predictive skill is currently limited by very few field-based, multi-factorial experiments, especially in tropical systems.
- Projected range shifts of mammal species will be limited by fragmented habitats and human pressures, and declines in species richness are *likely*, especially in protected areas.
- Because of the important control by disturbance, management options exist to develop adaptive strategies for carbon sequestration and species conservation goals.

V.3.c. Productivity

Forest productivity

Climate strongly influences forest productivity and species composition. Research reviewed in the North America IPCC chapter indicates that forest growth appears to have increased slightly in the previous decade (less than 1% per decade) in regions where growth has historically been limited by low temperatures and short growing seasons (Caspersen et al., 2000; McKenzie et al., 2001; Joos et al., 2002; Boisvenue and Running, 2006 in Field et al., 2007). However, as noted by Ryan et al. (2008), it is difficult to separate the role of climate from other potentially influencing factors particularly because these interactions vary by location. Other potentially influential factors include increases in precipitation (observed in the Midwest and Lake States), increases in nitrogen deposition, temperature increases and a lengthened growing season in the northern United States, changing age structure of forests (greater percentage of forests in young age classes), and evolving management practices.

Research reviewed by the IPCC (Field et al., 2007) indicates the vegetation growing season has increased by an average of 2 days per decade since 1950 in Canada and the conterminous United States, with most of the increase resulting from earlier spring warming (Bonsal et al., 2001; Easterling 2002; Bonsal and Prowse, 2003; Feng and Hu, 2004). While this allows a greater period of growth and, thus, potential to increase productivity, earlier warming can also contribute to dryer conditions and increased potential for disturbance, both of which may act to offset the increases. As above, the effects of these trends vary by region. Field et al. (2007) found that in temperature-limited areas, productivity had increased, while in areas subject to drought it had decreased. For example, height growth had increased since the 1 970s in black spruce at the forest–tundra transition in eastern Canada (Gamache and Payette, 2004). On the other hand, radial growth of white spruce on dry south-facing slopes in Alaska has decreased over the last 90 years (Barber et al., 2000), while growth rates in semi-arid forests of the southwestern United States have decreased since 1895, which is correlated with drought linked to warming temperatures (McKenzie et al., 2001). For a widespread species like lodgepole pine, a 3 °C temperature increase would increase growth in the northern part of its range, decrease growth in the middle, and decimate southern forests (Field et al., 2007).

Additional research presented by Field et al. (2007) provided evidence that these relationships can also occur within the same region based on topographical differences. Research in subalpine forests in the Pacific Northwest from 1895 to 1991 found that growth of subalpine fir and mountain hemlock was negatively correlated with spring snowpack depth and positively correlated with summer temperatures. This indicates temperature limitations on the growing season on high-elevation, north-facing slopes. At lower elevations, however, growth was negatively correlated with summer temperature, suggesting water limitations. (Peterson and Peterson, 2001; Peterson et al., 2002 in Field et al., 2007). There is also evidence of shifting species ranges. For example, aspen have advanced into the more cold-tolerant spruce–fir forests in Colorado over the past 100 years (Elliott and Baker, 2004 in Field et al., 2007) and lodgepole pine has advanced northward into areas previously dominated by the more cold-tolerant black spruce in the Yukon (Johnstone and Chapin, 2003 in Field et al., 2007).

Overall, productivity gains in one area may be offset by losses elsewhere, both through direct effects of climate changes and through important secondary effects on frequency and intensity of natural disturbances such as fire, insect outbreaks, ice storms, and windstorms. For example, Easterling et al. (2007) cited research projecting short-term productivity increases in California forests, in the area available for productive softwood growth, through 2020 with reductions in the long run (up to 2100) (Mendelsohn, 2003). Recent years have also seen a substantial increase in the area affected by wildfires and insect outbreaks in U.S. forests. These disturbances are further discussed in the following section. Easterling et al. (2007) also noted the changing climate will have substantial impacts on non-timber goods (e.g., seeds, nuts, hunting, resins, and plants used in pharmaceutical and botanical medicine and in the cosmetics industry) and services (e.g., wildlife habitat and recreation opportunities) offered by forest ecosystems.

Moreover, the commercial U.S. forest sector is expected to be affected by changing forest productivity in different regions of the world. Sohngen and Sedjo (2005) show two climate change scenarios where North American forests undergo more dieback in general than forests in other regions of the world, and where certain North American forest yields increase but less so compared to other regions (in Easterling et al., 2007). The implication is that forests in other parts of the world (including tropical forests with shorter rotations) could have a competitive advantage within the global forestry sector under a changing climate.

Grassland and shrubland productivity

Fischlin et al. (2007) found the ecosystem function and species composition of grasslands and savanna are *likely* to respond mainly to precipitation change and warming in temperate systems. In tropical systems, however, CO_2 fertilization and emergent responses to herbivory and fire regime will also exert strong control. Specific conclusions included the following:

- Rainfall change and variability is *very likely* to affect vegetation in tropical grassland and savanna systems with, for example, a reduction in cover and productivity simulated along an aridity gradient in southern African savanna in response to the observed drying trend of about 8 mm per year since 1970 (Woodward and Lomas, 2004a).
- Changing amounts and variability of rainfall may also strongly control the responses of temperate grassland to future climate change (Novick et al., 2004; Zha et al., 2005). For example, a Canadian grassland fixed roughly five times as much carbon in a year with 30% higher rainfall, while a 15% rainfall reduction led to a net carbon loss (Flanagan et al., 2002). Similarly, Mongolian steppe grassland switched from

carbon sink to carbon source in response to seasonal water stress, although carbon balance was neutral on an annual basis (Li et al., 2005).

- Nonlinear responses to increasing rainfall variability may be expected, as ecosystem models of mixed C_3/C_4 grasslands show initially positive net primary productivity relationships with increasing rainfall variability, but greater variability ultimately reduces both net primary productivity and ecosystem stability even if the rainfall total is kept constant (Mitchell and Csillag, 2001).
- Empirical results for C_4 grasslands confirm a similar monotonic relationship between net primary productivity and rainfall variability (Nippert et al., 2006).
- Increased rainfall variability was more significant than rainfall amount for tall-grass prairie productivity (Fay et al., 2000, 2002), with a 50% increase in dry-spell duration causing a 10% reduction in net primary productivity (Fay et al., 2003) and a 13% reduction in soil respiration (Harper et al., 2005).
- Elevated CO_2 has important effects on production and soil water balance in most grassland types, mediated strongly by reduced stomatal conductance and resulting increases in soil water (Leakey et al., 2006) in many grassland types (Nelson et al., 2004; Niklaus and Körner, 2004; Stock et al., 2005).
- In short-grass prairie, elevated CO_2 and 2.6 °C warming increased production by 26 to 47%, regardless of grass photosynthetic type (J. Morgan et al., 2001). In C_4 tropical grassland, no relative increase in herbaceous C_3 success occurred with a doubling of the ambient CO_2 (Stock et al., 2005). Regional climate modeling indicates that CO_2 fertilization effects on grasslands may scale up to affect regional climate (Eastman et al., 2001).

V.3.d. Disturbance

In large measure, the composition, structure, function, and condition of our land resources have been shaped by their disturbance history (both human-induced and natural). Disturbances drive change at both small (e.g., local wind event resulting in minor windthrow) and large scales (e.g., the 2002 Biscuit fire covering approximately 2,000 square kilometers (500,000 acres) in southern Oregon and northern California). On an annual basis, more than 220,000 square kilometers (55 million acres) of forestland in the United States are impacted by disturbance annually (Dale et al., 2001).

Wildland fires

While in some cases a changing climate may have positive impacts on the productivity of forest systems, changes in disturbance patterns are expected to have a substantial impact on overall gains or losses. According to studies reviewed by the IPCC (Field et al., 2007), in the absence of dramatic increases in disturbance, the effects of climate change on the potential for commercial forest harvest in the 2040s ranged from mixed for a low emissions scenario to positive for a high emissions scenario (Perez-Garcia et al., 2002). However, the tendency for North American producers to suffer losses increases if climate change is accompanied by increased disturbance, with simulated losses averaging $1 to 2 billion per year over the 21st century (Sohngen and Sedjo, 2005).

More prevalent fire activity has recently been observed in the United States and other world regions (Fischlin et al., 2007). Wildfires and droughts, among other extreme events (e.g., hurricanes), can result in short-term losses and long-term shifts in forest ecosystems. Wildland fires can also have direct economic implications due to large fire suppression costs (OFCM, 2005). The frequency and severity of these extreme events are expected to be altered by climate change.

In a review of fire activity in the western United States from 1974 to 2004, Westerling et al. (2006) found that both the frequency of large wildfires and fire season length increased substantially after 1985, and that these changes were correlated with advances in the timing of spring snowmelt and increases in spring and summer air temperatures. They concluded that earlier spring snowmelt contributed to greater wildfire frequency by extending the period during which ignitions could potentially occur and by reducing water availability to ecosystems in midsummer, thus enhancing drying of vegetation and surface fuels (Westerling et al., 2006). These trends in increased fire size correspond with the increased cost of fire suppression (Calkin et al., 2005).

Fire activity also has increased in recent decades in boreal forests across North America as more than twice as much area burned during the 1980s and 1990s than during the 1960s and 1970s. As in the western United States, a key predictor of burned area in boreal North America is air temperature, with warmer summer temperatures causing an increase in burned area on both interannual and decadal time scales (Gillett et al., 2004, Duffy et al., 2005, Flannigan et al., 2005 in Field et al., 2007). In Alaska, for example, June air temperatures alone explained approximately 38% of the variance of the natural log of annual burned area during 1950 to 2003 (Duffy et al., 2005).

The IPCC (Field et al., 2007) noted a number of observed changes to U.S. wildfire size and frequency, often associating these changes with changes in average temperatures:

- Since 1980, an average of about 22,000 km^2 per year (13,700 mi^2 per year) has burned in wildfires, almost twice the 1920 to 1980 average of about 13,000 km^2 per year (8,080 mi^2 per year) (Schoennagel et al., 2004).
- The forested area burned in the western United States from 1987 to 2003 is 6.7 times the area burned from 1970 to 1986 (Westerling et al., 2006).
- Human vulnerability to wildfires has increased, with a rising population in the wildland– urban interface.
- In the last three decades, the wildfire season in the western United States increased by 78 days, and burn durations of fires greater than 1,000 hectares (2,470 acres) have increased from 7.5 to 37.1 days, in response to a spring/summer warming of 0.87 °C (Westerling et al., 2006).
- Earlier spring snowmelt has led to longer growing seasons and drought, especially at higher elevations, where the increase in wildfire activity has been greatest (Westerling et al., 2006).
- In the southwestern United States, fire activity is correlated with ENSO positive phases and higher Palmer Drought Severity Indices (Kitzberger et al., 2001; McKenzie et al., 2004).[46]

These findings led the IPCC to conclude (Field et al., 2007): "Disturbances such as wildfire and insect outbreaks are increasing and are *likely* to intensify in a warmer future with drier soils and longer growing seasons (*very high confidence*)."

This conclusion is supported by findings reported in SAP 4.3. Ryan et al. (2008) found that several lines of evidence suggest that large, stand-replacing wildfires are expected to increase in frequency over the next several decades because of climate. Historical records unequivocally show that warmer and drier periods during the last millennium are associated with more frequent and severe wildfires in western forests. General circulation model (GCM) projections of future climate during 2010 to 2029 suggest that the number of low-humidity days (and high fire danger days) will increase across much of the western United States, allowing for more wildfire activity with the assumption that fuel densities and land management strategies remain the same (Flannigan et al., 2000; Brown et al., 2004). Flannigan et al. (2000) used two GCM simulations of future climate to calculate a seasonal severity rating related to fire intensity and difficulty of fire control. Depending on the GCM used, forest fire seasonal severity rating in the Southeast is projected to increase from 10 to 30%, and from 10 to 20% in the Northeast by 2060. Other biome models used with a variety of GCM climate projections simulate a larger increase in fire activity and biomass loss in the Southeast, sufficient to convert the southernmost closed-canopy Southeast forests to savannas (Bachelet et al., 2001).

In SAP 4.3, Ryan et al. (2008) found that in grasslands, particularly arid lands, non-native plant invasions, promoted by enhanced nitrogen deposition (Fenn et al., 2003) and increased anthropogenic disturbance (Wisdom et al., 2005), will have a major impact on how arid land ecosystems respond to climate and climate change. Once established, non-native annual and perennial grasses can create a continuous load of fine fuels in areas where vegetation historically occurred in patches across the landscape. This results in a changed fire regime and predisposes arid lands to fires more frequent and intense than those with which they evolved.

Not only will an increase in fire activity result in a greater area of disturbance, fire emissions across North America will have important consequences for climate forcing agents, air quality, and ecosystem services. More frequent fire will increase emissions of greenhouse gases and aerosols (Amiro et al., 2001 in Ryan et al., 2008) and increase deposition of black carbon aerosols on snow and sea ice (Flanner et al., 2007 in Ryan et al., 2008). Even though many forests will regrow and sequester the carbon released in the fire, forests burned in the next few decades can be sources of CO_2 for decades and not recover the carbon lost for centuries (Kashian et al., 2006 in Ryan et al., 2008)

Insects and pathogens

Insects and diseases are a natural part of forested ecosystems; they are nearly always present at endemic levels within terrestrial ecosystems. However, when conditions are right, their populations increase dramatically to epidemic levels with a concomitant increase in the damage they produce. Their effects vary from defoliation and retarded growth, to reduction in timber quality, to massive forest diebacks. Insect lifecycles are sensitive to climate variables and can be a factor in determining when they increase to epidemic levels.

Ryan et al. (2008) noted that insects and pathogens affect more area annually than any other disturbance event. Outbreaks are estimated to result in $1.5 billion annually in losses to U.S. forest ecosystems (Dale et al., 2001). Extensive reviews of the effects of climate change on insects and pathogens have reported many cases where climate change has affected and/or will affect forest insect species range and abundance (Ayres and Lombardero, 2000;

Malmström and Raffa, 2000; Bale et al., 2002). Ryan et al. (2008) cite the following examples of major outbreaks in recent years:

- Two mountain pine beetle outbreaks affected more than 10 million hectares (Mha) of forest in British Columbia (Taylor et al., 2006) and another 267,000 ha in Colorado (Colorado State Forest Service, 2007).
- More than 1.5 Mha of forest was attacked by spruce beetle in southern Alaska and western Canada (Berg et al., 2006).
- Greater than 1.2 Mha of pinyon pine mortality occurred because of extreme drought, coupled with an ips beetle outbreak in the Southwest (Breshears et al., 2005).
- Ecologically important whitebark pine is being attacked by mountain pine beetle in the northern and central Rockies (Logan and Powell, 2001). For example, almost 70,000 ha, or 17%, of whitebark pine forest in the Greater Yellowstone Ecosystem is infested by mountain pine beetle (Gibson, 2006).

Climate plays a major role in driving, or at least influencing, infestations of these important forest insect species in the United States (e.g., Holsten et al., 1999; Logan et al., 2003; Carroll et al., 2004; Tran et al., in press in Ryan et al., 2008), and the evidence suggests these recent large outbreaks are influenced by observed increases in temperature. Specifically, temperature controls lifecycle development rates, influences synchronization of mass attacks required to overcome tree defenses, and determines winter mortality rates and suitable range (Hansen et al., 2001; Logan and Powell, 2001; Hansen and Bentz, 2003; Tran et al., in press in Ryan et al., 2008). In addition, warming trends in the United States have also resulted in a longer period of activity for insects and proliferation of some species, such as the mountain pine beetle (Easterling et al., 2007). Changing climatic conditions may also affect insect outbreaks by affecting the overall health and productivity of trees. Specifically, drought stress, resulting from decreased precipitation and/or warming, reduces the ability of a tree to mount a defense against insect attack (Carroll et al., 2004; Breshears et al., 2005 in Ryan et al., 2008).

Research reviewed by Ryan et al. (2008) suggests that warming temperatures are expected to result in an expansion of suitable range for mountain pine beetle (Logan and Powell, 2001) and southern pine beetle (Ungerer et al., 1999) and increase the probability of spruce beetle outbreak (Logan et al., 2003). Climate change also appears to be encouraging the expansion of non-native insects, including the hemlock woolly adelgid (Parker et al., 1999) and the gypsy moth (Logan et al., 2003).

Storm events

While less is known about the effects of storms, it is expected that tree mortality resulting from storms would reduce future carbon sequestration. Ryan et al. (2008) cited recent analyses that demonstrated that a single Category 3 hurricane or severe ice storm could each transfer to the decomposable pool the equivalent of 10% of the carbon the United States sequesters annually, with subsequent reductions in sequestration caused by damage to forest stands (McNulty, 2002; H. McCarthy et al., 2006 in Ryan et al., 2008). For example, Hurricanes Rita and Katrina together damaged a total of 2,200 ha and 63 million cubic meters of timber volume (Stanturf et al., 2007). When decomposed over the next several years, this will release a total of 105 Tg carbon into the atmosphere, roughly equal to the annual net sink for U.S. forests (Chambers et al., 2007).

The IPCC (Field et al., 2007) found with *high confidence* that, across North America, impacts of climate change on commercial forestry potential are *likely* to be sensitive to changes in disturbances from insects and diseases, as well as wildfires.

V.3.e. Species composition

Forest composition

Climate change and associated changes in disturbance regimes will result in shifts in the distributions of tree species and the composition of forest stands. With warming, forests will extend further north and to higher elevations. Over currently dry regions, increased precipitation may allow forests to displace grasslands and savannas. Changes in forest composition in turn can alter the frequencies, intensities, and impacts of disturbances such as fire, insect outbreaks, and disease.

In Alaska and neighboring Arctic regions, there is strong evidence of recent vegetation composition change, as outlined by the IPCC (Anisimov et al., 2007 and references therein):

- Aerial photographs show increased shrub abundance in 70% of 200 locations.
- Along the Arctic to subarctic boundary, the tree line has moved about 10 km (6 mi) northwards, and 2% of Alaskan tundra on the Seward Peninsula has been displaced by forest in the past 50 years.
- The pattern of northward and upward tree line advances is comparable with earlier Holocene changes.
- Analyses of satellite images indicate that the length of the growing season is increasing by three days per decade in Alaska.

Likely rates of migration northward and to higher elevations are uncertain and depend not only on climate change, but also on future land use patterns and habitat fragmentation, which can impede seed source and species migration. Moreover, given that many plants are long-lived, species are expected to continue to persist for some time in their previous ranges despite changing conditions. Bioclimatic modeling based on outputs from five GCMs suggests that, over the next century, tree species richness will decrease in most parts of the conterminous United States, even though long-term trends (millennia) ultimately favor increased richness in some locations (Field et al., 2007). The Arctic Climate Impact Assessment (ACIA, 2004) also concluded that vegetation zones are projected to migrate northward, with forests encroaching on tundra and tundra encroaching on polar deserts. Limitations in amount and quality of soils are expected to hinder these poleward shifts.

Grassland and shrubland composition

Generally, these ecosystems have a long history of land management including grazing, fire suppression, and preferential management for particular species. These activities have a strong influence on the current species composition. For example, in some cases fire suppression has resulted in encroachment by trees from surrounding forests or upland slopes (e.g., juniper encroachment in the sagebrush steppe in the western United States). Research suggests increasing atmospheric CO_2 levels may promote the conversion of savannas to greater tree dominance (Bond et al., 2003).

In many regions, particularly arid lands of the United States, species composition is being markedly affected by invasions of non-native grasses. As noted above, such changes can have marked effects on the disturbance regime. Indeed, SAP 4.3 (Ryan et al., 2008) found that in arid lands of the United States, non-native grasses often act as 'transformer species' (Richardson et al., 2000; Grice, 2006) in that they change the character, condition, form, or nature of a natural ecosystem over substantial areas. Land use and climate markedly influence the probability, rate, and pattern of alien species invasion, and future change in each of these drivers will interact to strongly affect scenarios of plant invasion and ecosystem transformation (Sala et al., 2000; Walther et al., 2002; Hastings et al., 2005). Plant invasions are strongly influenced by seed dispersal and resource availability, but disturbance and abrupt climatic changes also play key roles (Clarke et al., 2005). Changes in ecosystem susceptibility to invasion by non-native plants may be expected with changes in climate (Ibarra et al., 1995; Mau-Crimmins et al., 2006), CO_2 (Smith et al., 2000; Nagel et al., 2004), and nitrogen deposition (Fenn et al., 2003). Elevation gradients have also been shown to influence the spread of invasive species. For cheatgrass, a common exotic annual in the Great Basin, the rate of invasion is related to temperature at higher elevations and soil water availability at lower elevations. Increased variability in soil moisture and reductions in perennial herbaceous cover also increased susceptibility of low-elevation sites to cheatgrass invasion (Chambers et al., 2007).

As noted in SAP 4.3 (Ryan et al., 2008), non-native plant invasions, promoted by enhanced nitrogen deposition (Fenn et al., 2003) and increased anthropogenic disturbance (Wisdom et al., 2005), will have a major impact on how arid land ecosystems respond to climate and climate change. Once established, non-native annual and perennial grasses can generate massive, high- continuity fine-fuel loads that predispose arid lands to fires more frequent and intense than those with which they evolved. The result is the potential for desert scrub, shrub–steppe, and desert grassland/savanna biotic communities to be quickly and radically transformed into monocultures of invasive grasses over large areas. This is already well underway in the cold desert region (Knapp, 1998) and is in its early stages in hot deserts (Williams and Baruch, 2000; Kupfer and Miller, 2005; Salo, 2005; Mau-Crimmins et al., 2006). By virtue of their profound impact on the fire regime and hydrology, invasive plants in arid lands are expected to exceed direct climate impacts on native vegetation where they gain dominance (Clarke et al., 2005).

Chronic disturbance, such as grazing, will also affect rates of ecosystem change in response to changing climatic conditions because it reduces vegetation resistance to slow, long-term changes in climate (Cole, 1985; Overpeck et al., 1990). Plant communities dominated by long-lived perennials may exhibit considerable biological inertia, and changes in community composition may lag behind significant changes in climate.

V.4. Water Resources

Water is vital for all life and is essential to the health and welfare of ecosystems as well as human and social systems. Historically, the availability of water has played an important role in shaping plant, animal, and human communities. Most of the Earth's water is contained in the oceans; only 3% is freshwater. Of this 3%, a majority (68.7%) is frozen in ice caps and glaciers while the remainder is split between groundwater (approximately 30%) and surface water (approximately 0.3%). Given water's importance, plant, animal, and human communities are all sensitive to variations in the availability, storage, fluxes, and quality of surface and groundwater. These, in turn, are sensitive to climate change.

In SAP 4.3, Lettenmaier et al. (2008) found that water resource systems have been designed and operated to maintain a reliable supply despite the wide variability that can exist in water availability, primarily that of surface water, over days to months and years. However, these approaches are all based on a similar assumption known as stationarity, in which natural systems are assumed to fluctuate within a range of known conditions based on the historical record. However, as noted by Arnell (2002), Lettenmaier (2003), Milly et al. (2008), and others, in the era of climate change this assumption is no longer tenable. Kundzewicz et al. (2007) found that because water infrastructure (e.g., dikes and pipelines) has been designed for stationary climatic conditions, the global population is highly vulnerable to climate change impacts on freshwater resources.

The IPCC (Kundzewicz et al., 2007) found that climate change is one of many factors exerting pressure on existing freshwater systems. Other factors include water pollution, damming of rivers, wetland drainage, reduction in streamflow, and lowering of the groundwater table (e.g., due to irrigation). The authors conclude that while climate-related changes have been small compared to these other pressures to date, climate change is expected to result in increasing effects in the future. Ultimately, each of these factors influences the availability of and access to freshwater. In this section, we review effects of global change on water supply, water quality, and extreme events, and explore the implications for water use.

In regards to the hydrologic observing systems on which these sections are based, Lettenmaier et al. (2008) found that the current hydrologic observing system was not designed specifically for the purpose of detecting the effects of climate change on water resources. In many cases, the resulting data are unable to meet the predictive challenges of a rapidly changing climate.

Overall, the IPCC (Kundzewicz et al., 2007) made the following conclusions:

- The negative impacts of climate change on freshwater systems outweigh the benefits (*high confidence*).
- All IPCC regions show an overall net negative impact of climate change on water resources and freshwater ecosystems (*high confidence*).
- Areas in which runoff is projected to decline are *likely* to face a reduction in the value of the services provided by water resources (*very high confidence*).
- The beneficial impacts of increased annual runoff in other areas will be tempered by the negative effects of increased precipitation variability and seasonal runoff shifts on water supply, water quality, and flood risks (*high confidence*).
- The impacts of climate change on freshwater systems and their management are mainly due to the observed and projected increases in temperature, sea level, and precipitation variability (*very high confidence*).
- More than one-sixth of the world's population lives in glacier- or snowmelt-fed river basins and will be affected by the seasonal shift in streamflow, an increase in the

ratio of winter to annual flows, and possibly the reduction in low flows caused by decreased glacier extent or snow water storage (*high confidence*).

V.4.a. Water supply and precipitation

Surface water and precipitation

The primary driver of the land surface hydrologic system is precipitation. Current vulnerabilities of the system are strongly tied to precipitation variability. Observed trends in both are reviewed here. In the IPCC Fourth Assessment Report, Kundzewicz et al. (2007) found that climate- induced changes in river flows and lake and wetland levels depend on changes in the volume, timing, and intensity of precipitation and snowmelt, as well as the amount of precipitation falling as snow or rain (Chiew, 2007). Precipitation is expected to increase globally with important regional variations. Current observations show increased precipitation over land north of 30° N from 1901 to 2005 and decreases over land between 10° S and 30° N after the 1970s. Snow cover and glaciers are decreasing in most regions, particularly in spring. Permafrost is thawing between 0.02 m per year (Alaska) and 0.4 m per year (Tibetan Plateau). In many cases, precipitation may be more variable, with increases in one season followed by decreases in another. Intensified droughts have been observed in some drier regions since the 1970s. Changes in evapotranspiration driven by temperature, radiation, atmospheric humidity, increased atmospheric CO_2 concentration, and wind speed can offset small increases in precipitation and intensify the effects of decreased precipitation on surface waters.

As noted in Kundzewicz et al. (2007), effects vary across catchments depending on specific physical, hydrological, and geological characteristics. Current observations have identified highly variable streamflows globally, with increases in some basins and declines in others. Warming temperatures have resulted in earlier spring peak flows and increased winter base flows in North America and Eurasia. Catchments that are already stressed due to non-climatic drivers are highly vulnerable to additional impacts. In addition, vulnerability to precipitation variability is highest in semi-arid and arid low-income countries, where precipitation and streamflow are concentrated over a few months, and where year-to-year variations are high (Lenton, 2004). Without adequate storage infrastructure, these regions are highly vulnerable to current climate variability and expected increases in variability with future climate change.

Human water use has resulted in reduced water levels in many lakes worldwide. In some cases, declining precipitation was also a significant cause (e.g., in the case of Lake Chad, where both decreased precipitation and increased human water use account for the observed decrease in lake area since the 1960s) (Coe and Foley, 2001 in Kundzewicz et al., 2007). For the many lakes, rivers, and wetlands that have shrunk mainly due to human water use and drainage, climate change is expected to exacerbate the situation if it results in reduced net water availability (precipitation minus evapotranspiration). Observations have also identified warming lake temperatures and reduction in ice cover on lakes worldwide.

Surface water availability and precipitation differ greatly across the United States. Generally, conditions become increasingly dry from east to west. However, conditions in the upslope areas of the Cascade and coastal mountain ranges, especially in the Pacific Northwest, are much more humid (Lettenmaier et al., 2008). The driest climates occur in the

Intermountain West and the Southwest. Precipitation variability follows similar trends with less variability in the humid areas (eastern United States and Pacific Northwest) and the greatest variability in the arid and semiarid West (Lettenmaier et al., 2008). The IPCC (Kundzewicz et al., 2007) concluded with *high confidence* that semi-arid and arid areas are particularly exposed to the impacts of climate change on freshwater.

Specific to the United States, the IPCC identified the following trends in surface water supply, precipitation patterns, and snowpack (Field et al., 2007):

- Streamflow in the eastern United States has increased 25% in the last 60 years (Groisman et al., 2004), but over the last century has decreased by about 2% per decade in the central Rocky Mountain region (Rood et al., 2005).
- Since 1950, stream discharge in both the Colorado and Columbia River Basins has decreased, while over the same time period annual evapotranspiration from the conterminous United States increased by 55 mm (Walter et al., 2004).
- In regions with winter snow, warming has shifted the magnitude and timing of hydrologic events (Mote et al., 2005; Regonda et al., 2005; Stewart et al., 2005). The fraction of annual precipitation falling as rain (rather than snow) increased at 74% of the weather stations studied in the western mountains of the United States from 1949 to 2004 (Knowles et al., 2006).
- In Canada, warming from 1900 to 2003 led to a decrease in total precipitation as snowfall in the West and on the Prairies (Vincent and Mekis, 2006).
- Spring and summer snow cover has decreased in the U.S. West (Groisman et al., 2004). April snow water equivalent has declined 15 to 30% since 1950 in the western mountains of North America, particularly at lower elevations and primarily due to warming rather than changes in precipitation (see Mote et al., 2003, 2005; Lemke et al., 2007).
- Streamflow peaks in the snowmelt-dominated western mountains of the United States occurred one to four weeks earlier in 2002 than in 1948 (Stewart et al., 2005).
- Breakup of river and lake ice across North America has advanced by 0.2 to 12.9 days over the last 100 years (Magnuson et al., 2000).

As reported in Lettenmaier et al. (2008), Mauget (2003) analyzed data from 167 stream gauge stations over the period 1939 to 1998 and found increasing streamflows over time in the eastern United States with a more or less reverse pattern in the western United States, with an onset of dry conditions beginning in the 1980s. Similar to global observations, U.S. effects will vary from region to region.

As for the Arctic, research reviewed in the Arctic Climate Impact Assessment concluded that precipitation has increased by about 1% per decade over the past century although the results are highly variable spatially (ACIA, 2005). Much of the increase has fallen as rain, with the largest increases occurring in autumn and winter. The ice season has been reduced by one to three weeks in some areas from a combination of later freeze-up and earlier breakup of river and lake ice. Glaciers throughout North America are melting. Alaskan glaciers are melting particularly fast and represent about half of the estimated loss of glacial mass worldwide (ACIA, 2004). Permafrost plays an important role in providing an impermeable surface and forming lakes and ponds. The spatial pattern of lake disappearance strongly suggests that permafrost thawing is driving the changes. In the Arctic, river discharge to the ocean has increased during the past few decades, and peak flows in the spring are occurring earlier. These changes are projected to accelerate with future climate change. Snow cover extent in Alaska is projected to decrease by 10 to 20% by the 2070s, with greatest declines in spring (ACIA, 2004).

Projected impacts

Reviewing the research on these trends, Lettenmaier et al. (2008) drew the following conclusions in SAP 4.3:

- There is a trend toward reduced mountain snowpack and earlier spring snowmelt runoff peaks across much of the western United States. Evidence suggests this trend is very likely attributable, at least in part, to long-term warming, although some part may have been played by decadal-scale variability, including a shift in the PDO in the late 1970s. Where shifts to earlier snowmelt peaks and reduced summer and fall low flows have already been detected, continuing shifts in this direction are expected and may have substantial impacts on the performance of reservoir systems.
- The most recent (IPCC Fourth Assessment Report) climate model simulations project increased runoff over the eastern United States, gradually transitioning to little change in the Missouri and lower Mississippi, to substantial decreases in annual runoff in the interior of the West (Colorado and Great Basin). The projected drying in the interior of the West is quite consistent among models. These changes are, very roughly, consistent with observed trends in the second half of the 20th century, which show increased streamflow over most of the United States, but sporadic decreases in the West.
- Snowpacks in the mountainous headwaters regions of the western United States generally declined over the second half of the 20th century, especially at lower elevations and in locations where average winter temperatures are close to or above 0 °C. These trends toward reduced winter snow accumulation and earlier spring melt are also reflected in a tendency toward earlier runoff peaks in the spring, a shift that has not occurred in rainfall-dominated watersheds in the same region.
- Climate model projections of increased temperatures and slight precipitation increases indicate that modest streamflow increases are expected in the East, but that larger (in absolute value) declines are expected in the West, where the balance between precipitation and evaporative demand changes will be dominated by increased evaporative demand. However, because of the uncertainty in climate model projections of precipitation change, future projections of streamflow are highly uncertain across most of the United States. One exception is watersheds that are dominated by spring and summer snowmelt, most of which are in the western United States. In these cases, where shifts to earlier snowmelt peaks and reduced summer and fall low flows have already begun to be detected, continuing shifts in this direction are generally expected and may have substantial impacts on the performance of reservoir systems.

V.4.b. Groundwater

In SAP 4.3, Lettenmaier et al. (2008) found that relatively few studies have assessed the sensitivity of groundwater systems to a changing climate. Reviewing the available literature (e.g., Vaccaro, 1992; Loaiciga et al., 2000; Hanson and Dettinger, 2005; Scibek and Allen, 2006; Gurdak et al., 2007), the authors concluded that the ability to predict the effects of climate and climate change on groundwater systems is nowhere near as advanced as for surface water systems.

The available research suggests that groundwater systems generally respond more slowly to climate change than surface water systems. In general, groundwater levels correlate most strongly with precipitation, but temperature becomes more important for shallow aquifers, especially during warm periods. With climate change, availability of groundwater is expected to be influenced by changes in withdrawals (reflecting development, human and agricultural demand, and availability of other sources) and recharge (determined by temperature, timing, and amount of precipitation, and surface water interactions) (*medium confidence*) (Kundzewicz et al., 2007). In general, simulated aquifer levels respond to changes in temperature, precipitation, and the level of withdrawal.

According to the IPCC, base flows were found to decrease in scenarios that are drier or have higher pumping rates, and increase in wetter scenarios on average across world regions (Kundzewicz et al., 2007). The IPCC projects that efforts to offset declining surface water availability due to increasing precipitation variability will be hampered by the fact that groundwater recharge will decrease considerably in some already water-stressed regions (*high confidence*) where vulnerability is often exacerbated by the rapid increase in population and water demand (*very high confidence*). This is expected to be particularly acute in some water-stressed regions, such as the southwestern United States. Projections for the Ogallala aquifer region show that natural groundwater recharge decreases more than 20% in all simulations with different climate models and future warming scenarios of 2.5 °C or greater (Field et al., 2007). Groundwater resources can also be adversely impacted in coastal areas by saltwater intrusion from sea level rise. In addition, they conclude that sea level rise will extend areas of salinization of groundwater and estuaries, resulting in a decrease in freshwater availability for humans and ecosystems in coastal areas (*very high confidence*) (Kundzewicz et al., 2007).

V.4.c. Water quality

(adapted largely from Lettenmaier et al., 2008)

Water quality reflects the chemical inputs from the air and landscape and their biogeochemical transformation within the water (Murdoch et al., 2000). A warming climate may influence water quality through increased water temperatures and modification of regional patterns of precipitation. However, most water quality changes observed so far across the continental United States are likely attributable to causes other than climate change (Lettenmaier et al., 2008). Attribution of changes in water quality to climate impacts is difficult as water quality is sensitive to other human activities, particularly land use practices that alter landscapes and modify the flux of water as well as its thermal and nutrient characteristics.

Lettenmaier et al. (2008) found that two main factors that influence water quality are temperature and water quantity. Higher temperatures enhance rates of biogeochemical transformation and physiological processes of aquatic plants and animals. As temperatures increase, the ability of water to hold dissolved oxygen declines, with potential negative impacts on aquatic organisms. High nutrient loads can also contribute to anoxic conditions. Increased streamflow can dilute nutrient concentrations and thus diminish excessive biological production. However, higher flows can flush excess nutrients from sources of origin in a stream. The overall balance of these competing effects in a changing climate is not yet known.

Most studies examining the responses of water quality over time have focused on nutrient loading. Ramstack et al. (2004) reconstructed water chemistry before European settlement for 55 Minnesota lakes. They found that lakes in forested regions showed very little change in water quality since 1800. By contrast, about 30% of urban lakes and agricultural lakes showed significant increases in chloride (urban) or phosphorus (agricultural). These results indicate the strong influence of land use on water quality indicators.

Recent historical assessments of changes in water quality due to temperature trends have largely focused on salmonid fishes in the western United States (e.g., Bartholow, 2005; Crozier and Zabel, 2006 in Lettenmaier et al., 2008). Increases in water temperature can influence salmon by negatively affecting different life stages. This research suggests that smaller snowpacks that reduce autumn flows and cause higher water temperatures are expected to reduce salmon survival. Petersen and Kitchell (2001) found that warmer water temperatures are also associated with an expected higher mortality rate of young salmon due to fish predators.

Studies reviewed by the IPCC (Field et al., 2007; Kundzewicz et al., 2007) provide the following conclusions. In lakes and reservoirs, climate change effects are primarily caused by water temperature variations (caused by climate change or thermal pollution as a result of higher demand for cooling water in the energy sector) affecting dissolved oxygen regimes, redox potentials,[47] lake stratification, mixing rates, and the development of aquatic biota. In addition, increasing water temperature reduces the self-purification capacity of rivers by decreasing the amount of dissolved oxygen available for biodegradation. Simulations of future North American surface and bottom water temperatures of lakes, reservoirs, rivers, and estuaries consistently show increases of 2 °C to 7 °C, with summer surface temperatures exceeding 30 °C in midwestern and southern lakes and reservoirs. The IPCC projects that warming is *likely* to extend and intensify summer thermal stratification in surface waters, further contributing to oxygen depletion (Field et al., 2007 and references therein).

Higher water temperatures, increased precipitation intensity, and longer periods of low flows exacerbate many forms of water pollution and can affect ecosystems, human health, and water system reliability and operating costs. Pollutants of concern in this case include sediment, nutrients, organic matter, pathogens, pesticides, salt, and thermal pollution (Kundzewicz et al., 2007). Elevated surface water temperatures will promote algal blooms and increases in bacteria and fungi levels. The frequency of heavy precipitation events in the United States have increased through the 1990s (Field et al., 2007). Increases in intense rain events result in the introduction of more sediment, nutrients, pathogens, and toxics into water bodies from non-point sources. Increasing nutrient and sediment loads (due to more intense runoff events) will negatively affect water quality, possibly rendering a source unusable unless special treatment is introduced. Intense rainfall will lead to increases in suspended solids (turbidity) and pollutant levels in water bodies due to soil erosion (Kundzewicz et al., 2007). Even with enhanced phosphorus removal in wastewater treatment plants, algal growth in water bodies may increase with warming over the long term. Conditions are exacerbated during low flow conditions where small water quantities result in less dilution and greater concentrations of pollutants. In addition, fluctuating levels of lakes and reservoirs can reduce water quality through resuspension of bottom sediments and liberating compounds (Field et al., 2007).

The IPCC reviewed a number of region-specific studies on U.S. water quality and made the following projections (Field et al., 2007; Kundzewicz et al., 2007):

- Changes in precipitation may increase nitrogen loads from rivers in the Chesapeake and Delaware Bay regions by up to 50% by 2030 (Kundzewicz et al., 2007).
- Decreases in snow cover and increases in winter rain on bare soil will *likely* lengthen the erosion season and enhance erosion intensity. This will increase the potential for sediment- related water quality impacts in agricultural areas (Field et al., 2007).
- Increased rainfall amounts and intensities will lead to greater rates of erosion within the United States and in other regions unless protection measures are taken (Kundzewicz et al., 2007). Soil management practices (e.g., crop residue, no-till) in some regions (e.g., the Corn Belt) may not provide sufficient erosion protection against future intense precipitation and associated runoff (Field et al., 2007).
- For the Midwest, in simulated low flows used to develop pollutant discharge limits (Total Maximum Daily Loads) flows decrease over 60% with a 25% decrease in mean precipitation, reaching up to 100% with the incorporation of irrigation demands (Eheart et al., 1999).
- Restoration of beneficial uses (e.g., to address habitat loss, eutrophication, beach closures) under the Great Lakes Water Quality Agreement will *likely* be vulnerable to declines in water levels, warmer water temperatures, and more intense precipitation (Mortsch et al., 2003).
- Based on simulations, phosphorus remediation targets for the Bay of Quinte (Lake Ontario) and the surrounding watershed could be compromised as 3 to 4 °C warmer water temperatures contribute to 77 to 98% increases in summer phosphorus concentrations in the bay (Nicholls, 1999), and as changes in precipitation, streamflow, and erosion lead to increases in average phosphorus concentrations in streams of 25 to 35% (Walker, 2001).

The IPCC (Kundzewicz et al., 2007) concluded that, globally, higher water temperatures, increased precipitation intensity, and longer periods of low flows exacerbate many forms of water pollution, affecting ecosystems, human health, and water system reliability and operating costs (*high confidence*). For North America, they also concluded that climate change is *likely* to make it more difficult to achieve existing water quality goals for sediment (*high confidence*) (Field et al., 2007).

V.4.d. Extreme events

Higher temperatures increase the water-holding capacity of the atmosphere and encourage greater evaporation, resulting in conditions that favor increased climate variability, with more intense precipitation and more droughts (Trenberth et al., 2003; Milly et al., 2008). A changing climate can influence the occurrence of floods and droughts. Many climatic and non-climatic drivers influence flood and drought impacts, including intense and/or long-lasting precipitation events, rapid snowmelt, dam failure, or reduced conveyance due to ice jams or landslides. Flood magnitude and spatial extent depend on the intensity, volume, and time of precipitation, and the antecedent conditions of rivers and their drainage basins (e.g., presence of snow and ice, soil composition, level of human development, existence of prevention infrastructure) (Kundzewicz et al., 2007).

Research reviewed by the IPCC (Kundzewicz et al., 2007) suggested that while research to date has not provided clear evidence for a climate-related trend in floods during the last decades, there is suggestive evidence that floods may have been affected by the observed increase in precipitation intensity and other observed climate changes (e.g., an increase in westerly weather patterns during winter over Europe) that lead to very rainy low-pressure systems that often trigger floods (Kron and Berz, 2007). Globally, the number of great inland flood catastrophes during the last 10 years (between 1996 and 2005) is twice as large, per decade, as between 1950 and 1980, while economic losses have increased by a factor of five (Kron and Berz, 2007). The increase in flood damage is primarily driven by an increase in population and wealth in vulnerable areas and changes in land use. Floods have been the most reported natural disaster events in Africa, Asia, and Europe, and have affected more people across the globe (140 million per year on average) than all other natural disasters (WDR, 2003, 2004). In Bangladesh, three extreme floods have occurred in the last two decades, and in 1998, about 70% of the country's area was inundated (Mirza, 2003; Clarke and King, 2004). In some river basins (e.g., the Elbe River Basin in Germany), increasing flood risk drives the strengthening of flood protection systems by structural means, with detrimental effects on riparian and aquatic ecosystems (Wechsung et al., 2005).

Globally, increased intensity and variability of precipitation is projected to increase the risks of flooding and drought in many areas (*high confidence*). Many of these areas (e.g., Mediterranean Basin, western United States, southern Africa, and northeastern Brazil) will suffer a decrease in water resources due to climate change (*very high confidence*) (Kundzewicz et al., 2007).

In the United States, precipitation intensity will increase across the country, particularly at middle and high latitudes where mean precipitation also increases. This will affect the risk of flash flooding and urban flooding in these regions (Kundzewicz et al., 2007). In general, projected changes in precipitation extremes are larger than changes in mean precipitation (Field et al., 2007).

Some studies project widespread increases in extreme precipitation with greater risks of not only flooding from intense precipitation, but also droughts from greater temporal variability in precipitation (Christensen et al., 2007). Droughts affect rain-fed agricultural production as well as water supply for domestic, industrial, and agricultural purposes. Globally, some semi-arid and sub-humid regions (e.g., Australia, western United States, southern Canada, and the Sahel) have suffered from more intense and multi-annual droughts, highlighting the vulnerability of these regions to the increased drought occurrence that is expected in the future due to climate change (Nicholson, 2005).

Across North America, vulnerability to extended drought is increasing as population growth and economic development create more demands from agricultural, municipal, and industrial uses, resulting in frequent over-allocation of water resources (Alberta Environment, 2002; Morehouse et al., 2002; Postel and Richter, 2003; Pulwarty et al., 2005 in Field et al., 2007). While much of the United States has experienced increases in precipitation and streamflow, there is some evidence of long-term drying and increases in drought severity and duration in the West and Southwest. This is probably due to a combination of decadal-scale climate variability and longterm change (Lettenmaier et al., 2008). However, the eastern regions are also vulnerable to droughts and attendant reductions in water supply, changes in water quality and ecosystem function, and challenges in allocation (Dupigny-Giroux, 2001; Bonsal et al., 2004; Wheaton et al., 2005).

Projections for the western mountains of the United States suggest that warming, and changes in the form, timing, and amount of precipitation will *very likely* lead to earlier melting and significant reductions in snowpack by the middle of the 21st century (*high confidence*; Field et al., 2007). In mountainous snowmelt-dominated watersheds, projections suggest advances in the timing of snowmelt runoff, increases in winter and early spring flows (raising flooding potential), and substantially decreased summer flows. Heavily utilized water systems of the western United States that rely on capturing snowmelt runoff, such as the Columbia River system, will be especially vulnerable (Field et al., 2007). Reduced snowpack has been identified as a major concern for the State of California (California Energy Commission, 2006b).

In SAP 4.3, Lettenmaier et al. (2008) found that U.S. consumptive use of water per capita has declined over the last two decades, primarily as a result of various improvements in water use efficiency related both to legal mandates and to water pricing, as well as some changes in water laws that have facilitated reallocation of water, particularly in the western United States and during droughts. Trends toward increased water use efficiency seem likely to continue in the coming decades. Pressures for reallocation of water will be greatest in areas of highest population growth, such as the Southwest. Declining water consumption, if it continues, will help mitigate the impacts of climate change on water resources.

There is evidence that much more severe droughts have occurred in North America prior to the instrumental record of roughly the last 100 years (Lettenmaier et al., 2008). For instance, Woodhouse and Overpeck (1998), using paleoclimatic indicators (primarily tree rings), find that many droughts over the last 2,000 years have eclipsed the major U.S. droughts of the 1930s and 1 950s, with much more severe droughts occurring as recently as the 1 600s. Although the nature of future drought stress remains unclear, for those areas where climate models suggest drying, such as the Southwest (e.g., Seager et al., 2007), droughts that are more severe than those encountered in the instrumental record may become increasingly likely.

In addition to the effects on water supply, extreme events, such as floods and droughts, are expected to reduce water quality. Increased erosion and runoff rates during flood events will wash pollutants (e.g., organic matter, fertilizers, pesticides, and heavy metals) from soils into water bodies, subsequently affecting species and ecosystems. During drought events, the lack of precipitation and subsequent low flow conditions will impair water quality by reducing the amount of water available to dilute pollutants. These effects from floods and droughts will make it more difficult to achieve pollutant discharge limits and water quality goals (Kundzewicz et al., 2007).

V.4.e. Implications for water users

On global water use, the IPCC (Kundzewicz et al., 2007) drew the following conclusions:

- Climate change affects the function and operation of existing water infrastructure as well as water management practices (*very high confidence*).
- Adverse effects of climate on freshwater systems aggravate the impacts of other stresses, such as population growth, changing economic activity, land use change, and urbanization (*very high confidence*).
- Globally, water demand will grow in the coming decades, primarily due to population growth and increased affluence. Regionally, large changes in irrigation water demand as a result of climate change are *likely* (*high confidence*).
- Current water management practices are *very likely* to be inadequate to reduce the negative impacts of climate change on water supply reliability, flood risk, health, energy, and aquatic ecosystems (*very high confidence*). Improved incorporation of current climate variability into water-related management would make adaptation to future climate change easier (*very high confidence*).

In the United States, many competing water uses will be adversely affected by climate change impacts on water supply and quality. The IPCC reviewed a number of studies describing the impacts of climate change on water uses in the United States that showed the following:

- Decreased water supply and lower water levels are *likely* to exacerbate challenges relating to navigation in the United States (Field et al., 2007). Some studies have found that low flow conditions may restrict ship loading in shallow ports and harbors (Kundzewicz et al., 2007). However, navigational benefits from climate change exist as well. For example, the navigation season for the Northern Sea Route is projected to increase from the current 20 to 30 days per year to 90 to 100 days by 2080 (ACIA, 2004).
- Climate change impacts on water supply and quality will affect agricultural practices, including the increase in irrigation demand in dry regions and the aggravation of non- point source water pollution (e.g., pollution from urban areas, roads, or agricultural fields) problems in areas susceptible to intense rainfall events and flooding (Field et al., 2007).
- The U.S. energy sector, which relies heavily on water for generation (hydropower) and cooling capacity, will be adversely impacted by changes in water supply and quality in reservoirs and other water bodies (Wilbanks et al., 2007a).

Less reliable supplies of water are expected to create challenges for managing urban water systems as well as for industries that depend on large volumes of water. U.S. water managers currently anticipate local, regional, or statewide water shortages over the next 10 years. Threats to reliable supply are complicated by high population growth rates in western states where many resources are at or approaching full utilization. The IPCC (Field et al., 2007) reviewed several regional-level studies of climate change impacts on U.S. water management that showed the following:

- In the Great Lakes–St. Lawrence Basin, many, but not all, assessments project lower net basin supplies and lake water levels. Lower water levels are *likely* to influence many sectors, with multiple, interacting impacts. These impacts are projected with *high confidence*, in which atmosphere–lake interactions contribute to the uncertainty in assessment.

- Urban water supply systems in North America often draw water from considerable distances, so climate impacts need not be local to affect cities. By the 2020s, 41% of the water supply to southern California is *likely* to be vulnerable due to snowpack loss in the Sierra Nevada mountains and the Colorado River Basin.
- The New York area will *likely* experience greater water supply variability. New York City's system can *likely* adapt to future changes, but the region's smaller systems may be vulnerable, leading to a need for enhanced regional water distribution plans.

Drawing on these studies, the IPCC concluded that climate change will constrain North America's over-allocated water resources, increasing competition among agricultural, municipal, industrial, and ecological uses (*very high confidence*) (Field et al., 2007). Rising temperatures will diminish snowpack and increase evaporation, affecting seasonal availability of water. Higher demand from economic development, agriculture, and population growth will further limit surface and groundwater availability. In the Great Lakes and major river systems across the United States, lower water levels are *likely* to exacerbate challenges relating to water quality, navigation, recreation, hydropower generation, water transfers, and bi-national relationships.

In the Arctic, river discharge to the ocean has increased during the past few decades, and peak flows in the spring are occurring earlier. These changes are projected to accelerate with future climate change. Snow cover extent in Alaska is projected to decrease by 10 to 20% by the 2070s, with greatest declines in spring (ACIA, 2004).

The IPCC concluded with *high confidence* that under most climate change scenarios, water resources in small islands around the globe are *likely* to be seriously compromised (Mimura et al., 2007). Most small islands have a limited water supply, and water resources in these islands are especially vulnerable to future changes and distribution of rainfall. Reduced rainfall typically leads to decreased surface water supply and slower recharge rates of the freshwater lens,[48] which can result in prolonged drought impacts. Many islands in the Caribbean (which include the U.S. territories of Puerto Rico and U.S. Virgin Islands) are *likely* to experience increased water stress as a result of climate change. Under all SRES scenarios, reduced rainfall in summer is projected for the Caribbean, making it *unlikely* that the demand for water resources will be met. Increased rainfall in winter is *unlikely* to compensate for these water deficits due to lack of storage capacity (Mimura et al., 2007).

V.5. Social Systems and Settlements

In their review of industry, settlement, and society, the IPCC found (Wilbanks et al., 2007a) that human systems include social, economic, and institutional structures and processes. These systems are influenced by multiple factors and stresses (e.g., access to financial resources, urbanization, demographic shifts) and weather and climate are often viewed as secondary benefits (Ocampo and Martin, 2003; Thomas and Twyman, 2005). Effects from a changing climate will be felt in settlements through interactions with these other factors, easing or aggravating multiple stresses and, potentially, pushing stressed systems across a threshold of sustainability (Wilbanks, 2003b in Wilbanks et al., 2007a). On the other hand, some U.S. settlements may find opportunities in climate change. For example, extended growing seasons may allow for greater agricultural productivity in some areas, while warmer winters could reduce cold exposure in northern states.

Climate sensitivity varies across settlements and industrial sectors. As noted by the IPCC, while it may appear that industrialized countries are well equipped to cope with gradual climate change at a national level, at a local level there may be substantial variability in climate effects and capacities to adapt (Environment Canada, 1997; Kates and Wilbanks, 2003; London Climate Change Partnership, 2004; O'Brien et al., 2004; Kirshen et al., 2006). According to Wilbanks et al. (2007a), "Industries, settlements and human society are accustomed to variability in environmental conditions, and in many ways they have become resilient to it when it is a part of their normal experience. Environmental changes that are more extreme or persistent than that experience, however, can lead to vulnerabilities, especially if the changes are not foreseen and/or if capacities for adaptation are limited." Moreover, vulnerability is greater when changes occur rapidly, resulting in a limited ability to plan for or cope with impacts (Gamble et al., 2008). However, effects from a changing climate can be positive as well as negative.

The IPCC identified three primary ways that climate and climate change affect human systems (Wilbanks et al., 2007a):

- Climate provides a context for a variety of climate-sensitive human activities ranging from agriculture to tourism. For instance, rivers fed by rainfall enable irrigation and transportation and can enrich or damage landscapes.
- Climate affects the cost of maintaining climate-controlled internal environments for human life and activity. Clearly, higher temperatures increase costs of cooling and reduce costs of heating. See additional discussion of energy effects in Section V.7.
- Climate interacts with other types of stresses on human systems, in some cases reducing stresses but in other cases exacerbating them. For example, drought can contribute to rural– urban migration, which, combined with population growth, increases stress on urban infrastructures and socioeconomic conditions.

To date, relatively limited research has been conducted specifically on the effects of climate change on U.S. settlements. SAP 4.6 drew on the available research to begin to examine emerging findings and identify areas and issues that merit further discussion. Accordingly, this section draws heavily from their analysis. Findings from the IPCC are also illustrated in the sections below.

V.5.a. U.S. population trends

An overview of population trends and settlement patterns provides a framework to understand impacts of global change on social systems, settlements, and welfare. In SAP 4.6, Gamble et al. (2008) found that a changing climate, interacting with changes in land use and demographics, has the capacity to affect important human activities. This section begins by briefly exploring demographic trends expected to be affected by changes in temperature, precipitation, and extreme climate-related events.

The U.S. Census Bureau (2000a,b, 2002, 2005, 2007) provides the following description of national population trends:

- Since 1980, the U.S. population has grown by more than 40 million.
- This growth has been unevenly distributed around the Nation. More than 500 U.S. counties lost population over this time frame (with a total reduction of more than 2 million people). In addition, a small number of counties (40) accounted for more than half the growth (from either migration or natural increase).
- Over the next 25 years the United States is expected grow by more than 60 million people. This growth is also expected to be distributed unevenly across the country. Seven states are expected to account for more than two-thirds of this growth (Florida, Texas, California, Arizona, North Carolina, Georgia, and Washington).
- The majority of the total growth is also expected to take place in large urban areas and coastal counties, while many rural and urban fringe counties are expected to experience rapid growth in percentage terms.
- The U.S. population age 65 and over is expected to double in size within the next 25 years. By 2030, almost 1 out of 5 Americans—some 72 million people—will be 65 years or older. The age group 85 and older is now the fastest-growing segment of the U.S. population.

The greatest population growth between 2000 and 2005 occurred in western and southern states. The quickest growing states were Nevada (20.8%), Arizona (15.8%), Florida (11.3%), Georgia (10.8%), and Utah (10.6%). The growth in western states is expected to continue. Indeed, the 11 western states are expected to account for one-third of all U.S. population growth over the next 25 years (U.S. Census Bureau, 2005). One of the most pronounced trends over the past few decades has been movement toward coastal areas. This trend is also expected to continue in the future. In the following sections we explore interactions between these demographic trends and changing climatic variables.

V.5.b. Vulnerable communities

Vulnerability to climate changes varies across population segments and settlement areas. Some will be more sensitive to climate effects or ill equipped to undertake the necessary steps to adapt. This section addresses these issues from the community level, while Section V.6 emphasizes potential impacts on human health.

Vulnerability to or opportunities from climate change are related to three factors (Clark et al., 2000):

- Exposure: What effects will a place be exposed to (e.g., temperature or precipitation changes, changes in storm exposure and/or intensity, and changes in the sea level)?
- Sensitivity: If primary climate changes were to occur, how sensitive are the activities and populations of a settlement to those changes? For instance, a city dependent substantially on a regional agricultural or forestry economy might be considered more sensitive than a city whose economy is based mainly on an industrial sector less sensitive to climate variations.

- Adaptive capacity: If effects occur, what is the capability of a settlement to handle those impacts without disabling damages, perhaps even with new opportunities?

Vulnerability to the impacts of climate change is expected to vary across regions and settlements. Indeed, in many cases, vulnerability will vary based on specific characteristics of the settlement (Wilbanks et al., 2008). For example, some settlements are particularly vulnerable because of their location on or near coasts subject to storms and sea level rise or in areas at risk of other extreme events (e.g., wildfires). Some are already stressed by other forces that might interact with climate change effects, such as rapid population growth, aging physical infrastructures, poverty, and social friction. Some are considerably dependent on ecosystem services (e.g., snowpack as a source for water resources) or linkage systems, such as bridges or electric power lines, which could be vulnerable to impacts of climate change.

However, in SAP 4.6 Wilbanks et al. (2008) identify the following key concerns that transcend location:

- *Effects on health.* Higher temperatures in urban areas are related to higher concentrations of ground-level ozone, which has the potential to cause respiratory problems if not controlled. There is also some evidence that combined effects of heat stress and air pollution may be greater than simple additive effects (Patz and Balbus, 2001). Moreover, historical data show relationships between mortalities and temperature extremes (Rozenzweig and Solecki, 2001). Other health concerns include changes in exposure to water- and food-borne diseases, vector- borne diseases, concentrations of plant species associated with allergies, and exposures to extreme weather events such as storms, floods, and fires (discussed further in Section V.6).

- *Effects on water and other urban infrastructures.* Changes in precipitation patterns may lead to reductions in snowpack, river flows, groundwater levels, and in coastal areas to saline intrusion in rivers and groundwater, affecting water supply. In addition, warming may increase water demands (Kirshen, 2002; IPCC, 2007b). Moreover, storms, floods, and other severe weather events may affect other infrastructures, such as sanitation, transportation, supply lines for food and energy, and communication. Exposed structures such as bridges and electricity transmission networks are especially vulnerable. In many cases, infrastructures are interconnected—an impact on one can also affect other sectors and locations (Kirshen et al., forthcoming). For example, an interruption in energy supply in particular locations may result in increased heat stress for vulnerable populations who rely on electricity for cooling (Ruth et al., 2006).

- *Effects on energy requirements.* Warming is *virtually certain* to increase energy demand in U.S. cities for cooling in buildings, while it reduces demands for heating in buildings (see section V.7). Demands for cooling during warm periods could jeopardize the reliability of service in some regions by exceeding the supply capacity, especially during periods of unusually high temperatures. Higher temperatures also affect costs of living and business operation by increasing costs of climate control in buildings (Amato et al., 2005; Ruth and Lin, 2006; Kirshen et al., forthcoming).

- *Effects on the urban metabolism.* An urban area is a living, complex mega-organism, associated with a host of inputs, transformations, and outputs: heat, energy, materials, and others (Decker et al., 2000). An example is the Urban Heat Index, which measures the degree to which built and paved areas are associated with higher temperatures than surrounding rural areas. Imbalances in the urban metabolism can aggravate climate change impacts, such as the role of the Urban Heat Index in the formation of smog in cities.

- *Effects on economic competitiveness, opportunities, and risks.* Climate change has the potential not only to affect settlements directly, but also to affect them through impacts on other areas linked to their economies at regional, national, and international scales (Solecki and Rosenzweig, 2006). In addition, it can affect a settlement's economic base if it is sensitive to climate, as in areas where settlements are based on agriculture, forestry, water resources, or tourism (IPCC, 2007b). Climate can also be a factor in an area's comparative advantage for economic production and growth and can influence the costs of doing business (e.g., differences in costs for climate control for buildings). It is possible that regions exposed to risks from climate change will see movement of population and economic activity to other locations. One reason is public perceptions of risk, but a more powerful driving force may be the availability of insurance. The insurance sector is one of the most adaptable of all economic sectors, and its exposure to costs from severe storms and other extreme weather events is likely to lead it to withdraw (or to make much more expensive) private insurance coverage from areas vulnerable to climate change impacts (IPCC, 2007b).

- *Effects on social and political structures.* Climate change can add to stress on social and political structures by increasing management and budget requirements for public services such as public health care, disaster risk reduction, and even public security. As sources of stress grow and combine, the resilience of social and political structures that are already somewhat unstable is expected to suffer, especially in areas with relatively limited resources (Sherbinin et al., 2006). Where climate change adds to stress levels in settlements, it is expected to be especially problematic for vulnerable parts of the population: the poor, the elderly, those already in poor health, the disabled, those living alone, those with limited rights and power (e.g., recent immigrants with limited English skills), and/or indigenous populations dependent on one or a few resources. As one example, warmer temperatures in urban summers have a more immediate impact on populations who live and work without air conditioning.

Overall, climate change effects on human settlements in the United States are expected to occur as a result of interaction with other processes. Driving forces and stresses—technological, economic, and institutional—will have more impact on the sustainability of most settlements than climate change *per se* (Wilbanks et al., 2007a).

According to the IPCC (2007b):

> The most vulnerable industries, settlements and societies are generally those in coastal and river flood plains, those whose economies are closely linked with climate-sensitive resources, and those in areas prone to extreme weather events, especially where rapid urbanization is occurring (high confidence). Poor communities can be especially vulnerable, in particular those concentrated in high-risk areas. They tend to have more limited adaptive capacities, and are more dependent on climate-sensitive resources such as local water and food supplies (high confidence).

In the United States, the most vulnerable areas are *likely* to be Alaska, coastal and river basin locations susceptible to flooding, arid areas where water scarcity is a pressing issue, and areas whose economic bases are climate-sensitive (Field et al., 2007).

In Alaska and elsewhere in the Arctic, indigenous communities are facing major economic and cultural impacts. Many indigenous peoples depend on hunting polar bear, walrus, seals, and caribou; herding reindeer; fishing; and gathering, not only for food and to support the local economy, but also as the basis for cultural and social identity. Changes in species' ranges and availability, access to these species, a perceived reduction in weather predictability, and travel safety in changing ice and weather conditions present serious challenges to human health and food security, and possibly even the survival of some cultures (ACIA, 2004). In addition, thawing ground is beginning to destabilize transportation, buildings, and other facilities, posing needs for rebuilding with ongoing warming adding to construction and maintenance costs. One recent estimate of the value of Alaska's public infrastructure at risk from climate change by 2080 set the value at tens of billions of today's dollars. The largest estimated public costs were replacement of buildings, bridges, and other structures with long lifetimes (Larsen et al., 2007).

As highlighted in Section V.5.a, population growth is generally shifting toward areas more likely to be vulnerable to the effects of climate change. For example, many rapidly growing places in the intermountain West rely heavily on winter snowpack to provide municipal water resources. Projections of decreased snowpack and earlier spring melting suggest lower stream flows in the future, particularly during the high-demand period of summer. Moreover, as discussed in the next section, coastal areas are particularly at risk to climate change effects.

V.5.c. Sea level rise

Sea levels are rising and the IPCC concluded with high confi dence that the rate of sea level rise increased from the 19th to the 20th centuries (IPCC, 2007a). As described in section IV. 1.d, the causes for observed sea-level rise over the past century include thermal expansion of seawater as it warms and changes in land ice (e.g., melting of glaciers and snow caps). Over the 20th century, sea level rose about 1.7 ± 0.5 mm/yr (Bindoff et al., 2007). For the period 1993 to 2003, the rate was nearly twice as fast, at 3.1 ± 0.7 mm/yr. Some of this recent increase may be due to the observed acceleration in the rate of Greenland ice melting over the past decade (Rignot, 2006). However, there is considerably decadal variability in the tide gauge record so that it is unknown whether the higher rate in 1993 to 2003 is due to decadal variability or an increase in the longer-term trend. (Bindoff et al., 2007).

Over this same time period, land use intensity in coastal areas increased dramatically, a trend that is expected to continue (Nicholls et al., 2007). Regarding the effects of this use, the IPCC concluded (Nicholls et al., 2007):

- Coastal population growth in many of the world's deltas, barrier islands, and estuaries has led to widespread conversion of natural coastal landscapes to agriculture, aquaculture, silviculture, and industrial and residential uses (Valiela, 2006).
- It has been estimated that 23% of the world's population lives both within 100 km of the coast and below 100 m above sea level, and population densities in coastal regions are about three times higher than the global average (Small and Nicholls, 2003)
- Migration of people to coastal regions is common in both developed and developing nations. Sixty percent of the world's 39 metropolises with a population of over 5 million are located within 100 km of the coast, including 12 of the world's 16 cities with populations greater than 10 million.

Growth in coastal population has kept pace with population growth in other parts of the country, but given the small land area of the coasts, the density of coastal communities has been increasing (Crossett et al., 2004). Over 50% of the U.S. population now lives in the coastal zone, and coastal areas are projected to continue to increase in population, with associated increases in population density, over the next several decades. The overlay of this migration pattern with climate change projections has several implications. Perhaps the most obvious is the increased exposure of people and property to the effects of sea level rise and hurricanes (Kunkel et al., 1999). With rapidly growing communities near coastlines, property damages would be expected to increase even without any changes in storm frequency or intensity (Changnon et al., 2003).

The IPCC characterized some of the implications of this coastal population growth and associated human activities as follows (Nicholls et al., 2007):

- Enlargement of natural coastal inlets and dredging of waterways for navigation, port facilities, and pipelines exacerbate saltwater intrusion into surface and groundwater.
- Increasing shoreline retreat and risk of flooding of coastal cities in Thailand (Durongdej, 2001; Saito, 2001), India (Mohanti, 2000), Vietnam (Thanh et al., 2004), and the United States (Scavia et al., 2002) have been attributed to degradation of coastal ecosystems by human activities, illustrating a widespread trend.
- The major direct impacts include drainage of coastal wetlands, deforestation and reclamation, and discharge of sewage, fertilizers, and contaminants into coastal waters. Extractive activities include sand mining and hydrocarbon production, harvests of fisheries and other living resources, introductions of invasive species, and construction of seawalls and other structures.
- Engineering structures (such as damming, channelization, and diversions of coastal waterways) harden the coast, change circulation patterns, and alter freshwater, sediment, and nutrient delivery. Natural systems are often altered, even by soft

engineering solutions, such as beach nourishment and foredune construction (Nordstrom, 2000; Hamm and Stive, 2002).

- Ecosystem services on the coast are often disrupted by human activities. For example, tropical and subtropical mangrove forests and temperate salt marshes provide goods and services (they accumulate and transform nutrients, attenuate waves and storms, bind sediments, and support rich ecological communities), which are reduced by large-scale ecosystem conversion for agriculture, industrial and urban development, and aquaculture.

According to the IPCC (Scavia et al., 2002; Lotze et al., 2006), the impacts of these other human activities on the coastal zone have been more significant over the past century than effects that can be attributed to observed climate change. The cumulative effect of these non-climate, anthropogenic impacts increases the vulnerability of coastal systems to climate-related stressors.

For example, the degradation of natural coastal systems due to climate change, such as wetlands, beaches, and barrier islands, removes the natural defenses of coastal communities against extreme water levels during storms. Accordingly, Nicholls et al. (2007) found that many coastal cities are heavily dependent upon artificial coastal defenses (e.g., Tokyo, Shanghai, Hamburg, Rotterdam, and London). These urban systems are vulnerable to low-probability extreme events, with rising sea levels and increased extreme water levels exceeding prior defense standards and leading to systemic failures (domino effect). The IPCC found that the ports, roads, and railways along the U.S. Gulf and Atlantic Coasts are especially vulnerable to coastal flooding (Nicholls et al., 2007).

In the Great Lakes where sea level rise is not a concern, both extremely high and low water levels resulting from changes in the hydrologic cycle have been damaging and disruptive to shoreline communities (Nicholls et al., 2007). High lake water levels increase storm surge flooding, accelerate shoreline erosion, and damage industrial and commercial infrastructure located on the shore. Conversely, low lake water levels can pose problems for navigation, expose intake/discharge pipes for electrical utilities and municipal water treatment plants, and cause unpleasant odors.

The IPCC analyzed flood risk in 2080 from rising sea levels and storm surges under different climate scenarios (Nicholls et al., 2007). Under each scenario, the population at risk increases substantially. However, results show that upgrading coastline defenses can dramatically reduce projected impacts.

Projected impacts

Globally, the IPCC (Nicholls et al., 2007) made the following conclusions regarding sea level rise:

- Coasts will be exposed to increasing risks, including coastal erosion, over coming decades due to climate change and sea level rise (*very high confidence*).
- The impact of climate change on coasts is exacerbated by increasing human-induced pressures (*very high confidence*).

- Adaptation for the coasts of developing countries will be more challenging than for coasts of developed countries, due to constraints on adaptive capacity (*high confidence*).
- Adaptation costs for vulnerable coasts are much less than the costs of inaction (*high confidence*).
- The unavoidability of sea level rise, even in the longer term, frequently conflicts with present-day human development patterns and trends (*high confidence*).

The U.S. coastline is long and diverse with a wide range of coastal characteristics. Sea level rise changes the shape and location of coastlines by moving them landward along low-lying contours and exposing new areas to erosion (NRC, 2006a). Coasts subsiding due to natural or human- induced causes will experience larger relative rises in sea level. In some locations, such as deltas and coastal cities (e.g., the Mississippi Delta and surrounding cities), this effect can be significant (Nicholls et al., 2007). Rapid development, including an additional 25 million people in the coastal United States over the next 25 years, will further reduce the resilience of coastal areas to rising sea levels and increase the economic resources and infrastructure vulnerable to impacts (Field et al., 2007). Superimposed on the impacts of erosion and subsidence, the effects of rising sea level will exacerbate the loss of waterfront property and increase vulnerability to inundation hazards (Nicholls et al., 2007).

Climate change is expected to have a strong impact on saltwater intrusion into coastal sources of groundwater in the United States and other world regions. Sea-level rise and high rates of water withdrawal, promote the intrusion of saline water into the groundwater supplies, which adversely affects water quality. In some locations, reduced groundwater recharge, associated with decreases in precipitation and increases in evapotranspiration, will exacerbate the effects of sea level rise on salinization rates (Kundzewicz et al., 2007). This effect could impose enormous costs on water treatment infrastructure (i.e., costs associated with relocating infrastructure or building desalinization capacity), especially in densely populated coastal areas. Saltwater intrusion is also projected to occur in freshwater bodies along the coast. Estuarine and mangrove ecosystems can withstand a range of salinities on a short-term basis. However, they are not expected to survive permanent exposure to high-salinity environments. Saltwater intrusion into freshwater rivers has already been linked with the decline of bald cypress forests in Louisiana and cabbage palm forests in Florida. Given that these ecosystems provide a variety of ecosystem services and goods (e.g., spawning habitat for fish, pollutant filtration, sediment control, and storm surge attenuation), the loss of these areas could be significant (Kundzewicz et al., 2007).

Coastal indigenous communities are particularly vulnerable to impacts from sea level rise based on their geographic location, reliance on the local environment for aspects of everyday life such as diet and economy, and the current state of social, cultural, economic, and political change taking place in these regions (Anisimov et al., 2007). In northern areas of ice-rich permafrost, such as Alaska, coastal erosion rates are among the highest anywhere. These rates could be increased by rising sea levels, forcing the issue of relocation for threatened settlements. Adapting to these changes will be costly. For example, it has been estimated that relocating the village of Kivalina, Alaska (377 inhabitants in the 2000 census) to a nearby site would cost $54 million (Anisimov et al., 2007).

For small islands, particularly in the Pacific, some studies suggest that sea level rise could result in a reduction of island size, raising concerns for Hawaii and other U.S. territories

(Mimura et al., 2007). In some cases, accelerated coastal erosion may lead to island abandonment, as has been documented in the Chesapeake Bay. In the Caribbean and Pacific Islands, more than 50% of the population lives within 1.5 km of the shore. International airports, roads, capital cities, and other types of infrastructure are typically sited along the coasts of these islands as well. Therefore, the socioeconomic well-being of island communities will be threatened by inundation, storm surge, erosion, and other coastal hazards resulting from climate change (*high confidence*) (Mimura et al., 2007).

Regarding the effects of climate change on sea level rise and coastal areas in North America, the IPCC (Field et al., 2007) concluded:

- Rates of coastal wetland loss in the Chesapeake Bay and elsewhere will increase with accelerated sea level rise, in part due to 'coastal squeeze' (IPCC: *high confidence*).
- Coastal communities and habitats will be increasingly stressed by climate change impacts interacting with development and pollution (*very high confidence*). Sea level is rising along much of the coast, and the rate of change will increase in the future, exacerbating the impacts of progressive inundation, storm-surge flooding, and shoreline erosion.
- Storm impacts are *likely* to be more severe, especially along the Gulf and Atlantic Coasts. Salt marshes, other coastal habitats, and dependent species are threatened by sea level rise, fixed structures blocking landward migration, and changes in vegetation. Population growth and rising value of infrastructure in coastal areas increases vulnerability to climate variability and future climate change.

V.5.d. Extreme events

While individual extreme events cannot be directly linked to climate change, changing climate conditions are expected to result in increased incidence of extreme weather events. The combined effects of severe storms and sea level rise in coastal areas or increased risks of fire in drier arid areas are examples of how climate change may increase the magnitude of challenges already facing risk-prone communities. For example, the IPCC (Wilbanks et al., 2007a) estimated that, of the 131 million people affected by natural disasters in Asia in 2004, 97% were affected by weather-related disasters. Exposures in highly populated coastal and riverine areas and small island nations have been especially significant (ADRC et al., 2005).

Extreme events can also have substantial secondary effects on local economies. The United States ranks among the top 10 nations for international tourism receipts ($112 billion), with domestic tourism and outdoor recreation markets that are several times larger than most other countries. Climate variability affects many segments of this growing economic sector. For example, wildfires in Colorado (2002) caused tens of millions of dollars in tourism losses by reducing visitation and destroying infrastructure. Similar economic losses during that same year were caused by drought-affected water levels in rivers and reservoirs in the western United States and parts of the Great Lakes. The 10-day closure and clean-up following Hurricane George (September 1998) resulted in tourism revenue losses of approximately $32 million in the Florida Keys (Field et al., 2007 and references therein). Impacts of climate change on other recreational activities such as skiing, fishing, and hunting are highlighted in Saunders et al. (2008). While the North American tourism industry acknowledges the important influence of climate, its impacts have not been analyzed comprehensively.

Floods and storms

Nicholls et al. (2007) concluded that, globally, coasts are experiencing the adverse consequences of hazards related to climate and sea level (*very high confidence*). Although increases in mean sea level over the 21st century and beyond will inundate unprotected, low-lying areas, the most devastating impacts are *likely* to be associated with storm surge (Nicholls et al., 2007). For example, the Maryland Geological Survey estimated that more than 8 ha (20 acres) of the state's land was lost on the western shore of Chesapeake Bay in the wake of Tropical Storm Isabel in 2003, causing approximately $84 million in damages to shoreline structures (NRC, 2006a). Coasts are highly vulnerable to extreme events, such as storms, which impose substantial costs on coastal societies. Globally, about 120 million people are exposed to tropical cyclone hazards every year, and these storms killed 250,000 people from 1980 to 2000.

The IPCC (2007d) projects *likely* increases in intense tropical cyclones. Increases in tropical cyclone intensity are linked to increases in the risk of deaths, injuries, water- and food-borne diseases, as well as post-traumatic stress disorders (IPCC, 2007b). Drowning by storm surge, heightened by rising sea levels and more intense storms (as projected by the IPCC), is the major killer in coastal storms when there are large numbers of deaths (Confalonieri et al., 2007). High- density populations in low-lying coastal regions such as the U.S. Gulf of Mexico experience a high health burden from weather disasters, particularly among lower income groups.

In SAP 4.6, Wilbanks et al. (2008) provide the following description of vulnerability to storm activity in the southeast United States.

> Recent hurricanes striking the coast of the U.S. Southeast cannot be attributed clearly to climate change, but if climate change increases the intensity of storms as projected (IPCC, 2001; Emanuel, 2005) that experience suggests a range of possible impacts. As an extreme case, consider the example of Hurricane Katrina (Wilbanks, 2007a). In 2005, the city of New Orleans had a population of about half a million, located on the delta of the Mississippi River along the U.S. Gulf Coast. Urban development throughout the 20th Century has significantly increased land use and settlement in areas vulnerable to flooding, and a number of studies had indicated growing vulnerabilities to storms and flooding. In late August 2005, Hurricane Katrina moved onto the Louisiana and Mississippi coast with a storm surge, supplemented by waves, reaching up to 8.5 m above sea level. In New Orleans, the surge reached around 5 m, overtopping and breaching sections of the city's 4.5 m defenses, flooding 70 to 80% of New Orleans, with 55% of the city's properties inundated by more than 1.2 m and maximum flood depths up to 6 m. 1,101 people died in Louisiana, nearly all related to flooding, concentrated among the poor and elderly. Across the whole region, there were 1.75 million private insurance claims, costing in excess of $40 billion (Hartwig, 2006), while total economic costs are projected to be significantly in excess of $100 billion. Katrina also exhausted the federally backed National Flood Insurance Program (Hunter, 2006), which had to borrow $20.8 billion from the Government to fund the Katrina residential flood claims. In New Orleans alone, while flooding of residential structures caused $8-$ 10 billion in losses, $3-6 billion was uninsured. 34,000-35,000 of the flooded homes carried no flood insurance, including many that were not in a designated flood risk zone (Hartwig, 2006). Six months after Katrina, it was estimated that the population of New Orleans was 155,000, with the number projected to rise to 272,000 by September 2008—56% of its pre-Katrina level (K. McCarthy et al., 2006).

The recent severe tropical and extra-tropical storms, and resulting impacts, led the IPCC to conclude that North American urban centers with assumed high adaptive capacity remain vulnerable to extreme events (Field et al., 2007).

The vulnerability of some major urban centers in the United States is exacerbated by the fact that they are situated in low-lying flood plains. For example, areas of New Orleans and its vicinity are 1.5 to 3 m below sea level. Considering the rate of subsidence and using a mid-range estimate of 480 mm sea level rise by 2100, it is projected that this region could be 2.5 to 4.0 m or more below mean sea level by 2100 (Field et al., 2007). In this scenario, a storm surge from a Category 3 hurricane (estimated at 3 to 4 m without waves) could be 6 to 7 m above areas that were heavily populated in 2004 (Field et al., 2007).

Superimposed on accelerated sea level rise, storm intensity, wave height, and storm surge projections suggest more severe coastal flooding and erosion hazards (Nicholls et al., 2007). In New York City and Long Island, flooding from a combination of sea level rise and storm surge could be several meters deep (Field et al., 2007). Projections suggest that the return period of a 100-year flood event in this area might be reduced to 19 to 68 years, on average, by the 2050s, and to 4 to 60 years by the 2080s (Wilbanks et al., 2007a).

In addition, the IPCC projects a *very likely* increase in heavy precipitation event frequency over most areas. Heavy precipitation events are associated with increased risk of floods as well as infectious, respiratory, and skin diseases (IPCC, 2007b). Floods are low-probability, high-impact events that can overwhelm physical infrastructure, human resilience, and social organization (Confalonieri et al., 2007). Flood health impacts include deaths, injuries, infectious diseases, intoxications, and mental health problems (Greenough et al., 2001; Ahern et al., 2005 in Confalonieri et al., 2007). Flooding may also lead to contamination of waters with dangerous chemicals, heavy metals, or other hazardous substances, from storage or from chemicals already in the environment (Confalonieri et al., 2007). As the risk of flooding increases with climate change, so does the importance of major drainage systems. For example, new design approaches which explicitly design roads to act as drains can radically reduce the duration of flooding (OFCM, 2004).

Demand for waterfront property and land for building in the United States continues to grow, increasing the value of property at risk. Of the $19 trillion value of all insured residential and commercial property in the United States exposed to North Atlantic hurricanes, $7.2 trillion (4 1%) is located in coastal counties. This economic value includes 79% of the property in Florida, 63% of property in New York, and 61% of the property in Connecticut (AIR, 2002 in Field et al., 2007). The devastating effects of Hurricane Ivan in 2004 and Hurricanes Katrina, Rita, and Wilma in 2005 illustrate the vulnerability of North American infrastructure and urban systems that were not designed or not maintained to adequate safety margins. When protective systems fail, impacts can be widespread and multi-dimensional (Field et al., 2007).

Wildland fire

Wildland fires are discussed in greater detail in Section V.3.d. Disturbances such as forest fires, insect outbreaks, ice storms, and hurricanes can change forest productivity, carbon cycling, and species composition (Ryan et al., 2008). Forest fires and bushfires cause burns, smoke inhalation, and other injuries. Large fires are also accompanied by an increased number of patients seeking emergency services for inhalation of smoke and ash (Hoyt and Gerhart, 2004 in Confalonieri et al., 2007).

Wildfires have increased in extent and severity in recent years. At the same time, population has been growing rapidly, expanding beyond urban areas to at-risk areas known as the wildland– urban interface, where human settlements and development meet or intermix with wildland fuel (Hammer et al., 2007). These trends have resulted in increased vulnerability to both ecological and social impacts from increased fire activity, including financial losses from damaged property, lost revenue to local businesses, disruption of local social networks when residents are displaced, and a sense of devastation among citizens who have strong connections to the surrounding landscape (Toman et al., 2008). In some regions, changes in the mean and variability of temperature and precipitation are projected to increase the frequency and severity of fire events, including parts of the United States (Easterling et al., 2007).

Drought

Observed changes in drought are discussed in Sections IV.1.c and IV.1.d and projections of moisture availability and drought are discussed in Section IV.3.b. Areas affected by droughts are *likely* to increase, according to the IPCC (2007d). Specifically, it is *likely* that droughts will continue to be exacerbated by earlier and possibly lower spring snowmelt runoff in the mountainous West, which results in less water available in late summer (Karl et al., 2008). If total precipitation decreases or becomes more variable, extending the kinds of drought that have affected much of the interior West in recent years, water scarcity will be exacerbated, and increased water withdrawals from wells could affect aquifer levels and pumping costs (Wilbanks et al., 2008). Moreover, drying increases risks of fires, which have threatened urban areas in California and other western areas in recent years. Thus, projections of increased frequencies of drought combined with a greater fire risk put the increasing populations of desert Southwest cities at risk (Wilbanks et al., 2008). Health impacts associated with drought tend to mostly affect semi-arid and arid regions, poor areas and populations, and areas with human-induced water scarcity. Information about the effects of increasing drought on U.S. agriculture can be found in Section V.2.c.

V.6. Human Health

In SAP 4.6, Ebi et al. (2008) found that weather, climate variability, and climate change can affect health through the effects of environmental temperature on bodily functioning. Climate change can also have important secondary effects on heath. For example, climate change can alter or disrupt natural systems, making it possible for vector-, water- and food-borne diseases to spread or emerge in areas where they had been limited or not existed, or making it possible for such diseases to disappear by making areas less hospitable to the vector or pathogen (NRC, 2001b). Climate also can affect the incidence of diseases associated with air pollutants and aeroallergens. Ultimately, climate impacts on health are complex and will be influenced by multiple factors, including demographics; population and regional vulnerabilities; the future social, economic, and cultural context; availability of resources and technological options; built and natural environments; public health infrastructure; and the availability and quality of health and social services.

A comprehensive assessment of the potential impacts of climate variability and change on human health was published in 2000 as part of the First National Assessment of the Potential Impacts of Climate Variability and Change undertaken by the U.S. Global Change Research Program (NAST, 2001). These findings have been significantly updated in recent IPCC and CCSP reports.

In its Fourth Assessment Report, the IPCC produced a number of key findings summarizing the potential health effects of climate change in North America. The IPCC (2007b) concluded that, overall, it is expected that benefits will be outweighed by the negative health effects of rising temperatures worldwide, especially in developing countries. Gamble et al. (2008) concluded that there may be fewer cases of illness and death associated with climate change in the United States than in the developing world. Nevertheless, increased costs to human health and well being are anticipated for the United States.

Key conclusions drawn primarily from the IPCC Fourth Assessment Report (Field et al., 2007) for North America include:

- Increased deaths, injuries, infectious diseases, and stress-related disorders and other adverse effects associated with social disruption, migration, and loss of place from more frequent extreme weather.
- Increased frequency and severity of heat waves leading to more illness and death, particularly among the young, elderly, frail, poor, and outdoor workers and athletes.
- Expanded ranges of vector- and tick-borne diseases in North America but with modulation by public health measures and other factors.

SAP 4.6 (Gamble et al., 2008) concluded that greater wealth and a more developed public health system and infrastructure (e.g., water treatment plants, sewers, and drinking water systems; roads, rails and bridges; flood control structures) will continue to enhance our capacity to respond to climate change. Similarly, governments' capacities for disaster planning and emergency response are key assets that should allow the United States to adapt to many of the health effects associated with climate change. Other conclusions based in large part on SAP 4.6 (Gamble et al., 2008) regarding the health effects projected for the United States are summarized below.

- *It is very likely that heat-related morbidity and mortality will increase over the coming decades, however net changes in mortality are difficult to estimate because, in part, much depends on complexities in the relationship between mortality and global change.* According to the U.S. Census, the U.S. population is aging; the percent of the population over age 65 is projected to be 13% by 2010 and 20% by 2030 (over 50 million people). Older adults, very young children, and persons with compromised immune systems are vulnerable to temperature extremes. This suggests that temperature-related morbidity and mortality are likely to increase. Similarly, heat-related mortality affects poor and minority populations disproportionately, in part due to lack of air conditioning. The concentration of poverty in inner city neighborhoods leads to disproportionate adverse effects associated with urban heat islands. However, few studies have attempted to link the epidemiological findings to climate scenarios for the United States, and studies that have done so have focused on the effects of changes in average temperature, with mixed results.

- *The impacts of higher temperatures in urban areas and likely associated increases in tropospheric ozone concentrations may contribute to or exacerbate cardiovascular and pulmonary illness if current regulatory standards are not attained.* In addition, stagnant air masses related to climate change are likely to degrade air quality in some densely populated areas. It is important to recognize that the United States has a well-developed and successful national regulatory program for ozone, PM2.5, and other criteria pollutants. That is, the influence of climate change on air quality will play out against a backdrop of ongoing regulatory control that will shift the baseline concentrations of air pollutants. Studies to date have typically held air pollutant emissions constant over future decades (i.e., have examined the sensitivity of ozone concentrations to climate change rather than projecting actual future ozone concentrations). Physical features of communities, including housing quality and green space, social programs that affect access to health care, aspects of population composition (level of education, racial/ethnic composition), and social and cultural factors are all likely to affect vulnerability to air quality.

- *Hurricanes, extreme precipitation resulting in floods, and wildfires also have the potential to affect public health through direct and indirect health risks.* The health risks associated with such extreme events are likely to increase with the size of the population and the degree to which it is physically, mentally, or financially constrained in its ability to prepare for and respond to extreme weather events. For example, coastal evacuations prompted by imminent hurricane landfall are only moderately successful. Many of those who are advised to flee to higher ground stay behind in inadequate shelter. Surveys find that the public is either not aware of the appropriate preventive actions or incorrectly assesses the extent of their personal risk.

- *There will likely be an increase in the spread of several food- and water-borne pathogens among susceptible populations depending on the pathogens' survival, persistence, habitat range and transmission under changing climate and environmental conditions.* The primary climate-related factors that affect these pathogens include temperature, precipitation, extreme weather events, and shifts in their ecological regimes. Consistent with our understanding of climate change on human health, the impact of climate on food and water-borne pathogens will seldom be the only factor determining the burden of human injuries, illness, and death.

- *Health burdens related to climate change will vary by region.* For the continental United States, the northern latitudes are likely to experience the largest increases in average temperatures; they will also bear the brunt of increases in ground-level ozone and other airborne pollutants. Because Midwestern and Northeastern cities are generally not as well adapted to the heat as Southern cities, their populations are likely to be disproportionately affected by heat related illnesses as heat waves increase in frequency, severity, and duration. The range of many vectors is likely to extend northward and to higher elevations. For some vectors, such as rodents associated with Hantavirus, ranges are likely to expand, as the precipitation patterns under a warmer climate enhance the vegetation that controls the rodent population. Forest fires with their associated decrements to air quality and pulmonary effects are likely to increase in frequency, severity, distribution, and duration in the Southeast, the Intermountain West and the West.

- *Finally, climate change is very likely to accentuate the disparities already evident in the American health care system.* Many of the expected health effects are likely to fall disproportionately on the poor, the elderly, the disabled, and the uninsured. The most important adaptation to ameliorate health effects from climate change is to support and maintain the United States' public health infrastructure.

Valid projections of health effects require solid data on dose–response relationships between climate variables and health outcomes. These data are often lacking for several outcomes of concern. Additional research and development are needed to ensure that surveillance systems account for and anticipate the potential effects of climate change (NRC, 2001b). Surveillance systems will be needed in locations where changes in weather and climate may foster the spread of climate-sensitive pathogens and vectors into new regions (NRC, 2001b). Understanding associations between disease patterns and environmental variables can be used to develop early warning systems that warn of outbreaks before most cases have occurred. Increased understanding is needed of how to design these systems where there is limited knowledge of the interactions among climate, ecosystems, and infectious diseases (NAS, 2001 c). To date, few studies address the interaction of multi-sector climate impacts or of interactions between climate change health impacts and other kinds of local, regional, and global changes (Field et al., 2007). For example, climate change impacts on human health in urban areas will be compounded by aging infrastructure, maladapted urban form and building stock, urban heat islands, air pollution, population growth, and an aging population.

The remainder of this section describes the literature on the impacts of climate change on human health in four areas noted above: primary temperature effects, extreme events, climate-sensitive infectious diseases, and aeroallergens and air quality.

V.6.a. Temperature effects

According to the IPCC (2007a), it is *very likely* that there were fewer cold days and nights and warmer and more frequent hot days over most land areas during the late 20th century. It is *virtually certain* that these trends will continue during the 21st century.

As a result of the projected warming, the IPCC projects increases in heat-related mortality and morbidity globally (including in the United States) (Field et al., 2007; IPCC, 2007b). The projected warming is expected to result in fewer cold-related deaths due to reduced exposure to the cold. However, net changes in mortality are difficult to estimate because, in part, much depends on complexities in the relationship between mortality and the changes associated with global change.

Increased heat exposure

In addition to increases in average temperatures, heat waves have become more frequent (IPCC, 2007a). Heat waves are associated with marked short-term increases in mortality (Confalonieri et al., 2007). India has experienced 18 heat waves since 1980, each resulting in multiple deaths ranging from an estimated 29 deaths in 2000 to more than 3,000 deaths in

2003 (Mohanty and Panda, 2003; Government of Andhra Pradesh, 2004 in Confalonieri et al., 2007). Heat waves have also resulted in substantial mortality in more industrialized nations. The 2003 European heat wave is estimated to have caused 35,000 deaths (e.g., Hemon and Jougla, 2004; Martinez- Navarro et al., 2004; Johnson et al., 2005 in Confalonieri et al., 2007).

Hot temperatures have also been associated with increased morbidity, although there is less information available (Ebi et al., 2008). While effects have varied across studies, Schwartz et al. (2004) documented increased hospital admissions for cardiovascular disease and emergency room visits in North America during heat events (Schwartz et al., 2004).

In SAP 4.6, Ebi et al. (2008) identified the following heat-related impacts on U.S. health:

- Exposure to excessive natural heat caused a reported 4,780 deaths during the period 1979 to 2002, while hyperthermia was reported as a contributing factor in an additional 1,203 deaths (CDC, 2005a).
- Heat is expected to contribute to the exacerbation of chronic health conditions, such as cardiovascular, renal, respiratory, diabetes, and nervous system disorders. Several analyses have found associations between increased heat and increased all-cause mortality (Medina- Ramon et al., 2006).
- Vulnerability varies among populations. Particularly vulnerable groups include the elderly, the very young, city-dwellers, those with less education, people on medications such as diuretics, the socially isolated, the mentally ill, those lacking access to air conditioning, and outdoor laborers (e.g., McGeehin and Mirabelli, 2001; Diaz et al., 2002; Klinenberg, 2002). Among the most well-documented heat waves in the United States are those that occurred in 1980 (St. Louis and Kansas City, Missouri), 1995 (Chicago, Illinois), and 1999 (Cincinnati, Ohio; Philadelphia, Pennsylvania; and Chicago, Illinois). In all these episodes, the highest death rates occurred in people over 65 years of age.
- Less information exists on temperature-related morbidity, and those studies that have examined hospital admissions and temperature have not seen consistent effects, either by cause or by demonstrated coherence with mortality effects where both deaths and hospitalizations were examined simultaneously (Semenza et al., 1999; Kovats et al., 2004; Schwartz et al., 2004).
- In many cases, the urban heat island effect may increase heat-related mortality. This effect may cause air temperatures in urban areas to increase by 1 to 6 °C over the surrounding suburban and rural areas, due to absorption of heat by dark paved surfaces and buildings; lack of vegetation and trees; heat emitted from buildings, vehicles, and air conditioners; and reduced airflow around buildings (Pinho and Orgaz, 2000; Vose et al., 2004; Xu and Chen, 2004). However, in some regions, urban areas may not experience greater heat-related mortality than in rural areas (Sheridan and Dolney, 2003).
- High temperatures and high air pollution can interact to result in additional health impacts. The extent of interaction varies by location (Goodman et al., 2004; Bates, 2005; Ren et al., 2006).

Reviewing the research and emerging projections to date, the IPCC projected that heat-related morbidity and mortality will increase globally (Confalonieri et al., 2007). Estimates of heat- related mortality attributable to climate change are reduced but not eliminated when models include assumptions about acclimatization and adaptation (e.g., implementation of space cooling). Past research indicates that populations in the United States became less sensitive to high temperatures over the period 1964 to 1988 due, at least in part, to these factors (R. Davis et al., 2002, 2003, 2004 in Confalonieri et al., 2007).

However, demographic trends suggest that the vulnerable population in the United States will increase in the coming years. Across North America, the population over the age of 65 (the segment of the population most at risk of dying from heat waves) will increase slowly to 2010, and then grow dramatically as the 'baby boomers' age (Field et al., 2007). Moreover, much of the population growth in the next 50 years is expected to occur in cities (Cohen, 2003 in Confalonieri et al., 2007), increasing the total population exposed to the urban heat island effect.

The IPCC projects the following U.S. regional increases in heat and/or heat-related effects (Confalonieri et al., 2007; Field et al., 2007):

- Severe heat waves are projected to intensify in magnitude and duration over the portions of the United States where these events already occur (*high confidence*).
- By the 2080s, in Los Angeles, the number of heat wave days (at or above 32 °C or 90 °F) increases four-fold under the B 1 emissions scenario (low growth) and six- to eight-fold under the A1FI (high growth) emissions scenario (Hayhoe, 2004). The annual number of projected heat-related deaths in Los Angeles increases from about 165 in the 1990s to 319 to 1,182 in the 2080s for a range of emissions scenarios.
- Chicago is projected to experience 25% more frequent heat waves annually for the period spanning 2080 to 2099 for a business-as-usual (A1B) emissions scenario (Meehl and Tebaldi, 2004).

Reduced cold exposure

The IPCC found that exposure to cold conditions results in adverse health impacts particularly in northerly latitudes, where very low temperatures can be reached in a few hours and continue for extended periods (Confalonieri et al., 2007). Specific findings included the following:

- Adverse impacts occur mainly outdoors, among socially deprived people (e.g., alcoholics and the homeless), workers, and the elderly in temperate and cold climates (Ranhoff, 2000).
- Living in cold environments in polar regions is associated with a range of chronic conditions in the non-indigenous population (Sorogin et al, 1993) as well as with acute risk from frostbite and hypothermia (Hassi et al., 2005).
- In general, population sensitivity is greater in temperate countries with mild winters, as populations are less well adapted to cold (Eurowinter Group, 1997; Healy, 2003). However, in countries with populations well adapted to cold conditions, cold waves can still cause substantial increases in mortality if electricity or heating systems fail.
- Cold waves also affect health in warmer climates, such as in South East Asia (EM-DAT, 2006).

In SAP 4.6, Ebi et al. (2008) wrote about specific impacts of cold exposure in the United States. An average of 689 reported deaths annually (ranging from 417 to 1,021 per year) from 1979 to 2002 were attributed to exposure to excessive natural cold (16,555 total deaths) (Fallico et al., 2005). However, the mortality burden is probably underestimated because cold also contributes to deaths caused by respiratory and cardiovascular diseases. Because many factors contribute to winter mortality, it is highly uncertain how climate change could affect mortality. No projections have been published for the United States that incorporate critical factors, such as the influence of influenza outbreaks.

Vulnerability is associated with the following factors:

- Location of residence: living in Alaska, New Mexico, North Dakota, or Montana, or living in a state with a milder climate that experiences rapid temperature changes (North and South Carolina) or a western state with greater ranges in nighttime temperatures (e.g., Arizona) (Fallico et al., 2005);
- Race: African Americans show higher vulnerability (Fallico et al., 2005);
- Level of education: vulnerability is higher among those with lower education (O'Neill et al., 2003);
- Gender: females are more vulnerable (Wilkinson et al., 2004);
- Pre-existing respiratory illness (Wilkinson et al., 2004);
- Living in nursing homes (Hajat et al., 2007); and
- Other factors such as lack of protective clothing (Donaldson et al., 2001), income inequality, fuel poverty, and low residential thermal standards (Healy, 2003).

While the IPCC found that health projections have improved in recent years, there is still a lack of information on the effects of thermal stress on mortality outside the industrialized countries and limited specific results are available by country (Confalonieri et al., 2007). Moreover, the authors provide a caution that projections of reduced cold-related deaths can be overestimated unless they take into account the effects of influenza and season (Armstrong et al., 2004 in Confalonieri et al., 2007). Ultimately, Confalonieri et al. (2007) conclude that additional research is needed to understand how the balance of heat-related and cold-related mortality could change under different socioeconomic scenarios and climate projections.

While projections suggest the reductions in cold-related deaths will be greater than increases in heat-related deaths in the United Kingdom (Donaldson et al., 2001), limited research examines potential outcomes in the United States under different climate scenarios (Confalonieri et al., 2007).

V.6.b. Extreme events

In addition to the immediate effects of temperature on mortality and morbidity, a changing climate can also result in substantial secondary effects such as through extreme events. Vulnerability to weather disasters depends on the attributes of the people at risk (including where they live, age, income, education, and disability) and on broader social and environmental factors (level of disaster preparedness, health sector responses, and environmental degradation).

As Ebi et al. (2008) concluded in SAP 4.6, the United States experiences a wide range of extreme weather events, including hurricanes, floods, tornadoes, blizzards, windstorms, and drought. Other extreme events, such as wildfires, are strongly influenced by meteorological conditions. Morbidity and mortality due to an event increase with the intensity and duration of the event, and can decrease with advance warning and preparation. Health also can be affected by extreme events through secondary mechanisms such as carbon monoxide poisonings from portable electric generator use for electricity (CDC, 2006) or increased incidence of gastroenteritis cases among hurricane evacuees (CDC, 2005a). Moreover, these events can lead to substantial mental health impacts (e.g., post-traumatic stress disorder and depression) (Middleton et al., 2002; Russoniello et al., 2002; Verger et al., 2003; North et al., 2004; Fried et al., 2005; Weisler et al., 2006).

Climate effects on extreme events are discussed in greater detail in Section V.5.d above. Here the discussion is limited to additional information on health affects.

Floods and storms

Impacts of floods and storms include deaths, injuries, infectious diseases, intoxications, and mental health problems (Greenough et al., 2001; Ahern et al., 2005 in Confalonieri et al., 2007). Flooding may also lead to contamination of waters with dangerous chemicals, heavy metals, or other hazardous substances, from storage or from chemicals already in the environment (Confalonieri et al., 2007). In addition, increases in tropical cyclone intensity are linked to increases in the risk of deaths, injuries, water- and food-borne diseases, and post-traumatic stress disorders (IPCC, 2007b).

Major storms and floods can result in huge numbers of deaths. Confalonieri et al. (2007) cite the following examples:

- In 2003, 130 million people were affected by floods in China (EM-DAT, 2006).
- In 1999, 30,000 people died from storms followed by floods and landslides in Venezuela.
- In 2000 and 2001, 1,813 people died in floods in Mozambique (IFRC, 2002; Guha-Sapir et al., 2004).

High-density populations in low-lying coastal regions such as the U.S. Gulf of Mexico experience a high health burden from weather disasters, particularly among lower income groups. The 2005 hurricane season in the United States resulted in double the average number of lives lost over the previous 65 years. Overall, 2,002 lives were lost, and over 1,800 of those occurred during two storm events (Katrina and Rita) (NOAA, 2007a).

Following the storm event, lasting impacts may result particularly when sanitation systems have been overwhelmed. Indeed, in Confalonieri et al. (2007) the IPCC noted that populations with poor sanitation infrastructure and high burdens of infectious disease often experience increased rates of diarrheal diseases after flood events. For example, increases in cholera (Sur et al., 2000; Gabastou et al., 2002), cryptosporidiosis (Katsumata et al., 1998), and typhoid fever (Vollaard et al., 2004) have been reported in middle- and low-income nations. The impacts can be substantial; the above-cited 2001 floods in Mozambique are estimated to have resulted in an additional 447 deaths from increased incidence of diarrheal disease (Cairncross and Alvarinho, 2006).

While the IPCC (Confalonieri et al., 2007) noted that risks of infectious disease following floods are generally low in high-income countries, important exceptions have occurred. Most notably, following Hurricanes Katrina and Rita, the contamination of the drinking water supply with fecal bacteria resulted in increased incidence of diarrheal illness and some deaths (CDC, 2005b).

Wildfires

Regarding wildfires, the immediate risk of mortality is largely a function of the population in the affected area and the speed and intensity with which the wildfire moves through those areas. Although mortalities still occur in the United States on an annual basis (ranging from 14 to 28 from 1998 to 2002), a well-established warning and evacuation system exists and has resulted in substantial reductions. Wildfires can have substantial effects, including through increased eye and respiratory illnesses due to fire-related air pollution and mental health impacts from evacuations, lost property, and damage to resources. Large fires are also accompanied by an increased number of patients seeking emergency services for inhalation of smoke and ash (Hoyt and Gerhart, 2004). Further discussion of health effects from particulate emissions (including those from wildfires) is in Section V.6.e under 'Particulate Matter.'

Droughts

The IPCC (Confalonieri et al., 2007) concluded that drought can lead to the following health effects:

- deaths, malnutrition (under-nutrition, protein-energy malnutrition, and/or micronutrient deficiencies), infectious diseases, and respiratory diseases (Menne and Bertollini, 2000);
- diminished dietary diversity and reduced overall food consumption with subsequent negative health impacts (e.g., Hari Kumar et al., 2005; Aziz et al., 1990); and
- population movements, particularly rural to urban migration, which may result in displacement impacts from overcrowding and a lack of safe water, food, and shelter (Choudhury and Bhuiya, 1993, Menne and Bertollini, 2000; del Ninno and Lundberg, 2005).

V.6.c. Climate-sensitive infectious diseases

Ebi et al. (2008) observed that the distinct seasonal pattern exhibited by most vector-, water-, or food-borne and animal-associated diseases suggests that weather and/or climate influence their distribution and incidence throughout the United States. They concluded that evidence for the United States indicates that climate affects both the abundance of vectors and ticks and the distributions of vectors and ticks that can carry West Nile virus, Western Equine encephalitis, Eastern Equine encephalitis, Bluetongue virus, and Lyme disease—perhaps affecting disease risk, but sometimes in counter-intuitive ways that do not necessarily translate to increased disease incidence (Wegbreit and Reisen, 2000; Subak, 2003; McCabe and Bunnell, 2004; DeGaetano, 2005; Purse et al., 2005; Kunkel et al., 2006; Ostfeld et al., 2006; Shone et al., 2006). Field et al. (2007) concluded that climate change is *likely* to increase the risk and geographic spread of vector-borne infectious diseases, including Lyme disease and West Nile virus.

As for water- and food-borne diseases, Ebi et al. (2008) concluded that these cause significant morbidity in the United States. For example, there were 1,330 food-related disease outbreaks in 2002 (Lynch et al., 2006). In 2004, there were 34 outbreaks from recreational water and 30 outbreaks from drinking water (Dziuban et al., 2006; Liang et al., 2006). Water- and food-borne diseases primarily result in gastroenteritis. In 2003 and 2004, gastroenteritis was noted in 48 and 68% of reported recreational and drinking water outbreaks, respectively (Dziuban et al., 2006; Liang et al., 2006). Most pathogens of concern for food- and waterborne exposure are enteric and transmitted by the fecal–oral route. Climate may influence the pathogen by influencing its growth, survival, persistence, transmission, or virulence or through interactions with land use practices and climate variability. Storm events and flooding may result in the contamination of food crops (especially produce such as leafy greens and tomatoes) with feces from nearby livestock or feral animals.

Ebi et al. (2008) note that the incidence of many water- and food-borne diseases is associated with temperatures. Given these associations, Ebi et al. (2008) concluded that increasing temperatures may result in increased incidence of disease, including the following:

- Salmonellosis: annual peaks in salmonellosis cases occur within one to six weeks of the highest reported ambient temperatures in North America (United States and Canada) (Fleury et al., 2006; Naumova et al., 2007), Australia (D'Souza et al., 2004), and several countries across Europe (Kovats et al., 2004a).
- Leptospirosis: the most widespread zoonotic disease in the world, re-emerging in the United States (Meites et al., 2004), is linked to warm temperatures.
- *Vibrio* spp.: accounts for 20% of sporadic shellfish-related illnesses and over 95% of deaths (Lipp and Rose, 1997; Morris, 2003); infections are more frequently associated with warm temperatures (e.g., Janda et al., 1988; Lipp et al., 2002).
- Protozoan parasites, particularly *Cryptosporidium* and *Giardia*: case reports peak in late summer and early fall, particularly among younger age groups, generally associated with recreational water use (Dietz and Roberts, 2000; Furness et al., 2000).
- *Naegleria fowleri*: infections are relatively rare, but nearly always fatal (Lee et al., 2002); associated with naturally warm or thermally polluted bodies of water.

Such associations with temperature were not readily identified for the prevalence or infection rates of *Campylobacter* spp. (Fleury et al., 2006; Naumova et al., 2007 in Ebi et al., 2008). Temperature associations were also not readily identified for epidemiologically significant viruses including enteroviruses, rotaviruses, hepatitis A virus, and norovirus. Viruses account for 67% of food-borne disease, and the vast majority of these are due to norovirus (Mead et al., 1999). Norovirus infections typically peak in the winter (Cook et al., 1990; Lynch et al., 2006 in Ebi et al., 2008).

Research also suggests that climate variables, including regional variations and ENSO interactions, may be an important driver for influenza outbreaks (Ebi et al., 2008). However, available research has not identified general trends or mechanisms regarding the role of climate in infection (e.g., Ebi et al., 2001; Flahault et al., 2004; Viboud et al., 2004; Dushoff et al., 2005; Greene et al., 2006; Choi et al., 2006 in Ebi et al., 2008). No studies have been able to identify a clear role for temperature in viral infection patterns (Ebi et al., 2008).

Ebi et al. (2008) concluded that further expansion of ranges of insect- and rodent-borne diseases are expected in North America. However, major human epidemics of these diseases in the United States are *unlikely* if the public health infrastructure is maintained and improved as needed.

Water quality

Research reviewed in SAP 4.6 (Ebi et al., 2008) identified an association between precipitation intensity and outbreaks of waterborne illness in U.S. drinking water. An analysis of waterborne outbreaks associated with drinking water in the United States between 1948 and 1994 found that 51% of outbreaks occurred following a daily precipitation event in the 90th percentile while 68% occurred when precipitation levels reached the 80th percentile (Curriero et al., 2001). Thomas et al. (2006) found that risk of waterborne disease doubled when rainfall amounts surpassed the 93rd percentile. The relationship between rainfall and outbreaks exists for both surface and groundwater, although the association is strongest for surface water (Rose et al., 2000).

As noted above, floodwaters may increase the likelihood of contaminated drinking water and also lead to incidental exposure to contaminated flood waters. Severe flooding following Hurricane Floyd in 1999 resulted in twice the average rate of gastrointestinal illness in the affected areas (Setzer and Domino, 2004). Following the 2001 floods in the Midwest, contact with floodwater was shown to increase the rate and risk of gastrointestinal illness, especially among children. Tap water, however, continued to meet all regulatory standards (Wade et al., 2004).

Ebi et al. (2008) concluded that federal and state laws and regulatory programs protect much of the U.S. population from waterborne disease. However, if climate variability increases, current and future deficiencies in watershed protection, infrastructure, and storm drainage systems will tend to increase the risk of contamination events.

V.6.d. Aeroallergens

Exposure to allergens results in allergic illnesses in approximately 20% of the U.S. population (AAAAI, 1996–2006). Although there is substantial evidence suggesting a causal relationship between aeroallergens and allergic illnesses, it remains unclear which aeroallergens are most important for sensitization and subsequent disease development (Nielsen et al., 2002). Not only the type, but also the amount of aeroallergen to which an individual is exposed, influences the development of an allergic illness.

Climate change, including changes in atmospheric CO_2 concentrations, could affect the production, distribution, dispersion, and allergenicity of aeroallergens and the growth and distribution of weeds, grasses, and trees that produce them (Confalonieri et al., 2007). These changes in aeroallergens and subsequent human exposures could affect the prevalence and severity of allergy symptoms. However, the scientific literature does not provide definitive data or conclusions on how climate change might affect aeroallergens and subsequently the prevalence of allergic illnesses in the United States. In addition, there are numerous other factors that affect aeroallergen levels and the prevalence of associated allergic illnesses, such as changes in land use, air pollution, and adaptive responses, many of which are difficult to assess.

The IPCC concluded that pollens are *likely* to increase with elevated temperature and atmospheric CO_2 concentrations in North America (Field et al., 2007). Moreover, warming and climate extremes are *likely* to increase pollen and ozone levels, both of which have the potential to exacerbate symptoms in people with respiratory illness. Laboratory studies that used a doubling of atmospheric CO_2 stimulated a greater than 50% increase in ragweed pollen production (Wayne et al., 2002). A U.S.-based field study, which used existing temperature/CO_2 concentration differences between urban and rural areas as a proxy for climate change, found that ragweed grew faster, flowered earlier, and produced significantly greater aboveground biomass and ragweed pollen at urban locations than at rural locations (Ziska et al., 2003 in Field et al., 2007).

The IPCC (Confalonieri et al., 2007) noted that climate change has caused an earlier onset of the spring pollen season in North America (D'Amato et al., 2002; Weber, 2002; Beggs, 2004) and that there is limited evidence that the length of the pollen season has increased for some species. However, it is unclear whether the allergenic content of these pollens has changed. The IPCC concluded that introductions of new invasive plant species with high allergenic pollen present important health risks, noting that ragweed (*Ambrosia artemisiifolia*) is spreading in several parts of the world (Rybnicek and Jaeger, 2001; Huynen and Menne, 2003; Taramarcaz et al., 2005; Cecchi et al., 2006 in Confalonieri et al., 2007).

V.6.e. Air quality

The IPCC (2007b) projects with *virtual certainty* "declining air quality in cities" due to "warmer and fewer cold days and nights and/or warmer/more frequent hot days and nights over most land areas" across all world regions. Furthermore, the IPCC reports with *very high confidence* that climate change impacts on human health in U.S. cities will be compounded by population growth and an aging population (Field et al., 2007). Surface air concentrations of air pollutants are highly sensitive to winds, temperature, humidity, and precipitation (Denman et al., 2007). Climate change can be expected to influence the concentration and distribution of air pollutants through a variety of processes, including the modification of biogenic emissions, the change in chemical reaction rates, wash-out of pollutants by precipitation, and modification of weather patterns that influence pollutant buildup.

In summarizing the impact of climate change on ozone and particulate matter, the IPCC (Denman et al., 2007) found that "future climate change may cause significant air quality degradation by changing the dispersion rate of pollutants, the chemical environment for ozone and particulate matter generation and the strength of emissions from the biosphere, fires and dust." The IPCC also noted that large uncertainties limit the ability to provide a simple quantitative description of the interactions between biogeochemical processes and climate change (Denman et al., 2007).

In SAP 4.6, Ebi et al. (2008) concluded that currently millions of Americans live in areas that do not meet the health-based National Ambient Air Quality Standards for ozone and fine particulate matter, both of which have the potential to be influenced by climate change.

Tropospheric ozone

Ground-level ozone is formed mainly by reactions that occur in polluted air in the presence of sunlight. The key precursors for formation include nitrogen oxides (emitted mainly by the burning of fuels) and volatile organic compounds (emitted both by burning of fuels and by evaporation from vegetation and stored fuels). Ground-level ozone formation increases with greater sunlight and higher temperatures.

According to the IPCC (Denman et al., 2007), climate change is expected to lead to increases in regional ozone pollution in the United States and other countries. Breathing air containing ozone can reduce lung function, thereby aggravating asthma and other respiratory conditions. Ozone exposure has been associated with increases in respiratory infection susceptibility, medicine use by asthmatics, emergency department visits, and hospital admissions. Exposure may contribute to premature death in people with heart and lung disease (EPA, 2006). Vulnerability to ozone health effects is greater for persons who spend time, especially with physical exertion, outdoors during episode periods because this results in a higher cumulative dose to the lung. Thus, children, outdoor laborers, and athletes may be at greater risk than people who spend more time indoors and who are less active (Ebi et al., 2008). In contrast to human health effects, which are associated with short-term exposures, the most significant ozone-induced plant effects (e.g., biomass loss and yield reductions) result from the accumulation of ozone exposures over the growing season, with differentially greater impact resulting from exposures to higher concentrations and/or longer durations (EPA, 2006).

Tropospheric ozone is both naturally occurring and, as the primary constituent of urban smog, a secondary pollutant. As a secondary pollutant, it is formed through photochemical reactions involving nitrogen oxides (NO_x) and volatile organic compounds (VOCs) in the presence of sunlight. Biomass burning, which is known to make a large contribution to ozone in the tropical troposphere (Thompson et al., 1996), is expected to increase in the tropics and at high latitudes with climate change (Price and Rind, 1994; Stocks et al., 1998; Williams et al., 2001; Brown et al., 2004). As described below, climate change can affect ozone by modifying

- emissions of precursors,
- atmospheric chemistry, and
- transport and removal (Denman et al., 2007)

The IPCC (Denman et al., 2007) concluded that, for all world regions, climate change affects the sources of ozone precursors in the following ways:

- Physical response: NO_x emissions due to lightning are expected to increase in a warmer climate. General circulation models concur that climate warming could increase the influx of ozone from the stratosphere to the troposphere due to large-scale atmospheric circulation shifts (i.e., the Brewer–Dobson circulation).
- Biological response (soils, vegetation): biogenic VOC emissions increase with increasing temperature. Climate-induced changes in biogenic VOC emissions alone may be regionally substantial and cause significant increases in ozone concentrations (European Commission, 2003; Hogrefe et al., 2004; Hauglustaine et al., 2005). Sensitivity simulations for the 2050s, relative to the 1 990s (under the IPCC A2 scenario discussed in Section IV.3 .a), indicate that increased biogenic emissions alone add one to three parts per billion (ppb) to summertime average daily maximum 8-hour ozone concentrations[49] in the Midwest and along the Eastern Seaboard (Hogrefe et al., 2004). The IPCC (Meehl et al., 2007) reports that biogenic emissions

> are projected to increase by between 27 and 59%, contributing to a 30 to 50% increase in ozone formation over northern continental regions (under the IPCC A2 scenario for the 2090–2 100 timeframe, relative to 1990–2000) (Denman et al., 2007).

Climate change impacts on temperature and water vapor could affect ozone chemistry significantly (Denman et al., 2007). A number of studies in the United States have shown that summer daytime ozone concentrations correlate strongly with temperature. That is, ozone generally increases at higher temperatures. This correlation appears to reflect contributions of comparable magnitude from:

- temperature-dependent biogenic VOC emissions (as mentioned above);
- thermal decomposition of peroxyacetylnitrate (PAN), which acts as a reservoir for NOx, (as described immediately below); and
- the association of high temperatures with regional stagnation (also discussed below)

(Denman et al., 2007).

Climate change is projected to increase surface layer ozone concentrations in both urban and polluted rural environments (any world region of high emissions) due to decomposition of PAN at higher temperatures (Sillman and Samson, 1995; Liao et al., 2006). Warming enhances decomposition of PAN, releasing NO_x, an important ozone precursor (Stevenson et al., 2005). Model simulations for the year 2100 show that enhanced PAN thermal decomposition leads to an increase in ozone net production (Hauglustaine et al., 2005).

Atmospheric circulation can be expected to change in a warming climate and thus modify pollutant transport and removal. The more frequent occurrence of stagnant air events in urban or industrial areas could enhance the intensity of air pollution events, although the importance of these effects is not yet well quantified (Denman et al., 2007). The IPCC (2007a) concluded that "extra-tropical storm tracks are projected to move poleward, with consequent changes in wind, precipitation, and temperature patterns, continuing the broad pattern of observed trends over the last half-century."

The IPCC (Denman et al., 2007) cited the Mickley et al. (2004) study for the eastern United States, which found an increase in the severity and persistence of regional pollution episodes due to the reduced frequency of ventilation by storms tracking across Canada. This study found that surface cyclone (storm) activity decreased by approximately 10 to 20% in a future simulation (for 2050, under the IPCC A1B scenario), which is in general agreement with a number of observational studies over the northern mid-latitudes and North America (Zishka and Smith, 1980; Agee, 1991; Key and Chan, 1999; McCabe et al., 2001). In this study, summer pollution episodes in the northeastern United States were projected to increase in severity and duration; pollutant concentrations in episodes increase by 5 to 10% and episode durations increase from two to three to four days.

Regarding the role that water vapor plays in tropospheric ozone formation, the IPCC (Denman et al., 2007) reported that simulations for the 21st century indicate a decrease in the lifetime of tropospheric ozone due to increasing water vapor. The projected increase in water vapor both decelerates the chemical production and accelerates the chemical destruction of ozone (Meehl et al., 2007). Overall, the IPCC found that climate change is expected to decrease background tropospheric ozone due to higher water vapor and to increase regional and urban-scale ozone pollution due to higher temperatures and weaker air circulation (Denman et al., 2007; Confalonieri et al., 2007).

For North America, the IPCC (Field et al., 2007) reported that surface ozone concentration may increase with a warmer climate. Several studies predict increases in ozone concentrations due to climate change in the near future (2020s and 2030s) as well as for 2050 and beyond. The IPCC (Field et al., 2007; Wilbanks et al., 2007a) cited Hogrefe et al. (2004) who evaluated the effects of climate change on regional ozone in 15 U.S. cities, finding that average summertime daily 8- hour maximum ozone concentrations could increase by 2.7 ppb in the 2020s and by 4.2 ppb in the 2050s (under the IPCC A2 scenario). For the 2050s (under the IPCC A2 scenario), the IPCC (Field et al., 2007) cited Bell et al. (2007), who reported that the projected effects of climate change on ozone in 50 eastern U.S. cities increases the number of summer days exceeding the 8- hour EPA standard by 68% (Bell et al., 2007).

Mickley et al. (2004 in Confalonieri et al., 2007) projected that between 2000 and 2050 significant changes will occur at the high end of the pollutant concentration distribution (episodes) in the Midwest and Northeast. Bell and Ellis (2004) also found that the largest increases in ozone concentrations occur near peak values. While the summer average of daily maximum 8-hour ozone concentrations increases by 2.7, 4.2, and 5.0 ppb in the 2020s, 2050s, and 2080s (compared to the 1 990s), respectively, the fourth-highest summertime daily maximum 8-hour ozone concentrations (levels used to measure compliance with air quality standards for ozone) increase by 5.0, 6.4, and 8.2 ppb for the 2020s, 2050s, and 2080s, respectively when compared to the 1990s (Hogrefe et al., 2004). These findings raise particular health concerns.

The IPCC (Field et al., 2007) found that "warming and climate extremes are *likely* to increase respiratory illness, including exposure to pollen and ozone." Holding population, dose-response characteristics, and pollution prevention measures constant, ozone-related deaths from climate change in the United States are projected to increase by approximately 4.5% from the 1 990s to the 2050s (under the IPCC A2 scenario) (Bell et al., 2007; Field et al., 2007). According to the IPCC (Field et al., 2007), the "large potential population exposed to outdoor air pollution translates this small relative risk into a substantial attributable health risk." In New York City, health impacts could be further exacerbated by climate change interacting with urban heat island effects (Field et al., 2007). However, all of these models predict increasing ozone without considering expected emissions controls that would likely be required to ensure meeting the National Ambient Air Quality Standards.

The IPCC reported (Denman et al., 2007) that "the current generation of tropospheric ozone models is generally successful in describing the principal features of the present-day global ozone distribution." However, they also find "there are major discrepancies with observed longterm trends in ozone concentrations over the 20th century" and "resolving these discrepancies is needed to establish confidence in the models."

In addition to human health effects, tropospheric ozone has significant adverse effects on crop yields in the United States and other world regions, pasture and forest growth, and species composition (EPA, 2006; Easterling et al., 2007). Furthermore, the effects of air pollution on plant function may affect carbon storage. Recent research showed that tropospheric ozone resulted in significantly less enhancement of carbon sequestration rates under elevated CO_2 (Loya et al., 2003), due to negative effects of ozone on biomass productivity and changes in litter (organic ground cover) chemistry (Booker et al., 2005; Liu et al., 2005).

Particulate matter

Particulate matter is a far more complex pollutant than ozone. Particulate matter is a complex mixture of anthropogenic, biogenic, and natural materials suspended as aerosol particles in the atmosphere. When inhaled, the smallest of these particles can reach the deepest regions of the lungs. Scientific studies have found an association between exposure to particulate matter and significant health problems, including aggravated asthma, chronic bronchitis, reduced lung function, irregular heartbeat, heart attack, and premature death in people with heart or lung disease. Particle pollution also is the main cause of visibility impairment in the Nation's cities and national parks (EPA, 2004).

The overall directional impact of climate change on particulate matter levels in the United States remains uncertain. The body of literature specifically addressing the potential effects of climate change on particulate matter is limited, but there is a substantial body of literature describing how ambient particulate matter responds to a wide range of meteorological conditions. On the basis of this information, broad conclusions can be drawn concerning the behavior of ambient particulate matter concentrations given a set of the meteorological changes anticipated in a warming climate. Those meteorological changes are expected to affect particulate matter concentrations by modifying

- emissions of primary particulate matter and particulate matter precursor emissions,
- aerosol photochemistry, and
- transport and removal, as described below.

Particulate Emissions

Particulate matter and particulate matter precursor emissions are affected by climate change through physical response (wind-blown dust), biological response (forest fires and vegetation type/distribution), and human response (energy generation). Most natural aerosol sources are controlled by climatic parameters like wind, moisture, and temperature. Therefore, human-induced climate change is expected to affect the natural aerosol burden. Biogenic organic material is both directly emitted into the atmosphere and produced by VOCs. All biogenic VOC emissions are highly sensitive to changes in temperature, and are also highly sensitive to climate- induced changes in plant species composition and biomass distributions. Biogenic emission rates are predicted to increase, on average, across world regions by 10% per 1 °C increase in surface temperature (Guenther et al., 1993; Denman et al., 2007). The response of biogenic secondary organic carbon aerosol production to a temperature change, however, could be considerably lower than the response of biogenic VOC emissions, since aerosol yields can decrease with increasing temperature (Denman et al., 2007). Particulate matter emissions from wildland fires can contribute to acute and chronic illnesses of the respiratory system, particularly in children, including pneumonia, upper respiratory diseases, asthma, and chronic obstructive pulmonary diseases (WHO, 2002; Bowman and Johnston, 2005; Moore et al., 2006 in Confalonieri et al., 2007).

Photochemistry

Particulate matter chemistry is affected by changes in temperature brought about by climate change. Temperature is one of the most important meteorological variables influencing air quality in urban atmospheres because it affects gas and heterogeneous chemical reaction rates and gas-to-particle partitioning. The net effect that increased temperature has on airborne particle concentrations is a balance between increased production rates for secondary particulate matter (which increases particulate concentrations) and increased equilibrium vapor pressures for semi- volatile particulate compounds (which decreases particulate concentrations). Increased temperatures may either increase or decrease the concentration of semi-volatile secondary reaction products, such as ammonium nitrate, depending on ambient conditions. Regions with relatively warm initial temperatures (>17 °C) may experience a reduction in particulate ammonium nitrate concentrations as temperature increases, while regions with relatively cool initial temperatures (<17 °C) may experience minor reductions or even small increases in particulate ammonium nitrate concentrations as temperature increases.

Transport and Removal

The transport and removal of particulate matter is highly sensitive to winds and precipitation. Removal of particulate matter from the atmosphere occurs mainly by wet deposition (in which atmospheric pollutants mix with water vapor and fall to Earth as precipitation) (NRC, 2005a). Sulfate lifetime, for example, is estimated to be reduced from 4.7 days to 4.0 days as a result of increased wet deposition (Liao et al., 2006). Precipitation also affects soil moisture, with impacts on dust source strength and on stomatal opening/closure of plant leaves, hence affecting biogenic emissions (Denman et al., 2007). Precipitation has generally increased over land north of 30° N over the period 1900 to 2005 and it has become significantly wetter in eastern parts of North America (Trenberth et al., 2007). However, model parameterizations of wet deposition are highly uncertain and not fully realistic in their coupling to the hydrologic cycle (NRC, 2005a). For models to simulate accurately the seasonally varying pattern of precipitation, they must correctly simulate a number of processes (e.g., evapotranspiration, condensation, and transport) that are difficult to evaluate at a global scale (Randall et al., 2007).

V.7. Energy Production, Use, and Distribution

To date, most discussions on energy and climate change have focused on mitigation. However, "along with this role as a *driver* of climate change, the energy sector will be subject to *effects* of climate change" (Wilbanks et al., 2007b). Due to the pervasive role that energy plays in our society, even relatively small disruptions in energy supply can have major economic and social consequences. As discussed in earlier sections, a warming climate is expected to have certain effects in the United States, including rising average temperatures in most regions, changes in precipitation amounts and seasonal patterns in many regions, changes in the intensity and pattern of extreme weather events, and rising sea levels in coastal areas. These effects are expected to lead to changes in the production, consumption, and distribution of energy resources. For instance, average warming can be expected to increase energy requirements for cooling and reduce energy requirements for heating. Changes in precipitation could affect prospects for hydropower, positively or negatively. Increases in storm intensity could threaten further disruptions of the sort experienced in 2005 with Hurricane Katrina. Concerns about climate change impacts could change perceptions and valuations of energy technology alternatives. Any or all of these types of effects could have very real meaning for energy policies, decisions, and institutions in the United States, affecting discussions of courses of action and appropriate strategies for risk management (Wilbanks et al., 2007b).

In SAP 4.5, CCSP (Scott and Huang, 2007; Wilbanks et al., 2007b) cited the following potential effects:

- decreases in the amount of energy consumed in residential, commercial, and industrial buildings for space heating and water heating;
- increases in energy used in residential, commercial, and industrial buildings for space cooling;
- increases in energy consumed for residential and commercial refrigeration and industrial process cooling (e.g., in thermal power plants or steel mills);
- increases in energy used to supply other resources for climate-sensitive processes, such as pumping water for irrigated agriculture and municipal uses;
- changes in the balance of energy use among delivery forms and fuel types, as between electricity used for air conditioning and natural gas used for heating;
- changes in energy consumption in key climate-sensitive sectors of the economy, such as transportation, construction, agriculture, and others;
- positive or negative effects on prospects for hydropower from changes in precipitation patterns and amounts;
- further disruptions of the sort experienced in 2005 with Hurricanes Katrina and Rita; and
- changing perceptions and valuations of energy technology alternatives from concerns about climate change impacts.

Because of the emphasis on mitigation, limited information is available about climate change effects on the energy sector. SAP 4.5 is the first overview of vulnerability and adaptive response issues for the energy sector in the United States. This section relies heavily on this work.

V.7.a. Energy use

(adapted from Scott and Huang, 2007)

According to the IPCC, climate change is *likely* to affect U.S. energy use and energy production, physical infrastructures, and institutional infrastructures, and will *likely* interact with and possibly exacerbate ongoing environmental change and environmental pressures in settlements (Wilbanks et al., 2007a). As noted in SAP 4.5, overall research is relatively clear that climate warming will mean reductions in total U.S. heating requirements and increases in total cooling requirements for buildings (Wilbanks et al., 2007b). Changes will vary by region and season and will affect household and business energy demands and cost. In general, the changes imply increased demands for electricity, which supplies virtually all cooling energy services but only some heating services. Natural gas is the most common heating fuel, fuel oil is commonly used in the Northeast, while in parts of the country with relatively short, mild winters and/or inexpensive electricity, electricity is commonly used for heating.

With climate warming, less heating is required for industrial, commercial, and residential buildings in the United States. A review of the relevant literature in SAP 4.5 (Scott and Huang, 2007) found that projected reductions varied across studies depending on the amount of temperature change in the climate scenario, the calculated sensitivity of the building stock to warming, and the adjustments allowed in the building stock over time.

Heating

Studies on residential heating reductions have shown considerable variation. As reported in SAP 4.5 (Scott and Huang, 2007):

- Mansur et al. (2005) identified relatively modest reductions. When natural gas is available, the marginal impact of a 1 °C increase in January temperatures is predicted to reduce residential electricity consumption by 2.8% for electricity-only consumers and 2% for natural gas customers.
- Scott et al. (2005) projected about a 6 to 10% decrease in space heating per 1 °C increase by 2020.
- The decreases in Huang's (2006) studies varied considerably by location and building vintage; the overall average was about 9.2% per 1 °C increase.
- Regional level studies using building models found a sensitivity of about 6 to 10% per 1 °C in temperature change and only about 1% per 1 °C in studies using econometrics. This is possibly due in part to reactive increases in energy consumption (energy consumption 'take- backs') as heating energy costs decline with warmer weather as well as choice of region.
- Two studies particularly highlighted the potential for regional variation as results differed substantially despite the involvement of many of the same researchers using very similar methodologies. Amato et al. (2005) projected about a 7 to 33% decline in space heating in the 2020s in Massachusetts, which has a long heating season, while Ruth and Lin (2006) projected only a 2 to 3% decline space heating energy during the same time frame in Maryland, which has a much milder heating season.

Similar results are projected in the commercial sector (Scott and Huang, 2007):

- With building equipment and shell efficiencies maintained at 1990 baseline levels and a 3.9 °C temperature change, Belzer et al. (1996) predicted a decrease in annual space heating energy requirements of about 7.4 to 9.0% per 1 °C.
- Mansur et al. (2005) projected that a 1 °C increase in January temperatures would reduce electricity consumption by about 3%, natural gas consumption by 3%, and fuel oil demand by 12%.
- Huang (2006) showed that the impact of climate change on commercial building energy use varies greatly depending on climate and building type. For the entire U.S. commercial sector, the simulations showed a 9.2% decrease per 1 °C.
- As with the residential results, regional analyses suggest more dramatic decreases in energy demand in colder regions than in warmer ones. However, the differences are smaller in the commercial sector because commercial building energy demands are dominated by internal loads such as lighting and equipment. In Loveland and Brown

(1990), results vary from a 10.1 to 12.5% decrease per 1 °C. Findings for Massachusetts vary from a 7 to 8% decline in 2020 (Amato et al., 2005) and for Maryland a 2.7% decrease (Ruth and Lin, 2006).

Cooling

When considering demands for cooling, however, all studies reviewed in SAP 4.5 suggested increases in energy demand (primarily provided by electricity) in all regions with climate warming. Scott and Huang (2007) found that, in most cases, this effect is nonlinear with respect to temperature and humidity, such that the *percentage* impact increases more than proportionately with increases in temperature (Sailor, 2001). Overall, the commercial sector appears to be less sensitive to climate increases. Findings are summarized below:

- National studies projected residential increases in electricity demand by approximately 5 to 20% per 1 °C temperature increase (Rosenthal et al., 1995; Mansur et al., 2005; Scott et al., 2005; Huang, 2006). For the commercial sector, the projected increases varied from 3.5 to 15%.
- Two of these studies looked specifically at the residential sector cooling demand (rather than all electricity). Scott et al. (2005) projected a 12 to 20% increase in cooling demand per 1 °C in 2020, while Huang (2006) projected a 22.4% increase per 1 °C.
- Loveland and Brown (1990) found cooling energy consumption increased by between 55.7% (for Fort Worth, which started from a relatively high base) and 146% (for Seattle, which started from a very low base) for a temperature increase of 3.7 to 4.7 °C. This implies about a 17 to 31% increase in cooling energy consumption per 1 °C.
- In mild, cooler climates, relatively small increases in temperature can have a large impact on air-conditioning energy use by reducing the potential for relying on natural ventilation or night cooling. For example, in California, Mendelsohn (2003) projected that total energy expenditures for electricity used for residential cooling would increase nonlinearly and that net overall energy expenditures would increase with warming in the range of 1.5 °C. Sailor (2001) projected that per 1 °C increases would be 0.5% in New York and 5.8% in Florida.
- Amato et al. (2005) projected a 6.8% increase in Massachusetts by 2020 (increasing 10 to 40% by 2030) while results in Ruth et al. (2006) for the more southerly state of Maryland appear to be quite sensitive to electricity prices, ranging from a 2.5% increase at high prices (about 8 cents per kWh, 1990 dollars) to a 24% increase if prices were low (about 6 cents per kWh, 1990 dollars).
- State-level studies of commercial cooling generally follow national trends. Analyses performed with building energy models generally indicate a 10 to 15% electric energy increase for cooling per 1 °C. The econometric studies also show increases, although the percentage increases are much smaller as these models generally include the change in consumption of all electricity rather than just that used for space cooling.

These results are compounded by potential increases in market penetration of air conditioning, particularly in areas where temperature increases are expected to be greatest. Sailor and Pavlova (2003) have projected that the extra market penetration of air conditioning more than doubled the energy use due to temperature alone. Other trends that can influence future demand include demographic shifts (e.g., population currently expanding in southern and western states), increases in the floor space per building occupant in both the residential and commercial sectors, and improvements in building efficiency and building shell performance (Scott and Huang, 2007).

Combined heating and cooling

Many of these studies address both heating and cooling and attempt to come to a 'bottom line' net result for energy consumption. Results differ based on methods, time frame, scenario, and geography—with some showing slight annual increases in overall energy demand (e.g., Linder and Inglis, 1989; Mendelsohn, 2001; Mansur et al., 2005) and others slight reductions (e.g., Rosenthal et al., 1995; Scott et al., 2005). Results also differ by region. For example, Sailor's state-level econometric analyses (Sailor and Muñoz, 1997; Sailor, 2001; Sailor and Pavlova, 2003) projected a 5% increase per 1 °C warming in residential per capita electricity demand in Florida (a summer-peaking state dominated by air-conditioning demand) and a 3% decrease per 1 °C warming in Washington State (which uses electricity extensively for heating and is a winter-peaking system). Similarly, at the individual city level, Loveland and Brown (1990) projected lower residential energy load in northern cities such as Chicago, Minneapolis, and Seattle and increased energy loads in southern cities such as Charleston, Fort Worth, and Knoxville—while office buildings showed an overall increase in energy loads in all six cities.

Another important concern for energy use is timing of demand. In SAP 4.5, Scott and Huang (2007) indicate that research to date suggest temperature increases would increase peak demand for electricity in most regions of the country, except in the Pacific Northwest. For example, a multiregional study of regional electricity demand found that although annual electricity consumption increased slightly (3.4–5.1%), peak electricity demand would increase between 8.6 and 13.8%, and capacity requirements would increase 13.1 to 19.7%—costing tens of billions of dollars (Linder and Inglis, 1989).

In SAP 4.5, Scott and Huang (2007) found that across all these studies, results suggested only a slight change (increase or decrease) in net energy demand per 1 °C. However, even minor changes could have substantial economic results. For example, in California small increases of energy use of 0.6 to 2.6% would signify increases of 1,741 GWh to 7,516 GWh, while increases of 0.34 to 1.5 1% or 2.57 to 2.99% in electricity demand correspond to increased peak demand by 221 to 967 MW and 1,648 to 1,916 MW (Baxter and Calandri, 1992). To put these impacts in perspective, actual growth in non-coincident peak demand between 1990 and 2004 was 8,650 MW for total end-use load and 9,375 MW for gross generation (California Energy Commission, 2006a).

Drawing on the range of results covered here, Scott and Huang (2007) conclude that across all findings there is a robust result that, in the absence of a energy efficiency measures directed at space cooling, climate change would cause a significant increase in the demand for electricity in the United States, which would require the building of additional electricity generation (and probably transmission facilities) worth many billions of dollars.

These findings led the IPCC to the following conclusion for North America:

> Recent North American studies generally confirm earlier work showing a small net change (increase or decrease, depending on methods, scenarios, and location) in net demand for energy in buildings but a significant increase in demand for electricity for space cooling, with further increases caused by additional market penetration of air conditioning (*high confidence*). (Field et al., 2007)

V.7.b. Energy production and distribution

(adapted from Bull et al., 2007)

Climate change could affect U.S. energy production and supply in the following ways (Wilbanks et al., 2007a):

- immediate impacts from increased intensity of extreme weather events;
- reductions in water supplies in regions dependent on water resources for hydropower and/or thermal power plant cooling;
- influencing facility siting decisions (based on changing conditions); and
- positive or negative impacts for biomass, wind power, or solar energy production (where climatic conditions change).

Significant uncertainty exists about the potential impacts of climate change on energy production and distribution, in part because the timing and magnitude of climate impacts are uncertain. Nonetheless, every existing source of energy in the United States has some vulnerability to climate variability. Bull et al. (2007) found that U.S. energy production is dominated by fossil fuels: coal, petroleum, and natural gas. Renewable energy sources tend to be more sensitive to climate variables, but production from traditional fossil sources can also be adversely affected by air and water temperatures, while nuclear energy also requires a thermoelectric cooling process. Moreover, extreme weather events can affect all sources through adverse effects on production, distribution, and fuel transportation.

In SAP 4.5, Bull et al. (2007) reviewed the limited research literature available on these impacts to date and concluded that while effects on the existing infrastructure might be categorized as modest, local and industry-specific impacts could be large, especially in areas that may be prone to disproportional warming (Alaska) or weather disruptions (Gulf Coast and Gulf of Mexico). Potential impacts are highlighted in Table V.7.1.

Fossil and nuclear energy

As illustrated in Table V.7.1, climate change is expected to affect fossil fuel and nuclear power in similar ways. In both cases, the most direct climate impacts are related to power plant cooling and water availability. As currently designed, power plants require significant amounts of water, and they will be vulnerable to fluctuations in water supply. Regional-scale changes would probably mean that some areas would see significant increases in water availability, while other regions would see significant decreases. In those areas seeing a decline, the impact on power plant availability or even siting of new capacity could be significant (Bull et al., 2007). While new plant designs are flexible and new technologies for water reuse, heat rejection, and use of alternative water sources are being developed, situations where the development of new power plants is being slowed down or halted due to inadequate cooling water are becoming more frequent throughout the United States (SNL, 2006 in Bull et al., 2007).

In addition, Bull et al. (2007) found that average ambient temperatures can affect generation cycle efficiency, along with cooling water requirements in the electrical sector, and water requirements for energy production and refining. While these impacts may appear small based on the proportional reduction in system efficiency or reliability, the vast scale of the energy industry amplifies the impact. A net reduction in generation of 1% due to increased ambient temperature would represent a drop in supply of 25 billion kilowatt-hours (Maulbetsch and DiFilippo, 2006). In the Gulf of Mexico, which accounts for 20 to 30% of the total domestic oil and gas production in the United States, energy producers must adhere to guidelines regarding water temperature impacts as a result of the formation of related anoxic zones. Constraints on produced water discharges could increase costs and reduce production in the Gulf of Mexico region and elsewhere.

Regarding the requirements for the production of energy from fossil fuels, Bull et al. (2007) wrote in SAP 4.5:

> An October 2005 report produced by the National Energy Technology Laboratory stated, in part, that the production of energy from fossil fuels (coal, oil, and natural gas) is inextricably linked to the availability of adequate and sustainable supplies of water. While providing the United States with a majority of its annual energy needs, fossil fuels also place a high demand on the Nation's water resources in terms of both use and quality impacts (EIA, 2005). Thermoelectric generation is water intensive; on average, each kWh of electricity generated via the steam cycle requires approximately 25 gallons of water, a weighted average that captures total thermoelectric water withdrawals and generation for both once-through and recirculating cooling systems to produce. According to the United States Geological Survey (USGS), power plants rank only slightly behind irrigation in terms of freshwater withdrawals in the United States (USGS, 2004), although irrigation withdrawals tend to be more consumptive. Water is also required in the mining, processing, and transportation of coal to generate electricity all of which can have direct impacts on water quality. Surface and underground coal mining can result in acidic, metal laden water that must be treated before it can be discharged to nearby rivers and streams. In addition, the USGS estimates that in 2000 the mining industry withdrew approximately 2 billion gallons per day of freshwater. Although not directly related to water quality, about 10% of total U.S. coal shipments were delivered by barge in 2003 (USGS, 2004). Consequently, low river flows can create shortfalls in coal inventories at power plants.
>
> Freshwater availability is also a critical limiting factor in economic development and sustainability, which directly impacts electric-power supply. A 2003 study conducted by the Government Accountability Office indicates that 36 States anticipate water shortages in the next 10 years under normal water conditions, and 46 States expect water shortages under drought conditions (GAO, 2003). Water supply and demand estimates by the Electric Power Research Institute (EPRI) for the years 1995 and 2025 also indicate a high likelihood of local and regional water shortages in the United States (EPRI, 2003). The area that is expected to face the most serious water constraints is the arid southwestern United States.

Table V.7.1. Mechanisms of Climate Impacts on Various Energy Supplies in the United States

Energy Supplies		Climate Impact Mechanisms
Fossil Fuels (8 6%)	Coal (22%)	Cooling water quantity and quality (T), cooling efficiency (T, W, H), erosion in surface mining
	Natural Gas (23%)	Cooling water quantity and quality (T), cooling efficiency (T, W, H), disruptions of off-shore extraction (E)
	Petroleum (40%)	Cooling water quantity and quality, cooling efficiency (T, W, H), disruptions of off-shore extraction and transport (E)
	Liquefied Natural Gas (1%)	Disruptions of import operations (E)
Nuclear (8%)		Cooling water quantity and quality (T), cooling efficiency (T, W, H)
Renewables (6%)	Hydropower	Water availability and quality, temperature-related stresses, operational modification from extreme weather (floods/droughts), (T, E)
	Biomass	
	• Wood and forest products	Possible short-term impacts from timber kills or long-term impacts from timber kills and changes in tree growth rates (T, P, H, E, CO_2 levels)
	• Waste (municipal solid waste, landfill gas, etc.)	n/a
	• Agricultural resources (including derived biofuels)	Changes in food crop residue and dedicated energy crop growth rates (T, P, E, H, CO2 levels)
	Wind	Wind resource changes (intensity and duration), damage from extreme weather
	Solar	Insolation changes (clouds), damage from extreme weather
	Geothermal	Cooling efficiency for air-cooled geothermal (T)

Percentages Are of Total Domestic Consumption (T = Water/Air Temperature, W = Wind, H = Humidity, P = Precipitation, and E = Extreme Weather Events).

Source: EIA, 2004.

Accordingly, the competition for access to water is expected to increase between power producers and other needs.

Renewable energy

Because renewable energy depends directly on ambient natural resources (such as hydrologic resources, wind patterns and intensity, and solar radiation), it is expected to be more sensitive to climate variability than fossil or nuclear energy systems that rely on geological stores. Renewable energy systems are also vulnerable to damage from extreme weather events. At the same time, renewable energy production is a primary means for mitigating the impacts of potential climate change by reducing energy-related greenhouse gas emissions (Bull et al., 2007).

2005). While the season for transport by barge is *likely* to be extended, the season for ice roads will *likely* be compressed (Lonergan et al., 1993; Lemmen and Warren, 2004; Welch, 2006). Other types of roads are *likely* to require costly improvements in design and construction (Stiger, 2001; McBeath, 2003).

- Some of these risks are expected to be offset by improvements in technology and information systems (Andrey and Mills, 2004). Impacts of warming on paved roads can be ameliorated with altered road design, construction, and management (including changes in the asphalt mix) and the timing of spring load restrictions (Clayton et al., 2005; Mills et al., 2006).

Gulf Coast Case Study

To explore climate effects on transportation systems and infrastructure, SAP 4.7 used a case study approach focused on a segment of the U.S. central Gulf Coast. Given the limited nature of available information on this topic, the authors selected a case study approach to contribute to the development of research methodologies for application in other locations, while generating useful information for local and regional decisionmakers. The Gulf Coast was selected as the study region because of the presence of a full range of transportation modes and infrastructure and its importance as a transportation hub for vital goods and services for the country as a whole. Potter et al. (2008) found that the methods used in the case study can be applied to any region; however, the modeled climate projections and the specific implications of these scenarios for transportation facilities are specific to the Gulf Coast study area.

The following sections are excerpted from SAP 4.7.

V.8.a. Study region

(adapted from Potter et al., 2008)

In SAP 4.7, the authors describe the study area as follows:

- Includes 48 contiguous coastal counties in four states, running from Houston/Galveston, Texas, to Mobile, Alabama.
- The region is home to almost 10 million people living in a range of urban and rural settings and contains critical transportation infrastructure that provides vital services to its constituent states and the Nation as a whole.
- It is also highly vulnerable to sea level rise and storm impacts.
- The coastal geography of the region is highly dynamic due to a unique combination of geomorphic, tectonic, marine, and atmospheric forcings that shape both the shoreline and interior landforms.
- Due largely to its sedimentary history, much of the central Gulf Coast region is low-lying; the great majority of the study area lies below 30 m in elevation.
- Because of this low relief, much of the region is prone to flooding during heavy rainfall events, hurricanes, and lesser tropical storms.
- Land subsidence is a major factor in the region; specific rates of subsidence vary across the region, influenced both by the geomorphology of specific locations (e.g., sediments vary over time) as well as by human activities.

- Most of the coastline is also highly vulnerable to erosion and wetland loss, particularly in association with tropical storms and frontal passages. An estimated 56,000 ha (217 mi^2) of land were lost in Louisiana alone during Hurricane Katrina.
- Many Gulf Coast barrier islands are retreating and diminishing in size. The Chandeleur Islands, which serve as a first line of defense from approaching hurricanes for the New Orleans region, lost roughly 85% of their surface area during Hurricane Katrina.
- As barrier islands and mainland shorelines erode and submerge, onshore facilities in low- lying coastal areas become more susceptible to inundation and destruction.

V.8.b. Transportation infrastructure

(adapted from Potter et al., 2008)

The central Gulf Coast study area's transportation infrastructure is a robust network of multiple modes—critical both to the movement of passengers and goods within the region and to national and international transport as well:

- The region has 27,000 km (17,000 mi) of major highways—about 2% of the Nation's major highways—that carry 83.5 billion vehicle miles of travel annually. The area is served by 13 major transit agencies: over 136 providers offer a range of public transit services to Gulf Coast communities.
- Roughly two-thirds of all U.S. oil imports are transported through this region, and pipelines traversing the region transport over 90% of domestic Outer Continental Shelf oil and gas. Approximately one-half of all the natural gas used in the United States passes through or by the Henry Hub gas distribution point in Louisiana.
- The study area is home to the largest concentration of public and private freight handling ports in the United States, measured on a tonnage basis. These facilities handle a huge share—around 40%—of the Nation's waterborne tonnage. Four of the top five tonnage ports in the United States are located in the region: South Louisiana, Houston, Beaumont, and New Orleans. The study area also has four major container ports.
- Overall, more than half of the tonnage (54%) moving through study area ports is petroleum and petroleum products. Additionally, New Orleans provides the ocean gateway for much of the U.S. interior's agricultural production.
- The region sits at the center of transcontinental trucking and rail routes and contains one of only four major points in the United States where railcars are exchanged between the dominant eastern and western railroads.
- The study area also hosts the Nation's leading and third-leading inland waterway systems (the Mississippi River and the Gulf Intracoastal) based on tonnage. The inland waterways traversing this region provide 20 states with access to the Gulf of Mexico.
- The region hosts 61 publicly owned, public-use airports, including 11 commercial service facilities. Over 3.4 million aircraft takeoffs and landings take place at these airports annually, led by the major facilities at George Bush Intercontinental (IAH) (Houston), William P. Hobby (Houston), and Louis Armstrong New Orleans

International. IAH also is the leading airport in the study area for cargo, ranking 17th in the Nation for cargo tonnage.

Given the scale and strategic importance of the region's transportation infrastructure, it is critical to consider the potential vulnerabilities in the network that may be presented by climate change. A better understanding of these risks will help inform transportation managers as they plan future investments.

V.8.c. Trends in climate and coastal change in the Gulf Coast

(adapted from Savonis et al., 2008)

The central Gulf Coast is particularly vulnerable to climate variability and change because of the frequency with which hurricanes strike, because much of its land is sinking relative to mean sea level, and because much of its natural protection—in the form of barrier islands and wetlands— has been lost. While difficult to quantify, the loss of natural storm buffers is expected to intensify many of the climate impacts identified in this report, particularly in relation to storm damage.

- **Relative Sea Level Rise**—Much of the land in the Gulf Coast is sinking due to compaction of the sediment. As a result, this area is facing much higher increases in relative sea level rise (the combination of local land surface movement and change in mean sea level) than most other parts of the U.S. coast. Based on the output of an ensemble of GCMs, run with a range of IPCC emissions scenarios, relative sea level in the study area is *very likely* to increase by at least 0.3 m across the region and possibly as much as 2 m in some parts of the study area over the next 50 to 100 years. The analysis of even a middle range of potential sea level rise of 0.3 to 0.9 m indicates that a vast portion of the Gulf Coast from Houston to Mobile may be inundated in the future. The projected rate of relative sea level rise for the region during the next 50 to 100 years is consistent with historical trends, region-specific analyses, and the IPCC Fourth Assessment Report (2007a) findings, which assume no major changes in ice sheet dynamics. Protective structures, such as levees and sea walls, could mitigate some of these impacts, but considerable land area is still at risk of permanent flooding from rising tides, sinking land, and erosion during storms. Subsidence alone could account for a large part of the change in land area through the middle of this century, depending on the portion of the coast that is considered. Sea level rise induced by the changing climate will substantially worsen the impacts of subsidence on the region.

- **Storm Activity**—The region is vulnerable today to transportation infrastructure damage during hurricanes and, given the potential for increases in the number of hurricanes designated as Category 3 and above, this vulnerability is expected to increase. This preliminary analysis did not quantitatively assess the impact of the loss of protective barrier islands and wetlands, which will only serve to make storm effects worse. It also did not consider the possible synergistic impacts of storm activity with rising sea levels.

- **Average Temperature Increase**—All GCMs used by the IPCC in its Fourth Assessment Report (Christensen et al., 2007) indicate an increase in average annual Gulf Coast temperature through the end of this century. Based on GCM runs under three different IPCC emission scenarios (A1B, A2, and B1; discussed in Section IV.3.a), the average temperature in the Gulf Coast region appears likely to increase by at least 1.5 ± 1 °C during the next 50 years, with the greatest increase in temperature occurring in the summer.

- **Temperature Extremes**—Increases in average temperature will bring increases in extreme high temperature. Based on historical trends and model projections, it is expected that the number of days above 32.2 °C will increase significantly across the study area; this has implications for transportation operations and maintenance. The number of days above 32.2 °C could increase by as much as 50% during the next 50 years.

- **Precipitation Change**—Future changes in precipitation are much more difficult to model than temperature. Precipitation trends in the study area suggest increasing values with some regions in Mississippi and Alabama showing significant long-term trends. Yet while some GCM results indicate that average precipitation will increase in this region, others indicate a decline in average precipitation during the next 50 to 100 years. Because of this ambiguity, it is difficult to reach conclusions about what the future holds regarding change in mean precipitation. Even if average precipitation increases slightly, average annual runoff in the region could decline as temperature and evapotranspiration rates increase.

- **Extreme Rainfall Events**—Average annual precipitation increased at most recording stations within the study area since 1919, and the literature indicates that a trend toward more rainfall and more frequent heavy downpours is expected. At this stage, climate-modeling capacity is insufficient to quantify effects on individual precipitation events, but the potential for temporary flooding in this region is clear. In an area where flooding already is a concern, this tendency could be exacerbated by extreme rainfall events. This impact will become increasingly important as relative sea level rises, putting more and more of the study area at risk.

V.8.d. Impacts of changing climate on Gulf Coast transportation

(adapted from Savonis et al., 2008)

Based on the trends in climate and coastal change, transportation infrastructures and the services that require them are vulnerable to future climate changes as well as other natural phenomena. While more study is needed to specify how vulnerable they are and what steps could be taken to reduce that vulnerability, it is clear that transportation planners in this region should not ignore these impacts.

- **Inundation from Relative Sea Level Rise**—While greater or lesser rises in relative sea level are possible, this study analyzed the effects of relative sea level rise of 0.6

and 1.2 m as realistic scenarios. Based on these levels, an untenable portion of the region's road, rail, and port network is at risk of permanent flooding.

Twenty-seven percent of the major roads, 9% of the rail lines, and 72% of the ports are at or below 122 cm in elevation, although portions of the infrastructure are guarded by protective structures such as levees and dikes. While flood protection measures will continue to be an important strategy, rising sea levels in areas with insufficient protection may be a major concern for transportation planners. Furthermore, the crucial connectivity of the intermodal system in the area means that the services of the network can be threatened even if small segments are inundated.

While these impacts are very significant, they can be addressed and adaptive strategies developed if transportation agencies carefully consider them in their decisions. The effectiveness of such strategies will depend on the strategies selected and the magnitude of the problem because scenarios of lower emissions demonstrate lesser impacts. It may be that in some cases, the adaptive strategy may be wholly successful, while in others further steps may need to be taken

- **Flooding and Damage from Storm Activity**—As the central Gulf Coast already is vulnerable to hurricanes, so is its transportation infrastructure. This study examined the potential for short-term flooding associated with a 5.5- and a 7.0-m storm surge. Based on these levels, a great deal of the study area's infrastructure is subject to temporary flooding. More than half (64% of interstates; 57% of arterials) of the area's major highways, almost half of the rail miles, 29 airports, and virtually all of the ports are subject to flooding.

  The nature and extent of the flooding depends on where a hurricane makes landfall and its specific characteristics. Hurricanes Katrina and Rita demonstrated that that this temporary flooding can extend for miles inland.

  This study did not examine in detail the potential for damage due to storm surge, wind speeds, debris, or other characteristics of hurricanes since this, too, greatly depends on where the hurricane strikes. Given the energy associated with hurricane storm surge, concern must be raised for any infrastructure in its direct path that is not designed to withstand the impact of a Category 3 hurricane or greater.

  Climate change appears to worsen the region's vulnerability to hurricanes, as warming seas give rise to more energetic storms. The literature indicates that the intensity of major storms may increase by 5 to 20%. This indicates that Category 3 storms and higher may return more frequently to the central Gulf Coast and thus cause more disruptions of transportation services.

  The impacts of such storms need to be examined in greater detail; storms may cause even greater damage under future conditions not considered here. If the barrier islands and shorelines continue to be lost at historical rates and as relative sea level rises, the destructive potential of tropical storms is expected to increase.

- **Effects of Temperature Increase**—As the average temperature in the central Gulf Coast is expected to rise by 0.5 to 2.5 °C, the daily high temperatures, particularly in summer, and the number of days above 32.2 °C are also expected to increase. These combined effects will raise costs related to the construction, maintenance, and operations of transportation infrastructure and vehicles. Maintenance costs will increase for some types of infrastructure because they deteriorate more quickly at temperatures above 32 °C. An increase in daily high temperatures could increase the potential for rail buckling in certain types of track. Construction costs could increase because of restrictions on days above 32 °C, since work crews may be unable to be deployed during extreme heat events and concrete strength is affected by the temperature at which it sets. Increases in daily high temperatures would affect aircraft performance and runway length because runways need to be longer when daily temperatures are higher (all other things being equal). While potentially costly and burdensome, these impacts may be addressed by transportation agencies by absorbing the increased costs and increasing the level of maintenance for affected facilities.

- **Effects of Change in Average Precipitation**—It is difficult to determine how transportation infrastructure and services might be affected by changes in average precipitation, since models project either a wetter or a drier climate in the southeastern United States. In either case, the changes in average rainfall are relatively slight, and the existing transportation network may be equipped to manage this.

- **Effects of Increased Extreme Precipitation Events**—Of more concern is the potential for short-term flooding due to heavier downpours. Even if average precipitation declines, the intensity of those storms can lead to temporary flooding as culverts and other drainage systems are overloaded. Further, the Louisiana Department of Transportation and Development reports that prolonged flooding of one to five weeks can damage the pavement substructure and necessitate rehabilitation (Gaspard et al., 2007). The central Gulf Coast already is prone to temporary flooding, and transportation representatives struggle with the disruptions these events cause. As the climate changes, flooding will probably become more frequent and more disruptive as the intensity of these downpours is expected to increase. As relative sea level rises, it appears probable that even more infrastructure will be at risk because overall water levels already will be so much higher. While these impacts cannot be quantified at present, transportation representatives can monitor where flooding occurs and how the sea is rising as an early warning system about what facilities are at immediate risk and warrant high-priority attention. In a transportation system that already is under stress due to congestion, and with people and freight haulers increasingly dependent on just-in-time delivery, the economic, safety, and social ramifications of even temporary flooding may be significant.

V.8.e. Impacts by transportation mode

(adapted from Kafalenos et al., 2008)

Roads

Similar to most of the Nation, roads make up the backbone of the transportation network in the Gulf Coast. Highways are the chief mode for transporting people across the region, and, combined with rail, move freight throughout the region and to other parts of the United

States. While temperature and precipitation changes have some implications for highway design and maintenance, the key impacts on the highway network result from relative sea level rise and storm surge.

Temperature

- Projected changes in average temperatures appear to have moderate implications for highways, while increases in extreme heat may be significant.
- Maintenance and construction costs for roads and bridges are expected to increase as temperatures increase. Further, higher temperatures cause some pavement materials to degrade faster, requiring earlier replacement. Such costs are expected to grow as the number of days above 32 °C (90 °F)—projected to grow from the current average of 77 days to a range of 99 to 131 days over the next century—increases, as well as the projected maximum record temperatures in the region.
- Currently, the designs of steel and concrete bridges and of pavements in the study area typically are based on a maximum design temperature of 46 to 53 °C. The increase in maximum record temperatures implied by the climate model projections are less than these values, although under the climate scenarios they would approach those values over the next century. It may be prudent for future designers of highway facilities to ensure that joints in steel and concrete bridge superstructures and concrete road surfaces can adequately accommodate thermal expansion resulting from these temperatures.

Precipitation

The analysis indicates little change in mean annual precipitation through either 2050 or 2100, but the range of possible futures includes both reductions and increases in seasonal precipitation. In either case, the analysis points to potential reductions in soil moisture and runoff as temperatures and the number of days between rainfall events increase.

Scenarios of insignificant change or reduction in average precipitation, coupled with drier soils and less runoff, may result in the following:

- decline of slides in slopes adjacent to highways;
- less settling under pavements, with a decrease in cracking and undermining of pavement base courses;
- uniform decreases in runoff could reduce scouring of bridge piers in rivers and streams; and
- stresses on animal and plant populations brought about by higher temperatures and changes in rainfall patterns could make it more difficult and expensive to mitigate the impacts of highway development on the natural environment.

Modest increases in average annual rainfall may not significantly affect pavement settling, bridge scour, and ecosystem impacts because of the effects of increasing temperature on evaporation rates.

However, an increase in the frequency of extreme precipitation events may contribute to the following:

- increased accident rates;
- more frequent short-term flooding and bridge scour, as well as more culvert washouts;
- exceeding the capacity of storm-water management infrastructure; and
- more frequent slides, requiring increased maintenance.

However, some states, such as Louisiana, already address precipitation through pavement grooving and sloping, and thus may have adequate capacity to handle some increase in precipitation.

Relative Sea Level Rise

The effects of 61 and 122 cm (2 and 4 ft) changes were analyzed to assess the implications of relative sea level rise on highways. Currently, about 209 km (130 mi) or about 1% of major highways (interstates and arterials) in the study region are located on land that is at or below sea level. The presence or absence of protective structures was not considered in this baseline analysis, but would be an important factor in subsequent assessments.

Risk from 61-cm relative sea level rise:

- Twenty percent of the arterial miles and 19% of the interstate miles in the study area are at elevations below 61 cm (2 ft) and thus at risk from sea level rise unless elevated or protected by levees.
- Most of the highways at risk are located in the Mississippi River Delta near New Orleans.
- The most notable highways at risk are I-10 and U.S. 90, with 220 km (137 mi) and 235 km (146 mi), respectively, passing through areas that will be below sea level if sea levels rise by 61 cm (2 ft).

Risk from 122-cm relative sea level rise:

- Twenty-eight percent of the arterial miles and 24% of the interstate miles are at elevations below 122 cm (4 ft).
- The majority of the highways at risk from a 122-cm (4-ft) increase in relative sea level are similarly located in the Mississippi River Delta near New Orleans.
- The most notable highways at risk remain I-10 and U.S. 90, with the number of kilometers increasing to 684 km (425 mi) and 628 km (390 mi) passing through areas below sea level, respectively.

Many of the National Highway System (NHS) Intermodal Connectors, public roads that provide access to major transportation facilities, such as rail, ports, and airports, pass through low-lying areas concentrated in the Mississippi River Delta, where sea level rise is expected to have the most pervasive impact. Of the 1,041 km (647 mi) of Intermodal Connectors, 238 km (148 mi), or 23%, are at risk from a 61-cm (2-ft) increase in relative sea levels; and a total of 444 km (276 mi), or 43%, are at risk from a 122-cm (4-ft) increase. The following terminals are at risk:

- for the 61-cm (2-ft) increase scenario—the New Orleans International Airport, Port Fourchon, most rail terminals in New Orleans, ferry terminals in New Orleans, and ferry terminals outside of the Mississippi River Delta in Galveston and Houston; and
- for the 122-cm (4-ft) scenario—all those above, as well as port facilities in Lake Charles, Galveston, Pascagoula, and Gulfport.

It is worth noting that the loss of use of a small individual segment of a given highway may make significant portions of that road network impassable. Further, even if a particular interstate or arterial is passable, if the feeder roads are flooded, then the larger road becomes less usable.

Storm Activity

About half of the region's arterial miles and about three-quarters of the intermodal connectors are vulnerable to a storm surge of 5.5 m (18 ft), and these proportions are even higher for a 7-m (23-ft) storm surge.

Surge Wave Crests and Effects on Bridges

The wave energy during storm surge events is greatest at the crest of the wave. The facilities most at risk are bridge decks and supports that are constructed at the wave height levels reached during a storm. The impact of the 2005 hurricanes vividly illustrates some of the factors involved in infrastructure vulnerability. While only a small percentage of the study area's bridges are located at the shore and have bridge decks or structures at these heights, when storm waves meet those bridges, the effect is devastating; massive bridge spans were dislodged during Hurricane Katrina. Although these bridges are few in number compared to the over 8,000 bridges in the functionally classified system, over 24 bridges were hit by wave surges resulting from Hurricane Katrina and experienced serious damage.

Surge Inundation

A substantial portion of the highway system across the study area is vulnerable to surge inundation: 51% of all arterials and 56% of the interstates are in the 5. 5-m (18-ft) surge risk areas. At the 7-m (23-ft) level, these percentages increase only slightly: 57% of all arterials and 64% of the interstates are in 7-m (23 -ft) surge risk areas.

The risk from surge inundation for NHS Intermodal (IM) Connectors is even greater than that for all highways. Seventy-three percent of IM Connector miles are located in areas that would be inundated by a 5.5-m (18-ft) surge, and the proportion of IM connectors that is vulnerable at the 7-m (23-ft) level is only slightly higher.

While inundation from storm surges is a temporary event, during each period of inundation the highway is not passable, and after the surge dissipates, highways must be cleared of debris before they can function properly. Of particular concern is that a substantial portion of all of the major east–west highways in the study area, particularly I-10/I-12, are at risk to storm surge inundation in some areas, and during storm events and the recovery period from these events, all long-distance highway travel through the study area is expected to be disrupted.

Wind

Wind from storms may affect the highway signs, traffic signals, and lighting fixtures throughout the study area. The wind design speed for signs and supports in the study area is typically 160 to 200 km per hour (100 to 125 mph). These designs should accommodate all but the most severe storm events. More significant safety and operational impacts are expected from debris blown onto roadways and from crashes precipitated by debris or severe winds.

Freight and passenger rail

Rail lines in the region play a key role in transporting freight, and a minor role in intercity passenger traffic. Rail connectivity is vital to the functioning of many, if not most, of the marine freight facilities in the study area. Of the four main climate drivers examined in this study, storm surge could be the most significant for rail.

Temperature

Projected increases in average temperatures generally fall within the current standards for existing rail track and facilities. However, the increase in temperature extremes—very hot days—could increase the incidence of buckling or 'sun kinks' on all the rail tracks in the study area. This occurs when compressive forces in the rail, due to restrained expansion during hot weather, exceed the lateral stiffness of the track causing the track to become displaced laterally.

Precipitation

The primary impacts on rail infrastructure from precipitation are erosion of the track subgrade and rotting of wooden crossties. Erosion of the subgrade can wash away ballast and weaken the foundation, making the track unstable for passage of heavy locomotives and railcars. Without ballast, wood crossties would rot at a faster rate, leading to more buckling and unstable track.

The precipitation projections do not indicate that design changes are warranted to prevent increased erosion or moisture damage to railroad track. The runoff projections point to even fewer problems with erosion over the next century than are present today, due to possibly less precipitation and slightly higher temperatures. However, if the frequency and/or the intensity of extreme rainfall events increase, it could lead to higher rates of erosion and railroad bridge scour, as well as higher safety risks and increased maintenance requirements.

It should be noted that many existing facilities at low elevations are protected by levees and other physical structures, which provide some resistance to gradual changes in sea level and the impacts of storm surge.

Relative Sea Level Rise

The effects on rail lines and facilities of relative sea level of 61 and 122 cm (2 and 4 ft) over the next 50 to 100 years were analyzed. The obvious impacts for both of these sea level rise scenarios are water damage or complete submersion of existing rail track and facilities. Currently, about 80 km (50 mi) or about 2% of rail lines in the study region are located on land that is at or below sea level.

Most of the rail lines in and around New Orleans are expected to be affected by relative sea level rise. The heavily traveled CSX line between Mobile and New Orleans, which was damaged during Hurricane Katrina, also is at risk, as are several area short lines. The following rail lines would be affected if relative sea level rises 61 cm (2 ft):

- most rail lines in and around New Orleans;
- BNSF line between Lafayette and New Orleans;
- CN/IC line into New Orleans;
- CSX line between Mobile and New Orleans;
- CSX line north of Mobile;
- Louisiana and Delta Railroad west of New Orleans;
- portions of the MSE rail line in Mississippi;
- the New Orleans and Gulf Coast Railway line between New Orleans and Myrtle;
- NS line into New Orleans;
- portions of the Port Bienville Railroad;
- segments of the UP line west of New Orleans; and
- various segments of track around Lake Charles and Galveston.

Further degradation of these lines is expected to occur should relative sea level increase by 122 cm (4 ft), with additional problems on the KCS route into New Orleans, the NS line north of Mobile, and selected track segments around Beaumont and Houston.

Storm Activity

Hurricane Katrina provided a vivid example of the devastating impacts of severe storm events on the rail system in the Gulf Coast Study Area. Making landfall on 29 August 2005, Katrina caused damage to all of the major railroads in the region. BNSF, CN, KCS, and UP all suffered damage, mostly to yards in and around New Orleans. CSX track and bridges also were damaged. NS had nearly 8 km (5 mi) of track washed away from the Lake Pontchartrain Bridge, which is 9.3 km (5.8 mi) long. By 13 September 2005, most of these railroads had resumed operations into New Orleans, at least on a partial basis.

One-third of the rail lines in the study region are vulnerable to a storm surge of 5.5 m (18 ft), and 41% are vulnerable to a storm surge of 7.0 m (23 ft). These include the heavily traveled CSX line from New Orleans to Mobile, and the UP and BNSF lines from New Orleans to Houston. Cities at risk include Mobile, Gulfport, Biloxi, New Orleans, Baton Rouge, Lafayette, Lake Charles, Beaumont, Port Arthur, and Galveston. Facilities at less than 5.5 m (18 ft) of elevation have the highest risk of 5.5-m (1 8-ft) storm surge impacts. These include 43% of the rail facilities in the study region. An additional 11 facilities are between 5.5 and 7.0 m (18 and 23 ft) of elevation and are expected to be affected by a 7.0-m (23-ft) storm surge.

Marine

Marine facilities, including both freight and non-freight ports, marinas, and industry support facilities, are most vulnerable to storm surge and relative sea level rise. Marine facilities and waterways are vital to the region, and to the Nation as a whole. The study area is one of the Nation's leading centers of marine activity. Much of the region's economy is directly linked to waterborne commerce; and in turn, this waterborne commerce supports a substantial portion of the U.S. economy.

Higher Temperatures

Higher temperatures may affect port facilities in three key ways. First, higher temperatures will increase costs of terminal construction and maintenance, particularly of any paved surfaces that will deteriorate more quickly if the frequency of high temperatures increases. Second, higher temperatures will lead to higher energy consumption and costs for refrigerated warehouses or 'reefer slots' (electrical plug-ins for containers with on-board cooling units). Third, higher temperatures would probably lead to increased stress on temperature-sensitive structures.

Container-handling cranes, warehouses, and other marine terminal assets are made of metals. With increasing record temperatures and days over 32 °C (90 °F), it may be necessary to design for higher maximum temperatures in replacement or new construction. On the other hand, most dock and wharf facilities are made of concrete and lumber, which are generally less sensitive to temperature fluctuations. It is possible that lock and dam structures could be affected, although this will require further investigation.

Precipitation

The prospect of more intense precipitation events could require the capacity of some storm- water retention and treatment facilities to be increased. The handling of storm water can be a significant expense for container terminals, auto terminals, and other terminals with large areas of impervious surface. Increasing environmental regulatory requirements also may add to costs of adapting storm-water handling infrastructure.

Relative Sea Level Rise

Of freight facilities in the study area, about 72% are vulnerable to a 122-cm (4-ft) rise in relative sea level. Of the 994 freight facilities in the U.S. Army Corps of Engineers database, 638 (64%) are in areas with elevations between 0 and 61 cm (2 ft) above sea level, and another 80 (8%) are in areas with elevations between 61 and 122 cm (2 and 4 ft). More than 75% of facilities are potentially vulnerable in Beaumont, Chocolate Bayou, Freeport, Galveston, New Orleans, Pascagoula, Plaquemines, Port Arthur, Port Bienville, and Texas City. Between 50 and 75% of facilities are potentially vulnerable in Gulfport, Houston, Lake Charles, Mobile, South Louisiana, and the Tennessee–Tombigbee Waterway. Only Baton Rouge, with 6% of facilities potentially at risk, appears to be well positioned to avoid impacts of sea level rise.

A similar situation faces non-freight facilities. Seventy-three percent of study area marine non- freight facilities are potentially vulnerable to a 122-cm (4-ft) increase in relative sea level. Of the 810 non-freight facilities in the U.S. Army Corp of Engineers database, 547 (68%) are in areas with elevations between 0 and 61 cm (2 ft) above sea level, and another 47 (6%) are in areas with elevations between 61 and 122 cm (2 and 4 ft). More than 75% of facilities are potentially vulnerable in Beaumont, Chocolate Bayou, Freeport, Galveston, New Orleans, Pascagoula, Plaquemines, Port Arthur, the Tennessee–Tombigbee Waterway, and Texas City; between 50% and 75% of facilities are potentially vulnerable in Houston, Lake

Charles, Mobile, and South Louisiana. Twenty-seven percent of Gulfport facilities and no Baton Rouge facilities are potentially at risk.

It is important to note that many existing facilities at low elevations are protected by levees and other physical structures, which should provide resistance to gradual changes in sea levels. The specific effects of existing protections have not been considered in this study. For facilities that are not appropriately protected, either by elevation or by structures, rising water levels pose an increased risk of chronic flooding, leading, in the worst case, to permanent inundation of marine terminal facilities, either completely or in part, rendering them inoperable.

Navigable depths are expected to increase in many harbors and navigation channels as a result of rising sea levels. This could lead to reduced dredging costs, but higher costs where rising water levels require changes to terminals. The functionality and/or protections of lock and dam structures controlling the inland waterway system also may be affected by relative sea level rise.

Storm Activity: Water and Wind Damage

While the actual facilities that would be flooded depend on the particulars of a given storm—the landfall location, direction, tidal conditions, etc.—fully 99% of all study area facilities are vulnerable to temporary and permanent impacts resulting from a 7.0-m (23-ft) storm surge, while almost 98% are vulnerable to temporary and permanent impacts resulting from a 5.5-m (1 8-ft) storm surge. All facilities are vulnerable to wind impacts. Similar to sea level rise, storm surge impacts on highway and rail connections could affect the ability to utilize ports for transport of goods to and from affected ports.

Aviation

Severe weather events could have the greatest impacts on aviation. Ultimately, the impact on the operational aspects of aviation could potentially supersede the overall magnitude of combined effects on aviation due to other factors discussed below.

Temperature

Higher temperatures affect aircraft performance and the runway lengths that are required. Generally speaking, the higher the temperature, the longer the runway that is required. Temperature increases considered by this report would indicate a small increase in baseline runway length requirements, assuming other relevant factors are held constant. However, advances in engine technology and airframe materials are expected to offset the potential effects of the temperature increases analyzed in this report, so that current runway lengths are expected to be sufficient.

Precipitation

In general, airlines, airports, and aircraft operate more efficiently in dry weather conditions than wet. Weather is a critical influence on aircraft performance and the outcome of the flight operations while taking off, landing, and while aloft. Precipitation affects aircraft and airports in several ways, such as decreasing visibility, slowing air traffic by requiring greater separation between aircraft, and decreasing braking effectiveness. On the ground, effects include creating turbulence, increasing the risk of icing of wings, and affecting engine thrust.

Precipitation effects depend on whether annual precipitation increases or decreases (results were inconclusive). Less precipitation is expected to have the following effects:

- reduce aircraft and air traffic delays;
- reduce periods of wet surfaces on runways, taxiways, and apron;
- reduce the risk of wing icing in the winter months;
- possibly increase convective weather (turbulence), as well as increase the number and severity of thunderstorms; and
- increase water vapor in the atmosphere, particularly during the summer months, potentially resulting in increased haze and reduced visibility for pilots.

Increased precipitation is expected to have the following effects:

- periods of low visibility would increase; and
- general aviation pilots would either learn to fly in Instrument Flight Rules (IFR) conditions by becoming 'instrument rated' or not fly during periods of reduced visibility and precipitation.

In either scenario, an increased intensity of individual rainfall events is expected. Increased extreme precipitation events would also affect commercial service aircraft operations. During severe thunderstorm activity, it is not unusual for an airline to cancel flights or, at a minimum, experience delays in operations. Navigation in heavy precipitation is possible and currently occurs on a daily basis in the national air system. However, precipitation almost always creates delays, particularly at the most congested airports. In addition, 8 of the 61 airports in the study area are located within 100-year floodplains.

Relative Sea Level Rise

Analysis indicates that three airports in the study area would be below mean sea level if relative sea level increases by 122 cm (4 ft). Each of these airports currently is protected by preventive infrastructure such as dikes and levees, which will need to be maintained. If feeder roads in the area are inundated, however, access to these airports may be disrupted. All three airports are located in Louisiana, and include New Orleans International (elevation: 122 cm or 4 ft), one of the study area's large commercial service airports, South LaFourche (elevation: 30 cm or 1 ft), a very small general aviation facility, and New Orleans Naval Air Station Joint Reserve Base (elevation: 91 cm or 3 ft), a military airport.

Storm Activity

Both storm surge and hurricane force winds can damage airport facilities. A variety of airports in the region would be vulnerable to the impacts of storm surge of 5.5 and 7.0 m (18 and 23 ft), though this depends on the specific characteristics of each individual storm event, including landfall location, wind speed, direction, and tidal conditions. There are 22 airports with elevations in the 0- to 5.5-m (18-ft) category and an additional 7 airports in the 5.8-m to 7.0-m (19- to 23-ft) category. This list includes some major airports in the region, such as

New Orleans International. In addition, the commercial service airport in Lake Charles, Louisiana would be vulnerable.

Pipelines

There is a combined total of 42,520 km (26,427 mi) of onshore liquid (oil and petroleum product) transmission and natural gas transmission pipelines in the Gulf Coast study area. This includes 22,913 km (14,241 miles) of onshore natural gas transmission pipelines and 19,607 km (12,186 miles) of onshore hazardous liquid pipelines (PHMSA, 2007). This region is essential to the distribution of the Nation's energy supply through pipeline transportation, and historically the landside pipelines have been relatively secure from disruption by increased storm activity and intensity. A number of risks and vulnerabilities to climate-related impacts have been revealed, however, particularly for submerged or very low elevation pipelines.

Some historical weather events have resulted in only minor impacts on pipelines, with the notable exceptions of Hurricanes Andrew's, Ivan's, Katrina's, and Rita's fairly extensive damage to underwater pipelines and flooded distribution lines in areas where houses were destroyed. Storm surge and high winds historically have not had much impact on pipelines—either onshore transmission lines or offshore pipelines—since they are strong structures; well stabilized and/or buried underground. Yet offshore pipelines have been damaged in relatively large numbers on occasion, as during Hurricanes Andrew and Ivan. Temperature shifts resulting from climate scenario projections are not expected to have much effect on pipelines. Increases or decreases in precipitation—either long-term or in the frequency or extent of droughts or inundation—could affect soil structure. Sea level rise would probably have little direct effect, but could affect water tables, soil stability, and the vulnerability of pipelines to normal wave action as well as storm surge.

Changes in soil structure, stability, and subsidence—whether undersea, landside, or in wetlands or transition elevations—could play an important role in pipeline-related risks. However, there is little information on this topic outside of earthquake risks. Recently, concerns have been raised about how wave action could affect the seabed, either by liquefying/destabilizing the sand or silt surface above a buried pipeline or by gradually eroding away seabed that had been covering the pipeline. It is unclear at present whether a changing climate might lead to conditions that exacerbate these effects and cause additional damage.

Temperature

The great majority of the transmission pipeline system is buried under at least 0.9 m (3 ft) of soil cover, both onshore and offshore. Pipelines typically carry products at significant temperature variations (natural gas is under pressure; whereas petroleum products are heated considerably above ambient temperatures). There is not expected to be any significant effect on pipelines due to direct effects from increased (or decreased) temperatures.

Precipitation Changes

Sustained periods of increases or decreases in precipitation, whether over months or the cumulative effect across years, can cause substantial soil changes due to drought or saturation. Changes in water tables may occur both from local climate changes as well as from global effects such as sea level rise. An increase in water table level or increased surface water runoff can cause erosion or slumping (collapse) of the soil surface, thereby leading to potential for pipeline exposure. In the lowland and marsh areas particularly associated with the coastal regions of Louisiana, the soil is being washed away due to storm activity. With the disappearance of the soil, the pipelines in these regions are losing cover.

Detailed analysis of geology and pipeline-specific conditions are required to draw more precise conclusions regarding the potential for serious disruption of the transmission pipeline system by climate-related soil changes. Nonetheless, this is an area of concern, as a considerable and unpredictable portion of the pipeline system could be vulnerable to these climate change and sealevel-induced impacts.

Another vulnerability is from expected short-term changes (such as torrents and floods), where significant changes in rate and energy of water flow are a result of increased precipitation. Risk analysis of the impacts of extreme events is required to determine appropriate adaptation or mitigation actions.

Storm Activity

Wave action during storms may affect pipelines. For offshore pipelines, in instances where significant subsidence occurs and the pipeline segment is exposed, this section is exposed to wave action. High-energy waves may subject a pipeline to stress levels it was not designed to withstand, causing a fracture. An exposed offshore pipeline also could be vulnerable to lateral and vertical displacement, exposure to vessel traffic and fishing trawls, or rupture by currents.

Storm damage will vary depending on the particulars of a given storm—the landfall location, direction of the storm, and tidal conditions. For example, Hurricanes Andrew and Ivan caused substantial damage to pipelines throughout the storm-front region. After Ivan, oil refineries had ample products to supply, but the pipelines could not deliver due to damages. In contrast, damage to pipelines from Katrina and Rita was relatively minor; most pipelines were ready to take product, but were hampered by the lack of available product due to refinery damage and/or power shortages.

References

AAAAI, (1996-2006). *Allergy Statistics*. American Academy of Allergy Asthma & Immunology, Milwaukee, WI. <www. aaaai. org/ media>. Aaheim, A. & L. Sygna, (2000). *Economic Impacts of Climate Change on Tuna Fisheries in Fiji Islands and Kiribati*. Cicero Report 4, Cicero, Oslo, Norway, 21

ACIA, (2004). *Impacts of a Warming Arctic: Arctic Climate Impact Assessment*. Cambridge University Press, 139.

ACIA, (2005). *Arctic Climate Impact Assessment*. Cambridge University Press, 1042 ADRC, Japan, CRED-EMDAT, Université Catholique de Louvain, and UNDP, 2005: *Natural Disasters Data Book–2004*. <web>.

Aerts, M., Cockrell, P., Nuessly, G., Raid, R., Schueneman, T. & Seal, D. (1999). Crop profile for corn (sweet) in Florida. <www.ipmcenters.org/CropProfiles/docs/FLcorn-sweet.html>.

Agee, E. M. (1991). Trends in cyclone and anticyclone frequency and comparison with periods of warming and cooling over the Northern Hemisphere. *Journal of Climate*, *4*, 263-267.

Ahern, M. J., Kovats, R. S., Wilkinson, P., Fewand, R. & Matthies, F. (2005). Global health impacts of floods: epidemiological evidence. *Epidemiologic Reviews*, *27*, 36-45.

Ainsworth, E. A. & Long, S. P. (2005). What have we learned from 15 years of free-air CO_2 enrichment (FACE)? A meta-analytic review of the responses of photosynthesis, canopy properties and plant production to rising CO_2. *New Phytologist*, *165*, 351-372.

Ainsworth, E. A. & Rogers, A. (2007). The response of photosynthesis and stomatal conductance to rising [CO_2]: mechanisms and environmental interactions. *Plant, Cell & Environment*, *30*, 258-270.

Ainsworth, E. A., Davey, P. A., Bernacchi, C. J., Dermody, O. C., Heaton, E. A., Moore, D. J., Morgan, P. B., Naidu, S. A., Yoo Ra, H. S., Zhu, X. G., Curtis, P. S. & Long, S. P. (2002). A meta- analysis of elevated [CO_2] effects on soybean (*Glycine max*) physiology, growth and yield. *Global Change Biology*, *8*, 695-709.

AIR, (2002).*Ten Years after Andrew: What Should We Be Preparing for Now?* AIR Technical Document HASR 0208, Applied Insurance Research, Inc., Boston, MA, 9 <www.air>.

Alberta Environment, (2002). *South Saskatchewan River Basin Water Management Plan, Phase One – Water Allocation Transfers: Appendices*. Alberta Environment, Edmonton, Alberta, Canada.

Alig, R. J., Adams, D., Joyce, L. & Sohngen, B. (2004). Climate change impacts and adaptation in forestry: responses by trees and markets. *Choices*, Fall 2004, 7-11. <www.choicesmagazine.org/2004-3/climate/2004-3-07.htm>.

Allen, R. G., Gichuki, F.N. & C. Rosenzweig, (1991). CO_2-induced climatic changes and irrigation-water requirements. *Journal of Water Resources Planning and Management*, *117*, 157-178.

Allison, E. H., Adger, W. N., Badjeck, M. C., Brown, K., Conway, D., Dulvy, N. K., Halls, A., Perry, A. & Reynolds, J. D. (2005). *Effects of Climate Change on the Sustainability of Capture and Enhancement Fisheries Important to the Poor: Analysis of the Vulnerability and Adaptability of Fisherfolk Living in Poverty*. Final Technical Report, Project No. R4778J, Fisheries Management Science Programme, MRAG for Department for International Development, London, 167 <www. fmsp. org. uk/ Documents/r4778j/R4778J_FTR1.pdf>.

Amato, A., Ruth, M., Kirshen, P. & Horwitz, J. (2005). Regional energy demand responses to climate change: Methodology and application to the Commonwealth of Massachusetts. *Climatic Change*, *71(1)*, 175-201.

Amiro, B. D., Todd, J. B., Wotton, B. M., Logan, K. A., Flannigan, M. D., Stocks, B. J., Mason, J. A., Martell, D. L. & Hirsch, K. G. (2001). Direct carbon emissions from Canadian forest fires, 1959-1999. *Canadian Journal of Forest Research*, *31*, 512-525.

Amstrup, S. C., Marcot, B. G. & Douglas, D. C. (2007). *Forecasting the Range-Wide Status of Polar Bears at Selected Times in the 21st Century*. Administrative Report, U.S. Department of the Interior, U.S. Geological Survey, Washington, DC, USA, 126.

Amthor, J. S. (2001). Effects of atmospheric CO_2 concentration on wheat yield: review of results from experiments using various approaches to control CO_2 concentration. *Field Crops Research*, *73*, 1-34.

Amundson, J. L., Mader, T. L., Rasby, R. J. & Hu, Q. S. (2006). Environmental effects on pregnancy rate in beef cattle. *Journal of Animal Science, 84*, 3415-3420.

Anderson, P. J. & Piatt, J. F. (1999). Community reorganization in the Gulf of Alaska following ocean climate regime shift. *Marine Ecology Progress Series, 189*, 117-123.

Andrey, J. & Mills, B. (2003). Climate change and the Canadian transportation system: Vulnerabilities and adaptations. In: *Weather and Transportation in Canada* [J. Andrey, & C.K. Knapper, (eds.)]. University of Waterloo, Waterloo, Ontario, 235-279.

Andrey, J. & Mills, B. (2004). Transportation. In: *Climate Change Impacts and Adaptations: A Canadian Perspective* [D. S. Lemmen, & F. J. Warren, (eds.)]. Government of Canada, Ottawa, Ontario, 131-149. <adaptation>.

Angert, A., Biraud, S., Bonfils, C., Henning, C. C., Buermann, W., Pinzon, J., Tucker, C. J. & Fung, I. Y. (2005). Drier summers cancel out the CO_2 uptake enhancement induced by warmer springs. *Proceedings of the National Academy of Sciences, 102*, 10823-10827.

Anisimov, O. A., Vaughan, D. G. Callaghan, T. V., Furgal, C., Marchant, H., Prowse, T. D., Vilhjálmsson, H. & Walsh, J. E. (2007). Polar regions (Arctic and Antarctic). In: *Climate Change 2007: Impacts, Adaptation and Vulnerability. Contribution of Working Group II to the Fourth Assessment Report of the Intergovernmental Panel on Climate Change* [M. L., Parry, O. F. Canziani, J. P., Palutikof, P. J., van der Linden, & C. E. Hanson, (eds.)]. Cambridge University Press, Cambridge, United Kingdom and New York, NY, USA, 683-685.

Anon, (2006). Bluetongue confirmed in France. News and Reports, *Veterinary Record, 159*, 331.

Antle, J. M., Capalbo, S. M. Elliott, E. T. & Paustian, K. H. (2004). Adaptation, spatial heterogeneity, and the vulnerability of agricultural systems to climate change and CO_2 fertilization: an integrated assessment approach. *Climatic Change, 64*, 289-315.

Armstrong, B., Mangtani, P., Fletcher, A., Kovats, R. S., McMichael, A. J., Pattenden, S. & Wilkinson, P. (2004). Effect of influenza vaccination on excess deaths occurring during periods of high circulation of influenza: cohort study in elderly people. *Brit. Med. J., 329*, 660-663.

Arnell, N. (2002). *Hydrology and Global Environmental Change*. Pearson Education Ltd, Edinburgh, 346.

Atkinson, M. D., Kettlewell, P. S., Hollins, P. D., Stephenson, D. B. & Hardwick, N. V. (2005). Summer climate mediates UK wheat quality response to winter North Atlantic Oscillation. *Agricultural and Forest Meteorology, 130*, 27-37.

Austin, J., D. Shindell, S. R., Beagley, C., Brühl, M., Dameris, E., Manzini, T., Nagashima, P., Newman, S., Pawson, G., Pitari, E., Rozanov, C., Schnadt, & T. G. Shepherd, (2003). Uncertainties and assessments of chemistry-climate models of the stratosphere. *Atmospheric Chemistry and Physics, 3(1)*, 1-27.

Ayres, M. P. & Lombardero, M. J. (2000). Assessing the consequences of global change for forest disturbance from herbivores and pathogens. *Science of the Total Environment, 262*, 263- 286.

Aziz, K. M. A., Hoque, B. A., Huttly, S., Minnatullah, K. M., Hasan, Z., Patwary, M. K., Rahaman, M. M. & Cairncross, S. (1990). *Water Supply, Sanitation and Hygiene Education: Report of a Health Impact Study in Mirzapur, Bangladesh.* Water and Sanitation Report Series, No. 1, World Bank, Washington, DC, 99

Bachelet, D., Neilson, R. P., Lenihan, J. M. & Drapek, R. J. (2001). Climate change effects on vegetation distribution and carbon budget in the United States. *Ecosystems*, *4*, 164-185.

Bachelet, D., Neilson, R. P., Lenihan, J. M. & Drapek, R. J. (2004). Regional differences in the carbon source-sink potential of natural vegetation in the U.S.A. *Environmental Management*, *33*, S23-S43, doi: 10.1 007/s00267-003-9 115-4.

Backlund, P., Janetos, A. & Schimel, D. (2008). *The Effects of Climate Change on Agriculture, Land Resources, Water Resources, and Biodiversity.* Synthesis and Assessment Product 4.3 by the U.S. Climate Change Science Program and the Subcommittee on Global Change Research, Washington, DC, USA.

Baker, J. T. & Allen, L. H. Jr., (1993a). Contrasting crop species responses to CO_2 and temperature: rice, soybean, and citrus. *Vegetatio*, *104/105*, 239-260.

Baker, J. T. & Allen, L. H. Jr., (1993b). Effects of CO_2 and temperature on rice: A summary of five growing seasons. *Journal of Agricultural Meteorology*, *48*, 575-5 82.

Baker, J. T., Boote, K. J. & Allen, L. H. Jr., (1995). Potential climate change effects on rice: Carbon dioxide and temperature. 31-47. *In* C. Rosenzweig, J. W. Jones, & Allen, L. H. Jr. (eds.). *Climate Change and Agriculture: Analysis of Potential International Impacts*, ASA Spec. Pub. No. 59, ASA-CSSA-SSSA, Madison, Wisconsin, USA.

Baldwin, M., Dameris, M., Austin, J., Bekki, S., Bregman, B., Butchart, N., Cordero, E., Gillett, N., Graf, H. F., Granier, C., Kinnison, D., Lal, S., Peter, T., Randel, W., Scinocca, J., Shindell, D., Struthers, H., Takahashi, M. & Thompson, D. (2007). Climate-ozone connections. In: *Scientific Assessment of Ozone Depletion: 2006*. Global Ozone Research and Monitoring Project, Report No. 50, World Meteorological Organization, Geneva, Switzerland.

Bale, J. S., Masters, G. J., Hodkinson, I. D., Awmack, C., Bezemer, T. M., Brown, V. K. Butterfield, J., Buse, A., Coulson, J. C., Farrar, J., Good, J. E. G., Harrington, R., Hartley, S., Jones, T. H., Lindroth, R. L., Press, M. C., Symrnioudis, I., Watt, A. D. & Whittaker, J. B. (2002). Herbivory in global climate change research: direct effects of rising temperature on insect herbivores. *Global Change Biology*, *8*, 1-16.

Barber, R. (2001). Upwelling ecosystems. In: *Encyclopedia of Ocean Sciences* [J. H., Steele, S. A. Thorpe, & K. K. Turekian, (eds.)]. Academic Press, London, 3128.

Barber, V. A., Juday, G. P. & Finney, B. P. (2000). Reduced growth of Alaskan white spruce in the twentieth century from temperature-induced drought stress. *Nature*, *405*, 668-673.

Barnett, T .P., Pierce, D. W., AchutaRao, K. M., Gleckler, P. J., Santer, B. D., Gregory, J. M. & Washington, W. M. (2005a). Penetration of human-induced warming into the world's oceans. *Science*, *309*, 284-287.

Barnett, T. P., Adam, J. C. & Lettenmaier, D. P. (2005b). Potential impacts of a warming climate on water availability in snow-dominated regions. *Nature*, *438*, 303-309.

Bartholow, J. M. (2005). Recent water temperature trends in the Lower Klamath River, California. *Journal of Fisheries Management*, *25*, 152-162.

Bates, D. V. (2005). Ambient ozone and mortality. *Epidemiology*, *16(4)*, 427-429.

Baxter, L. W. & Calandri, K. (1992). Global warming and electricity demand: A study of California. *Energy Policy*, *20(3)*, 23 3-244.

Beaubien, E. G. & Freeland, H. J. (2000). Spring phenology trends in Alberta, Canada: Links to ocean temperature. *International Journal of Biometeorology*, *44(2)*, 53-59.

Beaugrand, G. (2004). The North Sea regime shift: evidence, causes, mechanisms and consequences. *Progress in Oceanography*, *60*, 245-262.

Beaulac, I. & Doré, G. (2005). *Impacts du Dégel du Pergélisol sur les Infrastructures de Transport Aérien et Routier au Nunavik et Adaptations - état des connaissances.*

Facultées Sciences et de Génie, Université Laval, Montreal, Quebec, 141 Beebee, T.J.C., 1995 Amphibian breeding and climate. *Nature*, *374*, 2 19-220.

Beebee, T. J. C. (2002). Amphibian phenology and climate change. *Conservation Biology*, *16(6)*, 1454-1454, doi: 10.1046/j.1523-1739.2002.02102.x.

Beever, E. A., Brussard, P. F. & Berger, J. (2003). Patterns of apparent extirpation among isolated populations of pikas (*Ochotona princeps*) in the Great Basin. *Journal of Mammology*, *84*, 37-54.

Beggs, P. J. (2004). Impacts of climate change on aeroallergens: past and future. *Clinical & Experimental Allergy*, *34*, 1507-1513.

Bell, J. L., Sloan, L.C. & Snyder, M. A. (2004). Changes in extreme climatic events: A future climate scenario. *Journal of Climate*, *17(1)*, 8 1-87.

Bell, M. & Ellis, H. (2004). Sensitivity analysis of tropospheric ozone to modified biogenic emissions for the Mid-Atlantic region. *Atmospheric Environment*, *38*, 1879-1889.

Bell, M. L., Goldberg, R., Hogrefe, C., Kinney, P., Knowlton, K., Lynn, B., Rosenthal, J., Rosenzweig, C. & Patz, J. A. (2007). Climate change, ambient ozone, and health in 50 U.S. cities. *Climatic Change*, *82*, 61-76.

Belzer, D. B., Scott, M. J. & Sands, R. D. (1996). Climate change impacts on U.S. commercial building energy consumption: an analysis using sample survey data. *Energy Sources*, *18(2)*, 177-201.

Bender, J., Hertstein, U. & Black, C. (1999). Growth and yield responses of spring wheat to increasing carbon dioxide, ozone and physiological stresses: a statistical analysis of 'ESPACE-wheat' results. *European Journal of Agronomy*, *10*, 185-195.

Berg, E. E., Henry, J. D., Fastie, C. L., De Volder, A. D. & Matsuoka, S. M. (2006). Spruce beetle outbreaks on the Kenai Peninsula, Alaska, and Kluane National Park and Reserve, Yukon Territory: Relationship to summer temperatures and regional differences in disturbance regimes. *Forest Ecology and Management*, *227*, 219-232.

Berthelot, M., Friedlingstein, P., Ciais, P., Monfray, P., Dufresen, J. L., Treut, H. L. & Fairhead, L. (2002). Global response of the terrestrial biosphere and CO_2 and climate change using a coupled climate-carbon cycle model. *Global Biogeochemical Cycles*, *16*, 1084, doi: 10.1029/2001 GB001 827.

Bertness, M. D. & Ewanchuk, P. J. (2002). Latitudinal and climate-driven variation in the strength and nature of biological interactions in New England salt marshes. *Oecologia*, *132*, 392- 401.

Bertness, M. D., Ewanchuk, P. J. & Silliman, B. R. (2002). Anthropogenic modification of New England salt marsh landscapes. *Proceedings of the National Academy of Sciences*, *99*, 1395-1398.

Bindoff, N. L., Willebrand, J., Artale, V., Cazenave, A., Gregory, J., Gulev, S., Hanawa, K., Le Quéré, C., Levitus, S., Nojiri, Y., Shum, C. K., Talley, L. D. & Unnikrishnan, A. (2007). Observations: Oceanic climate change and sea level. In: *Climate Change 2007: The Physical Science Basis. Contribution of Working Group I to the Fourth Assessment Report of the Intergovernmental Panel on Climate Change* [S., Solomon, D., Qin, M., Manning, Z., Chen, M., Marquis, K.B., Averyt, M. Tignor, & H. L. Miller, (eds.)]. Cambridge University Press, Cambridge, United Kingdom and New York, NY, USA, 385-432.

Birdsey, R. A. (2006). Carbon accounting rules and guidelines for the United States forest sector. *Journal of Environmental Quality*, *35*, 1518-1524.

Blaustein, A. R., Belden, L. K., Olson, D. H., Green, D. M., Root, T. L. & Kiesecker, J. M. (2001). Amphibian breeding and climate change. *Conservation Biology*, *15(6)*, 1804-1809

Blaustein, A. R., Root, T. L., Kiesecker, J. M., Belden, L. K., Olson, D. H. & Green, D. M. (2002). Amphibian phenology and climate change. *Conservation Biology*, *16(6)*, 1454-1455.

Bodeker, G. E., Waugh, D. W., Akiyoshi, H., Braesicke, P., Eyring, V., Fahey, D. W., Manzini, E., Newchurch, M. J., Portmann, R. W., Robock, A., Shine, K. P., Steinbrecht, W. & Weatherhead, E. C. (2006). The Ozone layer in the 21st century. In: *Scientific Assessment of Ozone Depletion: 2006*. Global Ozone Research and Monitoring Project, Report No. 50, World Meteorological Organization, Geneva, Switzerland.

Boisvenue, C. & Running, S. W. (2006). Impacts of climate change on natural forest productivity – evidence since the middle of the 20th century. *Global Change Biology*, *12*, 862-882.

Bond, W. J., Midgley, G. F. & Woodward, F. I. (2003). The importance of low atmospheric CO_2 and fire in promoting the spread of grasslands and savannas. *Global Change Biology*, *9*, 973-982.

Bond, W. J., Woodward, F. I. & Midgley, G. F. (2005). The global distribution of ecosystems in a world without fire. *New Phytol.*, *165*, 525-537.

Bonsal, B. R. & Prowse, T. D. (2003). Trends and variability in spring and autumn 0°C-isotherm dates over Canada. *Climatic Change*, *57*, 341-358.

Bonsal, B. R., Zhang, X., Vincent, L. A., & Hood, W. D. (2001). Characteristics of daily and extreme temperatures over Canada. *Journal of Climate*, *14*, 1959-1976.

Bonsal, B. R., Koshida, G., O'Brien, E. G., & Wheaton, E. (2004). Droughts. In: *Threats to Water Availability in Canada*. NWRI Scientific Assessment Report Series No. 3, ACSD Science Assessment Series No.1, Environment Canada, *National Water Resources Institute*, Burlington, Ontario, 19-25.

Booker, F. L., Prior, S. A., Torbert, H. A., Fiscus, E. L., Pursley, W. A. & Hu, S. (2005). Decomposition of soybean grown under elevated concentrations of CO_2 and O_3. *Global Change Biology*, *11*, 685-698.

Boote, K. J., Jones, J. W. & Pickering, N. B. (1996). Potential uses and limitations of crop models. *Agronomy Journal*, *88*, 704-7 16.

Boote, K. J., Pickering, N. B. & Allen, L. H. Jr., (1997). Plant modeling: Advances and gaps in our capability to project future crop growth and yield in response to global climate change. In: *Advances in Carbon Dioxide Effects Research* [L. H., Allen, Jr., M. B., Kirkham, D. M., Olszyk, & C. E. Whitman, (eds.)]. ASA Special Publication No. 61, American Society of Agronomy, Madison, WI, 179-228.

Boote, K. J., Allen, L. H. Jr., Prasad, P. V., Baker, J. T., Gesch, R. W., Snyder, A. M.. Pan, D. & Thomas, J. M. (2005). Elevated temperature and CO_2 impacts on pollination, reproductive growth, and yield of several globally important crops. *Journal of Agricultural Meteorology*, *60*, 469-474.

Both, C. (2006). Climate change and adaptation of annual cycles of migratory birds. *Journal of Ornithology*, *147*(5, Suppl. 1), 68-68.

Bowman, D. M. J. S. & Johnston, F. H. (2005). Wildfire smoke, fire management, and human health. *EcoHealth*, *2*, 76-80.

Breshears, D. D., Cobb, N. S., Rich, P. M., Price, K. P., Allen, C. D., Balice, R. G., Romme, W. H., Kastens, J. H., Floyd, M. L., Belnap, J., Anderson, J. J., Myers, O. B. & Meyer, C. W. (2005). Regional vegetation die-off in response to global-change-type drought. *Proceedings of the National Academy of Sciences*, *102*, 15144-15148.

Bretherton, C. S., Ferrari, R. & Legg, S. (2004). Climate Process Teams: a new approach to improving climate models. *U.S. CLIVAR Variations Newsletter*, *2(1)*, 1-5.

Bromirski, P. D., Flick, R. E. & Cayan, D. R. (2003). Storminess variability along the California coast: 1958-2000. *Journal of Climate*, *16*, 982-993.

Brown, J. L. Liand, S. H. & Bhagabati, B. (1999). Long-term trend toward earlier breeding in an American bird: A response to global warming? *Proceedings of the National Academy of Sciences*, *96*, 5565-5569.

Brown, K. J., Clark, J. S. Grimm, E. C., Donovan, J. J., Mueller, P. G., Hansen, B. C. S. & Stefanova, I. (2005). Fire cycles in North American interior grasslands and their relation to prairie drought. *Proceedings of the National Academy of Sciences*, *102*, 8865-8870.

Brown, T. J., Hall, B. L. & Westerling, A. L. (2004). The impact of twenty-first century climate change on wildland fire danger in the western United States: An applications perspective. *Climatic Change*, *62*, 365-388.

Bull, S. R., Bilello, D. E., Ekmann, J., Sale, M. J. & Schmalzer, D. K. (2007). Effects of climate change on energy production and distribution in the United States. In: *Effects of Climate Change on Energy Production and Use in the United States.* Synthesis and Assessment Product 4.5 by the U.S. Climate Change Science Program and the Subcommittee on Global Change Research, Washington, DC, USA, 45-80.

Burke, E. J., Brown, S. J. & Christidis, N. (2006). Modelling the recent evolution of global drought and projections for the twenty-first century with the Hadley Centre climate model. *Journal of Hydrometeorology*, *7*, 1113-1125.

Burkett, V. R. (2002). *The Potential Impacts of Climate Change on Transportation.* Workshop Summary and Proceedings, 1-2 October 2002, DOT Center for Climate Change and Environmental Forecasting - Federal Research Partnership Workshop, Washington, District of Columbia. <climate.dot.gov/publications/workshop1002/index.html>.

Burkett, V. R., Wilcox, D. A., Stottlemeyer, R., Barrow, W., Fagre, D., Baron, J., Price, J. L., Nielsen, Allen, C. D., Peterson, D. L., Ruggerone, G. & Doyle, T. (2005). Nonlinear dynamics in ecosystem response to climatic change: case studies and policy implications. *Ecological Complexity*, *2*, 357-394.

Butler, C. (2003). The disproportionate effect of climate change on the arrival dates of short-distance migrant birds. *Ibis*, *145*, 484-495.

Cahoon, D. R., Hensel, P. F., Spencer, T., Reed, D. J., McKee, K. L. & Saintilan, N. (2006). Coastal wetland vulnerability to relative sea-level rise: wetland elevation trends and process controls. In: *Wetlands and Natural Resource Management* [J., Verhoeven, D., Whigham, R. Bobbink, & B. Beltman, (eds.)]. Springer Ecological Studies Series Vol. *190*, Springer, 271-342.

Cairncross, S. & Alvarinho, M. (2006). The Mozambique floods of 2000: health impact and response. In: *Flood Hazards and Health: Responding to Present and Future Risks.* [Few, R. & F. Matthies (eds.)]. Earthscan, London, 111-127.

California Energy Commission, (2006a). *Historic State-Wide California Electricity Demand.* <energy>.

California Energy Commission, 2006b *Our Changing Climate: Assessing the Risks to California.* <www.energy. ca. gov/ 2006publications/ CEC-500-2006-077/CEC-500-2006- 077.PDF>.

Calkin, D. E., Gebert, K. M., Jones, J. G. & Neilson, R. P. (2005). Forest service large fire area burned and suppression expression trends, 1970-2002. *Journal of Forestry, 103*, 179-183.

Carbone, G. J., Kiechle, W., Locke, L., Mearns, L. O., McDaniel, L. & Downton, M. W. (2003). Response of soybean and sorghum to varying spatial scales of climate change scenarios in the southeastern United States. *Climatic Change*, *60*, 73-98.

Carroll, A. L., Taylor, S. W., Regniere, J. & Safranyik, L. (2004). Effects of climate change on range expansion by the mountain pine beetle in British Columbia. In: *Mountain Pine Beetle Symposium: Challenges and Solutions* [T. L., Shore, J. E. Brooks, & J. E. Stone, (eds.)]. 30-3 1 October, 2003, Kelowna, British Columbia. Information Report BC-X-399, Natural Resources Canada, Canadian Forest Service, Pacific Forestry Centre, Victoria, BC, 223-232.

Caspersen, J. P., Pacala, S. W., Jenkins, J. C., Hurtt, G. C. & Moorcraft, P. R. (2000). Contributions of land-use history to carbon accumulation in U.S. forests. *Science*, *290*, 1148-1152.

Cayan, D. R., Kammerdiener, S., Dettinger, M. D., Caprio, J. M. & Peterson, D. H. (2001). Changes in the onset of spring in the western United States. *Bulletin of the American Meteorological Society*, *82*, 399-415.

CCSP, (2003). *Strategic Plan for the U.S. Climate Change Science Program. A report by the Climate Change Science Program and the Subcommittee on Global Change Research*, Washington, DC, 202

CDC, (2005a) Heat-related mortality – Arizona, 1993-2002, and United States, 1979-2002. *Morbidity & Mortality Weekly Report, 54(25)*, 628-630.

CDC, (2005b). Norovirus outbreak among evacuees from Hurricane Katrina Houston, Texas, September 2005. *Morbidity & Mortality Weekly Report, 54(40)*, 1016-1018.

CDC, 2006 Carbon monoxide poisonings after two major hurricanes - Alabama and Texas, August-October 2005. *Morbidity & Mortality Weekly Report, 55(9)*, 236-239.

Cecchi, L., Morabito, M., Domeneghetti, P., Crisci, M. A., Onorari, M. & Orlandini, S. (2006). Long distance transport of ragweed pollen as a potential cause of allergy in central Italy. *Annals of Allergy, Asthma & Immunology*, *96*, 86-91.

Cesar, H., Burke, L., & Pet-Soede, L. (2003). *The Economics of Worldwide Coral Reef Degradation.* Cesar Environmental Economic Consulting, Arnhem, The Netherlands.

Cess, R.D., 2005: Water vapor feedback in climate models. *Science*, *310*, 795-796.

Chambers, J. Q., Fisher, J. I., Zeng, H., Chapman, E. L., Baker, D. B. & Hurtt, G. C. (2007. Hurricane Katrina's carbon footprint on U.S. Gulf Coast forests. *Science*, *318*, 1107.

Chavez, F., L. Ryan, S.E. Lluch-Cota, and M. Ñiguen, 2003 From anchovies to sardines and back: Multidecadal change in the Pacific Ocean. *Science*, *229*, 217-221.

Chen, F. J., Wu, G., & Ge, F. (2004). Impacts of elevated CO_2 on the population abundance and reproductive activity of aphid *Sitobion avenae* Fabricius feeding on spring wheat. *Journal of Applied Entomology*, *128*, 723-730.

Chiew, F. H. S. (2007). Estimation of rainfall elasticity of streamflow in Australia. *Hydrol. Sci. J.*, *51*, 613-625.

Choi, K. M., Christakos, G. & Wilson, M. L. (2006). El Niño effects on influenza mortality risks in the state of California. *Public Health, 120(6)*, 505-5 16.

Choudhury, A. Y. & Bhuiya, A. (1993). Effects of biosocial variable on changes in nutritional status of rural Bangladeshi children, pre- and post-monsoon flooding. *Journal of Biosocial Science*, *25*, 35 1-357.

Christensen, J. H., Hewitson, B., Busuioc, A., Chen, A., Gao, X., Held, I., Jones, R., Kolli, R. K. Kwon, W. T., Laprise, R., Magaña Rueda, V., Mearns, L., Menéndez, C. G., Räisänen, J., Rinke, A., Sarr, A. & Whetton, P. (2007). Regional climate projections. *Climate Change 2007: The Physical Science Basis. Contribution of Working Group I to the Fourth Assessment Report of the Intergovernmental Panel on Climate Change* [S., Solomon, D., Qin, M., Manning, Z., Chen, M., Marquis, K. B., Averyt, M. Tignor, & H. L. Miller, (eds.)] Cambridge University Press, Cambridge and New York, 847-940.

Christensen, N. S., Wood, A. W., Voisin, N., Lettenmaier, D. P. & Palmer, R. N. (2004). The effects of climate change on the hydrology and water resources of the Colorado River basin. *Clim. Change*, *62*, 337-363.

Church, J. A., White, N. J., Coleman, R., Lambeck, K. & Mitrovica, J. X. (2004). Estimates of the regional distribution of sea level rise over the 1950-2000 period. *Journal of Climate*, *17*, 2609-2625.

Ciais, P., Reichstein, M., Viovy, N., Granier, A., Ogee, J., Allard, V., Aubinet, M., Buchmann, N., Chr. Bernhofer, Carrara, A., Chevallier, F., De Noblet, N., Friend, A. D., Friedlingstein, P., Grünwald, T., Heinesch, B., Keronen, P., Knohl, A., Krinner, G., Loustau, D., Manca, G., Matteucci, G., Miglietta, F., Ourcival, J. M., Papale, D., Pilegaard, K., Rambal, S., Seufert, G., Soussana, J. F., Sanz, M. J., Schulze, E. D., Vesala, T. & Valentini, R. (2005). Europe-wide reduction in primary productivity caused by the heat and drought in 2003. *Nature*, *437*, 529-533.

Clark, W. C., Jaeger, J., Corell, R., Kasperson, R., McCarthy, J. J., Cash, D., Cohen, S. J., Desanker, P., Dickson, N., Epstein, M. P., Guston, D. H., Hall, J. M., Jaeger, C., Janetos, A., Leary, N., Levy, M. A., Luers, A., MacCracken, M., Melillo, J., Moss, R., Nigg, J. M., Parry, M. L., Parson, E. A., Ribot, J. C., Schellnhuber, H. J., Schrag, D. P., Seielstad, G. A., Shea, E., Vogel, C. & Wilbanks, T. J. (2000). *Assessing Vulnerability to Global Environmental Risks*. Report of the Workshop on Vulnerability to Global Environmental Change: Challenges for Research, Assessment and Decision Making. 22-25 May, Airlie House, Warrenton, Virginia. Discussion Paper 2000-12, Environment and Natural Resources Program, Kennedy School of Government, Harvard University, Cambridge, MA, USA.

Clarke, L., Edmonds, J., Jacoby, J., Pitcher, H., Reilly, J., Richels, R., Parson, E., Burkett, V., Fisher-Vanden, K., Keith, D., Mearns, L., Pitcher, H., Rosenzweig, C. & Webster, M. (2007). *Scenarios of Greenhouse Gas Emissions and Atmospheric Concentrations (Part A) and Review of Integrated Scenario Development and Application (Part B).* Synthesis and Assessment Product 2.1 by the U.S. Climate Change Science Program and the Subcommittee on Global Change Research, Washington, DC, USA, 260.

Clarke, P. J., Latz, P. K. & Albrecht, D. E. (2005). Long-term changes in semi-arid vegetation: Invasion of an exotic perennial grass has larger effects than rainfall variability. *Journal of Vegetation Science*, *16*, 237-248.

Clarke, R. & King, J. (2004). *The Atlas of Water*. Earthscan, London, 128.

Clayton, A., Montufar, J., Regehr, J., Isaacs, C. & McGregor, R. (2005). *Aspects of the Potential Impacts of Climate Change on Seasonal Weight Limits and Trucking in the Prairie Region*. Prepared for Natural Resources Canada. <www. adaptation>.

Coakley, S. M., Scherm, H. & Chakraborty, S. (1999). Climate change and plant disease management. *Annual Review of Phytopathology, 37*, 3 99-426.

Coe, M. T. & Foley, J. A. (2001). Human and natural impacts on the water resources of the Lake Chad basin. *Journal of Geophysical Research, 106(D4)*, 3349-3356.

Cohen, J. C. (2003). Human population: the next half century. *Science, 302*, 1172-1175.

Cole, K. (1985). Past rates of change, species richness and a model of vegetation inertia in the Grand Canyon, Arizona. *American Naturalist, 125*, 289-303.

Collins, W. D., Bitz, C. M., Blackmon, M. L., Bonan, G. B., Bretherton, C. S., Carton, J. A., Chang, P., Doney, S. C., Hack, J. J., Henderson, T. B., Kiehl, J. T., Large, W. G., McKenna, D. S., Santer, B. D. & Smith, R. D. (2006). The Community Climate System Model: CCSM3. *Journal of Climate, 19*, 2122-2143.

Colorado State Forest Service, 2007 *2006 Report on the Health of Colorado's Forests*. Colorado Department of Natural Resources, Division of Forestry, Denver, CO, USA, *27*.

Confalonieri, U., Menne, B., Akhtar, R., Ebi, K. L., Hauengue, M., Kovats, R. S., Revich, B. & Woodward, A. (2007). Human health. In: *Climate Change 2007: Impacts, Adaptation and Vulnerability. Contribution of Working Group II to the Fourth Assessment Report of the Intergovernmental Panel on Climate Change* [M. L., Parry, O. F. Canziani, J. P. Palutikof, P. J. van der Linden, & C. E. Hanson, (eds.)]. Cambridge University Press, Cambridge, United Kingdom and New York, NY, USA, 391-431.

Cook, E. R., Woodhouse, C. A., Eakin, C. M., Meko, D. M. & Stahle, D. W. (2004). Long-term aridity changes in the Western United States. *Science, 306*, 1015-1018.

Cook, S. M., Glass, R. I., LeBaron, C. W. & Ho, M. S. (1990). Global seasonality of rotavirus infections. *Bulletin of the World Health Organization, 58(2)*, 17 1-177.

Corn, P. S. (2003). Amphibian breeding and climate change: Importance of snow in the mountains. *Conservation Biology, 17*, 622-625.

Costanza, R., dArge, R., deGroot, R., Farber, S., Grasso, M., Hannon, B., Limburg, K., Naeem, S., Oneill, R.V., Paruelo, J., Raskin, R.G., Sutton, P. & vandenBelt, M. (1997). The value of the world's ecosystem services and natural capital. *Nature, 387*, 253-260.

Cotton, P. (2003). Avian migration phenology and global climate change. *Proceedings of the National Academy of Sciences, 100*, 12219-12222.

Cox, P. M., Betts, R. A., Jones, C. D., Spall, S. A. & Totterdell, I. J. (2000). Acceleration of global warming due to carbon-cycle feedbacks in a coupled climate model. *Nature, 408*, 184- 187.

Croier, L. (2003). Winter warming facilitates range expansion: Cold tolerance of the butterfly *Atalopedes campestris*. *Oecologia, 135*, 648-656.

Crossett, K. M., Culliton, T. J., Wiley, P. C. & Goodspeed, T. R. (2004). *Population Trends Along the Coastal United States: 1980-2008*. National Oceanic and Atmospheric Administration, Silver Spring, MD.

Crozier, L. & Zabel, R. W. (2006). Climate impacts at multiple scales: evidence for differential population responses in juvenile Chinook salmon, *Journal of Animal Ecology, 75*, 1100- 1109.

Currie, D. J. (2001). Projected effects of climate change on patterns of vertebrate and tree species in the conterminous United States. *Ecosystems*, *4*, 2 16-225.

Curriero, F. C., Patz, J. A., Rose, J. B. & Lele, S. (2001). The association between extreme precipitation and waterborne disease outbreaks in the United States, 1948-1994. *American Journal of Public Health*, *91(8)*, 1194-1199.

Curry, R. & Mauritzen, C. (2005). Dilution of the northern North Atlantic in recent decades. *Science*, *308*, 1772-1774.

D'Amato, G., Liccardi, G., D'Amato, M. & Cazzola, M. (2002). Outdoor air pollution, climatic changes and allergic bronchial asthma. *European Respiratory Journal*, *20*, 763 - 776.

D'Souza, R. M., Becker, N. G., Hall, G. & Moodie, K. B. A. (2004). Does ambient temperature affect foodborne disease? *Epidemiology*, *15(1)*, 86-92.

Dai, A., Trenberth, K. E. & Qian, T. (2004). A global data set of Palmer Drought Severity Index for 1870–2002 Relationship with soil moisture and effects of surface warming. *Journal of Hydrometeorology*, *5*, 1117-1130.

Daily, G. C., Söderqvist, T., Aniyar, S., Arrow, K., Dasgupta, P., Ehrlich, P. R., Folke, C., Jansson, A., Jansson, B., Kautsky, N., Levin, S., Lubchenco, J., Mäler, K., Simpson, D., Starrett, D., Tilman, D. & Walker, B. (2000). The value of nature and the nature of value. *Science*, *289*, 395-396.

Dale, V. H. (1997).. The relationship between land-use change and climate change. *Ecological Applications*, *7*, 753-769.

Dale, V. H., Joyce, L. A., McNulty, S., Neilson, R. P., Ayres, M. P., Flannigan, M. D., Hanson, P. J., Irland, L. C., Lugo, A. E., Peterson, C. J., Simberloff, D., Swanson, F. J., Stocks, B. J. & Wotton, B. M. (2001). Climate change and forest disturbances. *BioScience*, *51*, 723-734.

Daszak, P., Cunningham, A. A. & Hyatt, A. D. (2000). Emerging infectious diseases of wildlife: threats to biodiversity and human health. *Science*, *287*, 443- 448.

Davis, M. S., Mader, T. L., Holt, S. M. & Parkhurst, A. M. (2003). Strategies to reduce feedlot cattle heat stress: effects on tympanic temperature. *Journal of Animal Science*, *81*, 649- 661.

Davis, R., Knappenberger, P., Novicoff, W. & Michaels, P. (2002). Decadal changes in heat related human mortality in the eastern United States. *Climate Research*, *22*, 175-184.

Davis, R., Knappenberger, P., Michaels, P. & Novicoff, W. (2003). Changing heat-related mortality in the United States. *Environmental Health Perspectives*, *111*, 1712 -1718.

Davis, R., Knappenberger, P., Michaels, P. & Novicoff, W. (2004). Seasonality of climate-human mortality relationships in US cities and impacts of climate change. *Climate Research*, *26*, 61-76. F. C. de la Chesnaye, & J. P. Weyant, (eds.), 2006, Multi-gas mitigation and climate policy. *The Energy Journal*, Special Issue No.3.

Decker, E., Elliot, S., Smith, F., Blake, D. & Rowland, F. (2000). Energy and material flow through the urban ecosystem. *Annual Review of Energy and the Environment*, *25*, 685-740.

DeGaetano, A. T. (2005). Meteorological effects on adult mosquito (*Culex*) populations in metropolitan New Jersey. *International Journal of Biometeorology*, *49(5)*, 345-353.

del Ninno, C. & Lundberg, M. (2005). Treading water: the long term impact of the 1998 flood on nutrition in Bangladesh. *Economics & Human Biology*, *3*, 67-96.

Delworth, T. L., Broccoli, A. J., Rosati, A., Stouffer, R. J., Balaji, V., Beesley, J. A., Cooke, W. F., Dixon, K. W., Dunne, J., Dunne, K. A., Durachta, J. W., Findell, K. L., Ginoux, P., Gnanadesikan, A., Gordon, C. T., Griffies, S. M., Gudgel, R., Harrison, M. J., Held, I. M., Hemler, R. S., Horowitz, L. W., Klein, S. A., Knutson, T. R., Kushner, P. J., Langenhorst, A. R., Lee, H. C., Lin, S. J., Lu, J., Malyshev, S. L., Milly, P. C. D., Ramaswamy, V., Russell, J., Schwarzkopf, M. D., Shevliakova, E., Sirutis, J. J., Spelman, M. J., Stern, W. F., Winton, M., Wittenberg, A. T., Wyman, B., Zeng, F. & Zhang, R. (2006). GFDL's CM2 global coupled climate models. Part I: Formulation and simulation characteristics. *Journal of Climate*, *19*, 643-674.

Denman, K. L., Brasseur, G., Chidthaisong, A., Ciais, P., Cox, P. M., Dickinson, R. E., Hauglustaine, D., Heinze, C., Holland, E., Jacob, D., Lohmann, U., Ramachandran, S., da Silva Dias, P. L., Wofsy, S. C. & Zhang, X. (2007). Couplings between changes in the climate system and biogeochemistry. In: *Climate Change 2007: The Physical Science Basis. Contribution of Working Group I to the Fourth Assessment Report of the Intergovernmental Panel on Climate Change* [S., Solomon, D., Qin, M., Manning, Z., Chen, M., Marquis, K. B., Averyt, M. Tignor, & H. L. Miller, (eds.)]. Cambridge University Press, Cambridge, 499-588.

Derocher, A. E., Lunn, N. J. & Stirling, I. (2004). Polar bears in a warming climate. *Integrative and Comparative Biology*, *44*, 163-176.

Dettinger, M. D. (2005). *Changes In Streamflow Timing In The Western United States In Recent Decades*, USGS Fact Sheet 2005-3018, U.S. Geological Survey.

Diaz, H. F., Eischeid, J. K., Duncan, C. & Bradley, R. S. (2003). Variability of freezing levels, melting season indicators, and snow cover for selected high-elevation and continental regions in the last 50 years. *Climatic Change*, *59*, 33-52.

Diaz, J., Jordan, A., Garcia, R., Lopez, C., Alberdi, J. C., Hernandez, E. & Otero, A. (2002). Heat waves in Madrid 1986-1997: effects on the health of the elderly. *International Archives of Occupational & Environmental Health*, *75(3)*, 163-170.

Dietz, V. J. & Roberts, J. M. (2000). National surveillance for infection with *Cryptosporidium parvum*, 1995-1998: what have we learned? *Public Health Reports*, *115*, 358-363.

Diffenbaugh, N. S., Pal, J. S., Trapp, R. J. & Giorgi, F. (2005). Fine-scale processes regulate the response of extreme events to global climate change. *Proceedings of the National Academy of Sciences*, *102(44)*, 15774-15778, doi: 10.1 073/pnas.0506042 102.

Donaldson, G. C., Rintamaki, H. & Nayha, S. (2001) Outdoor clothing: its relationship to geography, climate, behaviour and cold-related mortality in Europe. *International Journal of Biometeorology*, *45(1)*, 45-51.

du Vair, P., Wickizer, D. & Burer, M. J. (2002). Climate change and the potential implications for California's transportation system. In: *The Potential Impacts of Climate Change on Transportation, Federal Research Partnership Workshop*, 1-2 October, 2002, Washington, DC, 125-135. <climate. dot. gov/ publications/ workshop1002/ index.html>.

Ducklow, H. W., Baker, K., Martinson, D. G., Quetin, L. B., Ross, R. M. Smith, R. C. Stammerjohn, S. E., Vernet, M. & Fraser, W. (2007). Marine pelagic ecosystems: The West Antarctic Peninsula. *Philosophical Transactions of the Royal Society B*, *362*, 67-94.

Duffy, P. A., Walsh, J. E., Graham, J. M., Mann, D. H. & Rupp, T. S. (2005). Impacts of large-scale atmospheric-ocean variability on Alaskan fire season severity. *Ecological Applications, 15*, 1317-1330.

Dunn, P. O. & Winkler, D. (1999). Climate change has affected the breeding date of tree swallows throughout North America. *Proceedings of the Royal Society of London Bulletin, 266*, 2487-2490.

Dupigny-Giroux, L. A. (2001). Towards characterizing and planning for drought in Vermont – Part I: A climatological perspective. *Journal of the American Water Resources Association, 37*, 505-525.

Durongdej, S. (2001). Land use changes in coastal areas of Thailand. *Proceedings of the APN/SUR VAS/LOICZ Joint Conference on Coastal Impacts of Climate Change and Adaptation in the Asia – Pacific Region*, 14-16 November 2000, Kobe, Japan. Asia Pacific Network for Global Change Research, 113-117.

Dushoff, J., Plotkin, J. B., Viboud, C., Earn, D. J. & Simonsen, L. (2005). Mortality due to influenza in the United States – an annualized regression approach using multiple-cause mortality data. *American Journal of Epidemiology, 163(2)*, 181-187.

Dvortsov, V. L. & Solomon, S. (2001). Response of the stratospheric temperatures and ozone to pat and future increases in stratospheric humidity. *Journal of Geophysical Research, 106(D7)*, 7505-75 14.

Dyke, A. S. & Peltier, W. R. (2000). Forms, response times and variability of relative sea-level curves, glaciated North America. *Geomorphology, 32*, 3 15-333.

Dziuban, E. J., Liang, J. L., Craun, G. F., Hill, V., Yu, P. A., Painter, J., Moore, M. R., Calderon, R. L., Roy, S. L. & Beach, M. J. (2006). Surveillance for waterborne disease and outbreaks associated with recreational water – United States, 2003 – 2004. *Morbidity & Mortality Weekly Report, 55(12)*, 1-31.

Easterling, D. R. (2002). Recent changes in frost days and the frost-free season in the United States. *Bulletin of the American Meteorological Society, 83*, doi: 10.1175/1520-0477.

Easterling, W. et al., (2007). Food, Fibre and Forest Products. In: *Climate Change 2007: Impacts, Adaptation and Vulnerability. Contribution of Working Group II to the Fourth Assessment Report of the Intergovernmental Panel on Climate Change* [(eds.)]. Cambridge University Press, Cambridge, United Kingdom and New York, NY, USA.

Eastman, J. L. Coughenour, M. B. & Pielke, R. A. (2001). The regional effects of CO_2 and landscape change using a coupled plant and meteorological model. *Global Change Biol., 7*, 797-815.

Ebi, K. L., Exuzides, K. A., Lau, E., Kelsh, M. & Barnston, A. (2001). Association of normal weather periods and El Nino events with hospitalization for viral pneumonia in females: California, 1983–1998. *American Journal of Public Health, 91(8)*, 1200–1208.

Ebi, K. L., et al., (2008). (in press): Effects of global change on human health. Ch.2 in: *Analyses of the Effects of Global Change on Human Health and Welfare and Human Systems.* Synthesis and Assessment Product (SAP) 4.6 by the U.S. Climate Change Science Program and the Subcommittee on Global Change Research, Washington, DC.

Edmonds, J. A. & Rosenberg, N. J. (2005). Climate change impacts for the conterminous USA: An integrated assessment summary. *Climatic Change, 69*, 15 1-162.

Eheart, J. W., Wildermuth, A. J. & Herricks, E. E. (1999). The effects of climate change and irrigation on criterion low streamflows used for determining total maximum daily loads. *Journal of the American Water Resources Association, 35*, 1365-1372.

Ehrlich, P. R., Murphy, D. D., Singer, M. C., Sherwood, C. B., White, R R. & Brown, I. L. (1980). Extinction, reduction, stability and increase: The responses of checkerspot butterfly populations to the California drought. *Oecologia*, *46*, 101-105.

EIA, (2002). *Annual Coal Report*. DOE/EIA-05 84, Energy Information Administration, Washington, DC, USA.

EIA, (2004). *Annual Energy Review*. Energy Information Administration, Washington, DC, USA. EIA, 2005: *Renewable Energy Trends 2004*, Table 18, Renewable Electric Power Sector Net Generation by Energy Source and State, 2003. Energy Information Administration, Washington, DC, USA.

EIA, (2006). *Annual Energy Outlook 2006, with Projections to 2030*. DOE/EIA-0383, Energy Information Administration, Washington, DC, USA.

Elliott, G. P. & Baker, W. L. (2004). Quaking aspen at treeline: A century of change in the San Juan Mountains, Colorado, USA. *Journal of Biogeography*, *31*, 733-745.

Emanuel, K. A. (2005). Increasing destructiveness of tropical cyclones over the past 30 years *Nature*, *436*, 686-688.

EM-DAT, 2006 *The OFDA/CRED International Disaster Database*. WHO Collaborating Centre for Research on the Epidemiology of Disasters (CRED). <www.em-dat.net>. Environment Canada, 1997: The Canada country study: climate impacts and adaptation. Adaptation and Impacts Research Group, Downsview, Ontario.

EPA. 2004 Air quality criteria for particulate matter. Volume 1. http://cfpub.epa.gov/ncea/cfm/recordisplay.cfm?deid=87903 [Accessed May 5, 2008].

EPA, (2006). *Excessive Heat Events Guidebook*. EPA-430-B-06-005, U.S. Environmental Protection Agency, Office of Atmospheric Programs, Washington, DC, USA. <www.epa.gov/heatisland/about/pdf/EHEguide_final.pdf>.

EPRI, (2003). *A Survey of Water Use and Sustainability in the United States With a Focus on Power Generation*. EPRI Report No. 1005474, Electric Power Research Institute, Washington, DC, USA.

Erickson, J. E., Megonigal, J. P., Peresta, G. & Drake, B. G. (2007). Salinity and sea level mediate elevated CO_2 effects on C3 and C4 plant interactions and tissue nitrogen in a Chesapeake Bay tidal wetland. *Global Change Biology*, *13*, 202-215, doi:10.1 11 1/j.1365- 2486.2006.0 1 285.x.

Euliss, N. H., Gleason, R. A., Olness, A., McDougal, R. L., Murkin, H. R., Robarts, R.D., Bourbonniere, R.A. & Warner, B.G. (2006). North American prairie wetlands are important nonforested land-based carbon storage sites. *Science of the Total Environment*, *361(1-3)*, 179-188.

European Commission, (2003). *Ozone-Climate Interactions*. Air Pollution Research Report 81, EUR 20623, European Commission, Luxembourg, 143.

Eurowinter Group, (1997). Cold exposure and winter mortality from ischaemic heart disease, cerebrovascular disease, respiratory disease, and all causes in warm and cold regions of Europe. *Lancet*, *349*, 1341-1346.

Fallico, F., Nolte, K., Siciliano, L. & Yip, F. (2005). Hypothermia-related deaths – United States, 2003-2004. *Morbidity & Mortality Weekly Report*, *54(7)*, 173-175.

Fan, Y., Miguez-Macho, G., Weaver, C., Walko, R. & Robock, A. (2007). Incorporating water table dynamics in climate modeling, part I: water table observations and the equilibrium water table. *Journal of Geophysical Research*, *112*, D10125, doi: 10. 1029/2006JD0081 11.

Fay, P. A., Carlisle, J. D., Knapp, A. K., Blair, J. M. & Collins, S. L. (2000). Altering rainfall timing and quantity in a mesic grassland ecosystem: design and performance of rainfall manipulation shelters. *Ecosystems*, *3*, 308-319.

Fay, P. A., Carlisle, J. D., Danner, B. T., Lett, M. S., McCarron, J. K., Stewart, C., Knapp, A. K., Blair, J. M. & Collins, S. L. (2002). Altered rainfall patterns, gas exchange, and growth in grasses and forbs. *International Journal of Plant Sciences*, *163*, 549-557.

Fay, P. A., Carlisle, J. D., Knapp, A. K., Blair, J. M. & Collins, S. L. (2003). Productivity responses to altered rainfall patterns in a C-4-dominated grassland. *Oecologia*, *137*, 245-251.

Federal Register, 2006 Rules and Regulations. Endangered and Threatened Species: Final listing determination for elkhorn coral and staghorn coral. *Federal Register*, *71*, 26852.

Feng, S. & Hu, Q. (2004). Changes in agro-meteorological indicators in the contiguous United States: 1951-2000. *Theoretical and Applied Climatology*, *78*, 247-264.

Fenn, M. E., Baron, J. S., Allen, E. B., Reuth, H. M., Nydick, K. R., Geiser, L., Bowman, W. D., Sickman, J.O., Meixner, T., Johnson, D. W. & Neitlich, P. (2003). Ecological effects of nitrogen deposition in the western United States. *BioScience*, *53*, 404-420.

Ferguson, S. H., Stirling, I. & McLoughlin, P. (2005). Climate change and ringed seal (*Phoca hispida*) recruitment in western Hudson Bay. *Marine Mammal Science*, *21*, 12 1-135.

Fidje, A. & Martinsen, T. (2006). Effects of climate change on the utilization of solar cells in the Nordic region. Extended abstract for *European Conference on Impacts of Climate Change on Renewable Energy Sources*, 5-9 June 2006, Reykjavik, Iceland.

Field, C. B., Mortsch, L. D., Brklacich, M., Forbes, D. L., Kovacs, P., Patz, J. A., Running, S. W. & Scott, M. J. (2007). North America. In: *Climate Change 2007: Impacts, Adaptation and Vulnerability. Contribution of Working Group II to the Fourth Assessment Report of the Intergovernmental Panel on Climate Change* [M. L., Parry, O. F., Canziani, J. P., Palutikof, P. J. van der Linden, & C. E. Hanson, (eds.)]. Cambridge University Press, Cambridge, United Kingdom and New York, NY, USA, 617-652.

Fischlin, A., Midgley, G. F., Price, J. T., Leemans, R., Gopal, B., Turley, C., Rounsevell, M. D. A., Dube, O. P., Tarazona, J. & Velichko, A. A. (2007). Ecosystems, their properties, goods, and services. In: *Climate Change 2007: Impacts, Adaptation and Vulnerability. Contribution of Working Group II to the Fourth Assessment Report of the Intergovernmental Panel on Climate Change.* [M. L., Parry, O. F. Canziani, J. P. Palutikof, P. J. van der Linden, & C. E. Hanson, (eds.)]. Cambridge University Press, Cambridge, United Kingdom and New York, NY, USA, 211-272.

Fisher, B. S., Nakicenovic, N., Alfsen, K., Corfee Morlot, J., de la Chesnaye, F., Hourcade, J. Ch., Jiang, K., Kainuma, M., La Rovere, E., Matysek, A., Rana, A., Riahi, K., Richels, R., Rose, S. van Vuuren, D. & Warren, R. (2007). Issues related to mitigation in the long-term context. In: *Climate Change 2007: Mitigation. Contribution of Working Group III to the Fourth Assessment Report of the Intergovernmental Panel on Climate Change* [B., Metz, O. R. Davidson, P. R. Bosch, R. Dave, & L. A. Meyer, (eds.)]. Cambridge University Press, Cambridge, United Kingdom and New York, NY, USA, 169-250.

Flahault, A., Viboud, C., Pakdaman, K., Boelle, P. Y., Wilson, M. L., Myers, M. & Valleron, A. J. (2004). Association of influenza epidemics in France and the USA with global climate variability. In: *Proceedings of the International Conference on Options for the Control of Influenza V* [Y. Kawaoka, (ed.)]. Elsevier Inc., San Diego, CA, USA, 73-77.

Flanagan, L. B., Wever, L. A. & Carlson, P. J. (2002). Seasonal and interannual variation in carbon dioxide exchange and carbon balance in a northern temperate grassland. *Global Change Biology*, *8*, 599-615.

Flanner, M. G., Zender, C. S., Randerson, J. T. & Rasch, P. J. (2007). Present day climate forcing and response from black carbon in snow. *Journal of Geophysical Research*, *112*, D1 1202, doi: 10.1 029/2006JD008003.

Flannigan, M. D., Stocks, B. J. & Wotton, B. M. (2000). Climate change and forest fires. *Science of the Total Environment*, *262*, 221-229.

Flannigan, M. D., Logan, K. A., Amiro, B. D., Skinner, W. R. & Stocks, B. J. (2005). Future area burned in Canada. *Climatic Change*, *72*, 1-16.

Fleury, M., Charron, D. F., Holt, J. D., Allen, O. B., & Maarouf, A. R. (2006). A time series analysis of the relationship of ambient temperature and common bacterial enteric infections in two Canadian provinces. *International Journal of Biometeorology*, *60*, 385-391.

Forbes, D. L., Manson, G. K., Chagnon, R., Solomon, S. M., van der Sanden, J. J. & Lynds, T. L. (2002). Nearshore ice and climate change in the southern Gulf of St. Lawrence. In: *Ice in the Environment.* Proceedings 16th IAHR International Symposium on Ice, Dunedin, New Zealand, *1*, 344-351.

Forbes, D. L., Parkes, G. S., Manson, G. K. & Ketch, L. A. (2004). Storms and shoreline retreat in the southern Gulf of St. Lawrence. *Marine Geology*, *210*, 169-204.

Forister, M. L. & Shapiro, A. M. (2003). Climatic trends and advancing spring flight of butterflies in lowland California. *Global Change Biology*, *9*, 1130-1135.

Forster, P., Ramaswamy, V., Araxo, P., Berntsen, T., Betts, R. A., Fahey, D. W., Haywood, J., Lean, J., Lowe, D. C., Myhre, G., Nganga, J., Prinn, R., Raga, G., Schulze, M. & Van Dorland, R. (2007). Changes in atmospheric constituents and in radiative forcing. In: *Climate Change 2007 The Physical Science Basis. Contribution of Working Group I to the Fourth Assessment Report of the Intergovernmental Panel on Climate Change* [S., Solomon, D., Qin, M., Manning, Z., Chen, M., Marquis, K. B., Averyt, M. Tignor, & H. L. Miller, (eds.)]. Cambridge University Press, Cambridge, 130-234.

Franco, A. M. A., Hill, J. K., Kitschke, C., Collingham, Y. C., Roy, D. B., Fox, R., Huntley, B. & Thomas, C. D. (2006). Impacts of climate warming and habitat loss on extinctions at species' low latitude range boundaries. *Global Change Biology*, *12*, 1545-1553.

Frank, K. L., Mader, T. L., Harrington, J. A., Hahn, G. L. & Davis, M. S. (2001). Climate change effects on livestock production in the Great Plains. In: *Proceedings of the 6th International Livestock Environment Symposium* [R. R., Stowell, R., Bucklin, & R. W. Bottcher, (eds.)]. *American Society of Agricultural Engineering*, St. Joseph, Michigan, 35 1-358.

Fried, B. J., Domino, M. E. & Shadle, J. (2005). Use of mental health services after hurricane Floyd in North Carolina. *Psychiatric Services*, *56(11)*, 1367-1373.

Friedland, K. D., Reddin, D. G. McMenemy, J. R. & Drinkwater, K. F. (2003). Multidecadal trends in North American Atlantic salmon (*Salmo salar*) stocks and climate trends relevant to juvenile survival. *Canadian Journal of Fisheries and Aquatic Science*, *60*, 563-5 83.

Fuhlendorf, S. D., Briske, D. D. & Smeins, F. E. (2001). Herbaceous vegetation change in variable rangeland environments: The relative contribution of grazing and climatic variability. *Applied Vegetation Science*, *4*, 177-188.

Fung, I. Y., Doney, S. C., Lindsay, K. & John, J. (2005). Evolution of carbon sinks in a changing climate. *Proceedings of the National Academy of Sciences, 102(32),* 11201-11206.

Furness, B. W., Beach, M. J. & Roberts, J. M. (2000). Giardiasis surveillance – United States, 1992-1997. *Morbidity & Mortality Weekly Report, 49(07),* 1-13.

FWS, (2007). *12-Month Petition Finding and Proposed Rule to List the Polar Bear (Ursus maritimus) as Threatened through Its Range.* Fish and Wildlife Service, 50 CFR Part 17, RIN 1018-AV19.

Gabastou, J. M., Pesantes, C., Escalante, S., Narvaez, Y., Vela, E., Garcia, L., Zabala, D. & Yadon, Z. E. (2002). Characteristics of the cholera epidemic of 1998 in Ecuador during El Niño (in Spanish). *Revista Panamericana de Salud Pública, 12,* 157-164.

Gamache, I. & Payette, S. (2004). Height growth response of tree line black spruce to recent climate warming across the forest-tundra of eastern Canada. *Journal of Ecology, 92,* 835-845.

Gamble, J. L., et al., (2008) (in press): Analyses and effects of global change on human health and welfare and human systems. Ch. 1 in: *Analyses of the Effects of Global Change on Human Health and Welfare and Human Systems.* Synthesis and Assessment Product (SAP) 4.6 by the U.S. Climate Change Science Program and the Subcommittee on Global Change Research, Washington, DC.

Gan, J. (2004). Risk and damage of southern pine beetle outbreaks under global climate change. *Forest Ecology and Management, 191,* 61-71.

GAO, (2003). *Freshwater Supply, States' Views of How Federal Agencies Could Help Them Meet the Challenges of Expected Shortages.* Government Accountability Office, Washington, DC, USA.

Gaspard, K., Martinez, M., Zhang, Z. & Wu, Z. (2007). *Impact of Hurricane Katrina on Roadways in the New Orleans Area.* Technical Assistance Report No. 07-2TA, LTRC Pavement Research Group, Louisiana Department of Transportation and Development, Louisiana Transportation Research Center, 73.

Gerber, S., Joos, F. & Prentice, I. C. (2004). Sensitivity of a dynamic global vegetation model to climate and atmospheric CO_2. *Global Change Biology, 10,* 1223-1239.

Gibbs, J. P. & Breisch, A. R. (2001). Climate warming and calling phenology of frogs near Ithaca, New York, 1900-1999. *Conservation Biology, 15,* 1175-1178.

Gibson, K. E. (2006). *Mountain Pine Beetle Conditions in Whitebark Pine Stands in the Greater Yellowstone Ecosystem, 2006.* R1Pub06-03, Forest Health Protection Report, USDA Forest Service, Northern Region, Missoula, MT, USA.

Gille, S. (2008). Decadal-scale temperature trends in the Southern Hemisphere Ocean. *Journal of Climate* (in press).

Gillett, N. P., Weaver, A. J., Zwiers, F. W. & Flannigan, M. D. (2004). Detecting the effect of climate change on Canadian forest fires. *Geophysical Research Letters, 31,* L1 8211, doi: 10.1 029/2004GL020876.

Gitay, H., Brown, S., Easterling, W. & Jallow, B. (2001). Ecosystems and their goods and services. In: *Climate Change 2001: Impacts, Adaptation, and Vulnerability. Contribution of Working Group II to the Third Assessment Report of the Intergovernmental Panel on Climate Change* [J. J., McCarthy, O. F., Canziani, N. A., Leary, D. J. Dokken, & K. S. White (eds.)]. Cambridge University Press, Cambridge, 237-342.

Gnanadesikan, A., Dixon, K. W., Griffies, S. M., Balaji, V., Barreiro, M., Beesley, J. A., Cooke, W. F., Delworth, T. L., Gerdes, R., Harrison, M. J., Held, I. M., Hurlin, W. J., Lee, H. C., Liang, Z., Nong, G., Pacanowski, R. C., Rosati, A., Russell, J., Samuels, B. L., Song, Q., Spelman, M. J., Stouffer, R. J., Sweeney, C. O., Vecchi, G., Winton, M., Wittenberg, A. T., Zeng, F., Zhang, R. & Dunne, J. P. (2006) GFDL's CM2 global coupled climate models. Part II: The baseline ocean simulation. *Journal of Climate*, *19*, 675-697.

Gobbi, M., Fontaneto, D. & De Bernardi, F. (2006). Influence of climate changes on animal communities in space and time: the case of spider assemblages along an alpine glacier foreland. *Global Change Biology*, *12*, 1985-1992.

Gobron, N., Pinty, B., Melin, F., Taberner, M., Verstraete, M. M., Belward, A., Lavergne, T. & Widlowski, J. L. (2005). The state of vegetation in Europe following the 2003 drought. *International Journal of Remote Sensing*, *26*, 2013-2020.

Goodman, P. G., Dockery, D. W. & Clancy, L. (2004). Cause-specific mortality and the extended effects of particulate pollution and temperature exposure. *Environmental Health Perspectives*, *112(2)*, 179-185. [erratum appears in *Environmental Health Perspectives* 2004, *112(13)*, A729.]

Government of Andhra Pradesh, (2004). *Report of the State Level Committee on Heat Wave Conditions in Andhra Pradesh State*. Revenue (Disaster Management) Department, Government of Andhra Pradesh, Hyderabad, India, 67.

Grebmeier, J. M., Overland, J. E., Moore, S. E., Farley, E. V., Carmack, E. C., Cooper, L.W., Frey, K. E., Helle, J. H., McLaughlin, F. A. & McNutt, S. L. (2006). A major ecosystem shift in the Northern Bering Sea. *Science*, *311*, 1461-1464.

Greene, S. K., Ionides, E. L. & Wilson, M. L. (2006). Patterns of influenza-associated mortality among US elderly by geographic region and virus subtype, 1968-1998. *American Journal of Epidemiology*, *163(4)*, 316-326.

Greenough, G., McGeehin, M., Bernard, S. M., Trtanj, J., Riad, J. & Engleberg, D. (2001). The potential impacts of climate variability and change on health impacts of extreme weather events in the United States. *Environmental Health Perspectives*, *109*, S191 S198.

Gregg, W. W., Conkright, M. E., Ginoux, P., O'Reilly, J. E. & Casey, N. W. (2003). Ocean primary production and climate: Global decadal changes. *Geophysical Research Letters*, *30*, 1809, doi: 10.1 029/2003GL01 6889.

Grice, A. C. (2006). The impacts of invasive plant species on the biodiversity of Australian rangelands. *The Rangeland Journal*, *28*, 1-27.

Groisman, P.Y., Knight, R. W., Karl, T. R., Easterling, D. R., Sun, B. & Lawrimore, J. M. (2004). Contemporary changes of the hydrological cycle over the contiguous United States: Trends derived from in-situ observations. *Journal of Hydrometeorology*, *5*, 64-85.

Guenther, A. B., Zimmerman, P., Harley, P., Monson, R. & Fall, R. (1993). Isoprene and monoterpene emission rate variability - model evaluations and sensitivity analyses. *Journal of Geophysical Research*, *98(D7)*, 12609-12617.

Guha-Sapir, P., Hargitt, D. & Hoyois, H. (2004). *Thirty Years of Natural Disasters 1974–2003: The Numbers.* UCL Presses, Universitaires de Louvrain, Louvrain la Neuve, 188.

Guo, L. & Macdonald, R.W. (2006). Source and transport of terrigenous organic matter in the upper Yukon River: Evidence from isotope (δ13C, Δ14C, δ15N) composition of dissolved, colloidal, and particulate phases. *Global Biogeochemical Cycles*, *20*, GB2011, doi: 10.1 029/2005GB002593.

Gurdak, J. J., Hanson, R. T., McMahon, P. B., Bruce, B. W., McCray, J. E., Thyne, G. D. & Reedy, R. C. (2007). Climate variability controls on unsaturated water and chemical movement, High Plains Aquifer, USA. *Vadose Zone Journal*, *6*, 533-547.

Hahn, G. L. & Mader, T. L. (1997). Heat waves in relation to thermoregulation, feeding behavior and mortality of feedlot cattle. In: *Proceedings 5th International Livestock Environment Symposium*, American Society of Agricultural Engineers, St. Joseph, MI, USA, 563- 571.

Hahn, G. L., Chen, Y. R., Nienaber, J. A., Eigenberg, R. A. & Parkhurst, A. M. (1992). Characterizing animal stress through fractal analysis of thermoregulatory responses. *Journal of Thermal Biology*, *17*, 115-120.

Hahn, G. L., Mader, T. L., Gaughan, J. B., Hu, Q. & Nienaber, J. A. (1999). Heat waves and their impacts on feedlot cattle. In: *Proceedings 15th International Congress of Biometeorology and the International Congress on Urban Climatology*, Sydney, Australia.

Hahn, G. L., Mader, T., Spiers, D., Gaughan, J., Nienaber, J., Eigenberg, R., Brown-Brandl, T., Hu, Q., Griffin, D., Hungerford, L., Parkhurst, A., Leonard, M., Adams, W. & Adams, L. (2001). Heat wave impacts on feedlot cattle: Considerations for improved environmental management. In: *Proceedings 6th International Livestock Environment Symposium*, American Society of Agricultural Engineers, St. Joseph, MI, USA, 129-130.

Hajat, S., Kovats, R. & Lachowycz, K. (2007). Heat-related and cold-related deaths in England and Wales: who is at risk? *Occupational Environmental Medicine*, *64*, 93-100.

Hamlet, A. F., Mote, P. W., Clark, M. P. & Lettenmaier, D. P. (2005). Effects of temperature and precipitation variability on snowpack trends in the western U.S. *Journal of Climate*, *18*, 4545-4561.

L. Hamm, & M. J. F. Stive (eds.), 2002 Shore nourishment in Europe. *Coastal Engineering*, *47*, 79-263.

Hammer, R. B., Radelogg, V. C., Fried, J. S. & Stewart, S. I. (2007). Wildland-urban interface housing growth during the 1 990s in California, Oregon, and Washington. *International Journal of Wildland Fire*, *16*, 255-265.

Han, J. & Roads, J. O. (2004). U.S. climate sensitivity simulated with the NCEP Regional Spectral Model. *Climatic Change*, *62*, 115-154.

Hanesiak, J. M. & Wang, X. L. L. (2005). Adverse-weather trends in the Canadian Arctic. *Journal of Climate*, *18*, 3140-3156.

Hansen, E. M. & Bentz, B. (2003). Comparison of reproductive capacity among univoltine, semivoltine, and re-emerged parent spruce beetles (Coleoptera: Scolytidae). *Canadian Entomologist*, *135*, 697-7 12.

Hansen, E. M., Bentz, B. J. & Turner, D. L. (2001). Temperature-based model for predicting univoltine brood proportions in spruce beetle (Coleoptera: Scolytidae). *Canadian Entomologist*, *133*, 827-841.

Hansen, J. & Sato, M. (2004). Greenhouse gas growth rates. *Proceedings of the National Academy of Sciences*, *101*, 16109-16114, doi:10.1073/pnas.0406982101.

Hanson, R. T. & Dettinger, M. D. (2005). Ground water/surface water responses to global climate simulations, Santa Clara-Calleguas basin, Ventura, California. *Journal of the American Water Resources Association*, *41*, 517-536.

Hari Kumar, R., Venkaiah, K., Arlappa, N., Kumar, S., Brahmam, G. & Vijayaraghavan, K. (2005). Diet and nutritional status of the population in the severely drought affected areas of Gujarat. *Journal of Human Ecology*, *18*, 319-326.

Harper, C. W., Blair, J. M., Fay, P. A., Knapp, A. K., & Carlisle, J. D. (2005). Increased rainfall variability and reduced rainfall amount decreases soil CO_2 flux in a grassland ecosystem. *Global Change Biology*, *11*, 322-3 34.

Harrington, R., Fleming, R., Woiwood. I. P. (2001).. Climate change impacts on insect management and conservation in temperate regions: can they be predicted? *Agricultural and Forest Entomology*, *3*, 233-240.

Hartwig, R. (2006) *Hurricane Season of 2005, Impacts on U.S. P/C Markets, 2006 and Beyond.* Presentation to the Insurance Information Institute, March 2006, New York, NY, USA. <www.iii.org/media>.

Harvell, C. D., Kim, K., Burkholder, J. M., Colwell, R. R., Epstein, P. R., Grimes, D. J., Hofmann, E. E., Lipp, E. K. & coauthors, 1999 Emerging marine diseases – climate links and anthropogenic factors. *Science*, *285*, 1505-1510.

R., Hassan, R. Scholes, & N. Ash, (eds.), 2005 *Ecosystems and Human Wellbeing: Volume 1: Current State and Trends*. Island Press, Washington, DC, USA, 917.

Hassi, J., Rytkonen, M., Kotaniemi, J. & Rintamaki, H. (2005). Impacts of cold climate on human heat balance, performance and health in circumpolar areas. *International Journal of Circumpolar Health*, *64*, 459-467.

Hastings, A., Cuddington, K., Davies, K. F., Dugaw, C. J., Elmendorf, S., Freestone, A., Harrison, S., Holland, M., Lambrinos, J., Malvadkar, U., Melbourne, B. A., Moore, K., Taylor, C. & Thomson, D. (2005). The spatial spread of invasions: new developments in theory and evidence. *Ecology Letters*, *8*, 91-101.

Hatfield, J. L. & Prueger, J. H. (2004). Impact of changing precipitation patterns on water quality. *Journal Soil and Water Conservation*, *59*, 5 1-58.

Hatfield, J. L., Boote, K. J., Kimball, B. A., Wolfe, D. W., Ort, D. R., Izaurralde, R. C., Thomson, A. M., Morgan, J. A., Polley, H. W., Fay, P. A., Mader, T. L. & Hahn, G. L. (2008). Agriculture. In: *The Effects of Climate Change on Agriculture, Land Resources, Water Resources, and Biodiversity*. Synthesis and Assessment Product 4.3 by the U.S. Climate Change Science Program and the Subcommittee on Global Change Research, Washington, DC, USA.

Hauglustaine, D. A., Lathière, J., Szopa, S. & Folberth, G. (2005). Future tropospheric ozone simulated with a climate-chemistry-biosphere model. *Geophysical Research Letters*, *32*, L24807, doi: 10.1 029/2005GL024031.

Hayhoe, K. (2004). Emissions pathways, climate change, and impacts on California. *Proceedings of the National Academy of Sciences*, *101*, 12422.

Hayhoe, K., Cayan, D., Field, C., Frumhoff, P., Maurer, E., Miller, N., Moser, S., Schneider, S., Cahill, K., Cleland, E., Dale, L., Drapek, R., Hanemann, R. M., Kalkstein, L., Lenihan, J., Lunch, C., Neilson, R., Sheridan, S. & Verville, J. (2004). Emissions pathways, climate change, and impacts on California. *Proceedings of the National Academy of Sciences*, *101*, 12422-12427.

Healy, J. D. (2003). Excess winter mortality in Europe: a cross country analysis identifying key risk factors. *Journal of Epidemiology and Community Health*, *57(10)*, 784-789.

Hegerl, G. C., Crowley, T. J., Hyde, W. T. & Frame, D. J. (2006). Climate sensitivity constrained by temperature reconstructions over the past seven centuries. *Nature*, *440*, doi: 10.103 8/nature04679.

Hegerl, G. C., Zwiers, F. W., Braconnot, P., Gillett, N. P., Luo, Y., Marengo Orsini, J. A., Nicholls, N., Penner, J. E. & Stott, P. A. (2007). Understanding and attributing climate change. In: *Climate Change 2007: The Physical Science Basis. Contribution of Working Group I to the Fourth Assessment Report of the Intergovernmental Panel on Climate Change* [S., Solomon, D., Qin, M., Manning, Z., Chen, M., Marquis, K. B., Averyt, M. Tignor, & H. L. Miller, (eds.)]. Cambridge University Press, Cambridge, United Kingdom and New York, NY, USA, 663-745.

Heim, R. R. (2002). A review of twentieth-century drought indices used in the United States. *Bulletin of the American Meteorological Society*, *83*, 1149-1165.

Heinz Center, (2002). *The State of the Nation's Ecosystems*. Cambridge University Press, 288

Held, I. & B. Soden, 2006 Robust responses of the hydrological cycle to global warming. *Journal of Climate*, *19*, 5686-5699.

Hemon, D. & Jougla, E. (2004). La canicule du mois d'aout 2003 en France [The heatwave in France in August 2003]. *Rev. Epidemiol. Santé*, *52*, 3-5.

Hersteinsson, P. & Macdonald, D. W. (1992). Interspecific competition and the geographical distribution of red and arctic foxes, *Vulpes vulpes* and *Alopex lagopus*. *Oikos*, *64*, 505-515.

Hicke, J. A. & Lobell, D. B. (2004). Spatiotemporal patterns of cropland area and net primary production in the central United States estimated from USDA agricultural information. *Geophysical Research Letters*, *31*, L20502, doi: 10. 1029/2004GL020927.

Hill, J. K., Thomas, C. D., Fox, R., Telfer, M. G., Willis, S. G., Asher, J. & Huntley, B. (2002). Responses of butterflies to twentieth century climate warming: Implications for future ranges. *Proceedings of the Royal Society Series B*, *269(1505)*, 2163-2171.

Hoerling, M. & Kumar, A. (2003). The perfect ocean for drought. *Science*, *299*, 69 1-694.

Hofmann, E., Ford, S., Powell, E. & Klinck, J. (2001). Modeling studies of the effect of climate variability on MSX disease in eastern oyster (*Crassostrea virginica*) populations. *Hydrobiologia*, *460*, 195-212.

Hogg, E. H., Brandt, J. P. & Kochtubajda, B. (2002). Growth and dieback of aspen forests in northwestern Alberta, Canada in relation to climate and insects. *Canadian Journal of Forest Research*, *32*, 823-832.

Hogrefe, C. B., Lynn, B., Civerolo, K., Ku, J. Y., Rosenthal, J., Rosenzweig, C., Goldberg, R., Gaffin, S., Knowlton, K. & Kinney, P. L. (2004). Simulating changes in regional air pollution over the eastern United States due to changes in global and regional climate and emissions. *Journal of Geophysical Research*, *109*, D22301.

Holland, M. M., Bitz, C. M. & Tremblay, B. (2006). Future abrupt reductions in the summer Arctic sea ice. *Geophysical Research Letters*, *33*, L23 503, doi: 10. 1029/2006GL028024.

Holsten, E. H., Thier, R. W., Munson, A. S. & Gibson, K.E. (1999). *The Spruce Beetle. Forest Insect and Disease Leaflet 127*, USDA Forest Service.

Horie, T., Baker, J. T., Nakagawa, H., Matsui, T. & Kim, H. Y. (2000). Crop ecosystem responses to climatic change: Rice. Chapter 5. In: *Climate Change and Global Crop Productivity* [K. R. Reddy, & Hodges, H. F. (eds.)]. CAB International, New York, NY, USA, 81-106.

Houghton, R. A. (2003). Revised estimates of the annual net flux of carbon to the atmosphere from changes in land use and land management 1850-2000. *Tellus B*, *55*, 378-3 90.

Hoyt, K. S. & Gerhart, A. E. (2004). The San Diego County wildfires: perspectives of health care. *Disaster Management & Response*, *2*, 46-5 2.

Huang, Y. J. (2006). *The Impact of Climate Change on the Energy Use of the U.S. Residential and Commercial Building Sectors*. LBNL-60754, Lawrence Berkeley National Laboratory, Berkeley, CA, USA.

Hunter, J. R. (2006). Testimony before the Committee on Banking, Housing and Urban Affairs of the United States Senate Regarding Proposals to Reform the National Flood Insurance Program.

Huynen, M. & Menne, B. (2003). *Phenology and Human Health: Allergic Disorders*. Report of a WHO meeting in Rome, Italy, 16-17 January 2003. Health and Global Environmental Series, EUR/03/5036791, World Health Organization, Copenhagen, 64

Ibarra, F. A., Cox, J. R., Martin, M. H., Crowl, T. A., & Call, C. A. (1995). Predicting buffelgrass survival across a geographical and environmental gradient. *Journal of Range Management*, *48*, 53-59.

IFRC, (2002). *World Disaster Report 2002*. International Federation of Red Cross and Red Crescent Societies, Geneva, 240

Inouye, D. W. (2007). Impacts of global warming on pollinators. *Wings*, *30(2)*, 24-27.

Inouye, D. W. (2008). Effects of climate change on phenology, frost damage, and floral abundance of montane wildflowers. *Ecology*, *89(2)*, 353-362.

Inouye, D. W. & Wielgolaski, F. E. (2003). High altitude climates. In: *Phenology: an Integrative Environmental Science* [Schwartz, M. D. (ed.)]. Kluwer Academic Publishers, Dordrecht, The Netherlands, 195-2 14.

Inouye, D. W. Calder, W. A. & Waser, N. M. (1991). The effect of floral abundance on feeder censuses of hummingbird abundance. *Condor*, *93*, 279-285.

Inouye, D. W., Barr, B., Armitage, K. B. & Inouye, B. D. (2000). Climate change is affecting altitudinal migrants and hibernating species. *Proceedings of the National Academy of Sciences*, *97*, 1630-1633.

Inouye, D. W., Morales, M. & Dodge, G. (2002). Variation in timing and abundance of flowering by *Delphinium barbeyi* Huth (Ranunculaceae): the roles of snowpack, frost, and La Niña, in the context of climate change. *Oecologia*, *139*, 543-550.

Inouye, D. W., Saavedra, F. & Lee, W. (2003). Environmental influences on the phenology and abundance of flowering by *Androsace septentrionalis* L. (Primulaceae). *American Journal of Botany*, *90*, 905-9 10.

IPCC, (1990). *Climate Change: The IPCC Scientific Assessment*. Report prepared for the Intergovernmental Panel on Climate Change by Working Group I. Cambridge University Press, Cambridge, UK, 365.

IPCC, (1996). *Climate Change 1995: The Science of Climate Change. Contribution of Working Group I to the Second Assessment Report of the Intergovernmental Panel for Climate Change*. Cambridge University Press, Cambridge, UK, 572.

IPCC, (2000). *Special Report on Emissions Scenarios* (SRES). A Special Report of Working Group III of the Intergovernmental Panel on Climate Change [N. Nakicenovic et al. (eds.)]. Cambridge University Press, Cambridge, United Kingdom and New York, NY, USA.

IPCC, (2001). *Climate Change 2001: Impacts, Adaptation, and Vulnerability. Contribution of Working Group II to the Third Assessment Report of the Intergovernmental Panel on Climate Change* [J. J., McCarthy, O. F., Canziani, N. A., Leary, D. J. Dokken, & White, K. S. (eds.)]. Cambridge University Press, Cambridge, 967.

IPCC, (2007a). *Climate Change 2007: The Physical Science Basis. Contribution of Working Group I to the Fourth Assessment Report of the Intergovernmental Panel on Climate Change* [S., Solomon, D., Qin, M., Manning, Z., Chen, M., Marquis, K. B. Averyt, M. Tignor, & H. L. Miller, (eds.)]. Cambridge University Press, Cambridge, United Kingdom and New York, NY, USA, 996.

IPCC, (2007b). Summary for Policymakers. In: *Climate Change 2007: Impacts, Adaptation and Vulnerability. Contribution of Working Group II to the Fourth Assessment Report of the Intergovernmental Panel on Climate Change* [M. L., Parry, O. F., Canziani, J. P., Palutikof, P. J., van der Linden & C. E. Hanson, (eds.)]. Cambridge University Press, Cambridge, United Kingdom and New York, NY, USA.

IPCC, (2007c). Summary for Policymakers. In: *Climate Change 2007: Mitigation. Contribution of Working Group III to the Fourth Assessment Report of the Intergovernmental Panel on Climate Change* [B., Metz, O. R., Davidson, P. R., Bosch, R., Dave, L. A., Meyer, (eds.)]. Cambridge University Press, Cambridge, United Kingdom and New York, NY, USA.

IPCC, (2007d). Summary for Policymakers. In: *Climate Change 2007: The Physical Science Basis. Contribution of Working Group I to the Fourth Assessment Report of the Intergovernmental Panel on Climate Change* [S., Solomon, D., Qin, M., Manning, Z., Chen, M., Marquis, K. B., Averyt, M. Tignor, & H. L. Miller, (eds.)]. Cambridge University Press, Cambridge, United Kingdom and New York, NY, USA.

IPCC, (2007e). *Climate Change 2007: Synthesis Report. Contribution of Working Groups I, II, and II to the Fourth Assessment Report of the Intergovernmental Panel on Climate Change* [Core Writing Team, Pachauri, R. K. & A. Reisinger, (eds.)]. IPCC, Geneva, Switzerland, 104.

IUCN Polar Bear Specialist Group, (2006). Status of the polar bear. In: *Polar Bears: Proceedings of the 14th Working Meeting of the IUCN/SSC Polar Bear Specialist Group* [N. J. Lunn, & A. E. Derocher, (eds.).]. IUCN, Gland, Switzerland and Cambridge, UK, 35-55.

Izaurralde, R. C., Rosenberg, N. J., Brown, R. A. & Thomson, A. M. (2003). Integrated assessment of Hadley Centre climate change projections on water resources and agricultural productivity in the conterminous United States. II. Regional agricultural productivity in 2030 and 2095. *Agricultural and Forest Meteorology*, *117*, 97-122.

Izaurralde, R. C., Rosenberg, N. J., Thomson, A. M. & Brown, R. A. (2005). Climate change impacts for the conterminous USA: An integrated assessment. Part 6: Distribution and productivity of unmanaged ecosystems. *Climatic Change*, *69*, 107-126.

Jacobs, K., Garfin, G. & Lenart, M. (2005). More than just talk, connecting science and decision- making. *Environment*, *47*, 6-21.

Janda, J. M., Powers, C., Bryant, R. G. & Abbott, S. L. (1988). Clinical perspectives on the epidemiology and pathogenesis of clinically significant *Vibrio* spp. *Clinical Microbiology Reviews*, *1(3)*, 245-267.

Janetos, A. C., Shaw, R., Meyerson, L., Peterson, W., Inouye, D., Kelly, B. P. & Hansen, L. (2008). Biodiversity. In: *The Effects of Climate Change on Agriculture, Land Resources,*

Water Resources, and Biodiversity. Synthesis and Assessment Product 4.3 by the U.S. Climate Change Science Program and the Subcommittee on Global Change Research, Washington, DC, USA.

Jansen E., Overpeck, J., Briffa, K. R., Duplessy, J. C., Joos, F., Masson-Delmotte, V., Olago, D., Otto-Bliesner, B.,. Peltier, W. R., Rahmstorf, S., Ramesh, R., Raynaud, D., Rind, D., Solomina, O., Villalba, R. & Zhang, D. (2007). Palaeoclimate. In: *Climate Change 2007: The Physical Science Basis. Contribution of Working Group I to the Fourth Assessment Report of the Intergovernmental Panel on Climate Change* [S., Solomon, D., Qin, M., Manning, Z., Chen, M., Marquis, K.B., Averyt, M., Tignor, & Miller, H. L. (eds.)]. Cambridge University Press, Cambridge, United Kingdom and New York, NY, USA, 433-497.

Jastrow, J. D., Miller, R. M., Matamala, R., Norby, R. J., Boutton, T. W., Rice, C. W. & Owensby, C. E. (2005). Elevated atmospheric carbon dioxide increases soil carbon. *Global Change Biology*, *11*, 2057-2064.

Jayawickreme, D. H. & Hyndman, D. W. (2007). Evaluating the influence of land cover on seasonal water budgets using NEXRAD rainfall and streamflow data. *Water Resources Research*, *43*, W02408, doi: 10.1 029/2005WR004460.

Johnson, H., Kovats, R. S., McGregor, G. R., Stedman, J. R., Gibbs, M., Walton, H., Cook, L. & Black, E. (2005). The impact of the 2003 heatwave on mortality and hospital admissions in England. *Health Statistics Quarterly*, *25*, 6-12.

Johnson, T. R. (1998). *Climate Change and Sierra Nevada Snowpack*. M.S. Thesis. University of California, Santa Barbara, CA, USA.

Johnstone, J. F. & Chapin III, F. S. (2003). Non-equilibrium succession dynamics indicate continued northern migration of lodgepole pine. *Global Change Biology*, *9*, 1401-1409.

Jonzen, N., Lindén, A., Ergon, T., Knudsen, E., Vik, J. O., Rubolini, D., Piacentini, D., Brinch, C., Spina, F., Karlsson, L., Stervander, M., Andersson, A., Waldenström, J., Lehikoinen, A., Edvardsen, E., Solvang, R. & Stenseth, N. C. (2006). Rapid advance of spring arrival dates in long-distance migratory birds. *Science*, *312*, 1959-1961.

Joos, F., Prentice, I. C. & House, J. I. (2002). Growth enhancement due to global atmospheric change as predicted by terrestrial ecosystem models: consistent with US forest inventory data. *Global Change Biology*, *8(4)*, 299-303.

Julius, S., West, J., Baron, J., Joyce, L., Kareiva, P., Keller, B., Palmer, M., Peterson, C., Scott, J. M. (2008). *Preliminary Review of Adaptation Options for Climate-Sensitive Ecosystems and Resources*. Synthesis and Assessment Product 4.4 by the U.S. Climate Change Science Program and the Subcommittee on Global Change Research, Washington, DC, USA, 784.

Kafalenos, R. S., Leonard, K. J., Beagan, D. M., Burkett, V. R., Keim, B. D., Meyers, A., Hunt, D. T., Hyman, R. C., Maynard, M. K., Fritsche, B., Henk, R. H., Seymour, E. J., Olson, L. E., Potter, J. R. & Savonis, M. J. (2008). What are the implications of climate change and variability for Gulf Coast transportation? In: *Impacts of Climate Change and Variability on Transportation Systems and Infrastructure: Gulf Coast Study, Phase I*. Synthesis and Assessment Product 4.7 by the U.S. Climate Change Science Program and the Subcommittee on Global Change Research [M. J., Savonis, V. R. Burkett, & J. R. Potter, (eds.)], Washington, DC, USA.

Karl, T. R., Hassol, S.J., Miller, C.D.& Murray W.L. (eds.), (2006). *Temperature Trends in the Lower Atmosphere: Steps for Understanding and Reconciling Differences*. Synthesis and Assessment Product 1.1 by the U.S. Climate Change Science Program and the Subcommittee on Global Change Research, Washington, DC, USA.

Karl, T. R., Meehl, G. A., Miller, C. D., Hassol, S. J., Waple, A. M. & Murray, W. L. (eds.), 2008 (in press): *Weather and Climate Extremes in a Changing Climate; Regions of Focus: North America, Hawaii, Caribbean, and U.S. Pacific Islands*. Synthesis and Assessment Product 3.3 by the U.S. Climate Change Science Program and the Subcommittee on Global Change Research. Washington, DC, USA, 164.

Karoly, D. J. & Wu, Q. (2005). Detection of regional surface temperature trends. *Journal of Climate*, *18*, 4337-4343.

Kashian, D. M., Romme, W. H., Tinker, D. B., Turner, M. G. & Ryan, M. G. (2006). Carbon storage on landscapes with stand-replacing fires. *BioScience*, *56*, 598-606.

Kasischke, E. S. & Turetsky, M. R. (2006). Recent changes in the fire regime across the North American boreal region: Spatial and temporal patterns of burning across Canada and Alaska. *Geophysical Research Letters*, *33*, doi: 10.1 029/2006GL025677.

Kates, R. & Wilbanks, T. (2003). Making the global local: responding to climate change concerns from the bottom up. *Environment*, *45*, 12-23.

Katsumata, T., Hosea, D., Wasito, E. B., Kohno, S., Hara, K., Soeparto, P. & Ranuh, I. G. (1998). Cryptosporidiosis in Indonesia: a hospital-based study and a community-based survey. *American Journal of Tropical Medicine and Hygiene*, *59*, 628-632.

Kennish, M. J. (2001). Coastal saltmarsh systems in the US: a review of anthropogenic impacts. *Journal of Coastal Research*, *17*, 731-748.

Kenny, A. & Mollmann, C. (2006). Towards integrated ecosystem assessments for the North and Baltic Seas: synthesizing GLOBEC research. *GLOBEC International Newsletter*, *12(2)*, 64-65.

Kettlewell, P. S., Easey, J., Stephenson, D. B. & Poulton, P. R. (2006). Soil moisture mediates association between the winter North Atlantic Oscillation and summer growth in the Park Grass Experiment. *Proceedings of the Royal Society of London B*, *273*, 1149-1154.

Key, J. R. & Chan, A. C. K. (1999). Multidecadal global and regional trends in 1000 mb and 500 mb cyclone frequencies. *Geophysical Research Letters*, *26*, 2053-2056.

Kiehl, J. T. & Shields, C. A. (2005). Climate simulation of the latest Permian: Implications for mass extinction. *Geology*, *33*, 757-760.

Kirshen, P. H. (2002). Potential impacts of global warming in eastern Massachusetts. *Journal of Water Resources Planning and Management*, *128(3)*, 216-226.

Kirshen, P., Ruth, M. & Anderson, W. (2006). Climate's long-term impacts on urban infrastructures and services: the case ofMetro Boston. *Regional Climate Change and Variability: Impacts and Responses*, M. Ruth, K. Donaghy and P.H. Kirshen, Eds., Edward Elgar Publishers, Cheltenham, 190-252.

Kirshen, P., Ruth, M. & Anderson, W. forthcoming: Interdependencies of urban climate change impacts and adaptation strategies: A case study of metropolitan Boston. *Climatic Change*.

Kitzberger, T., Swetnam, T. W. & Veblen, T. T. (2001). Inter-hemispheric synchrony of forest fires and the El Niño-Southern Oscillation. *Global Ecology and Biogeography*, *10*, 315- 326.

Klinenberg, E. (2002). *Heat Wave: A Social Autopsy of Disaster in Chicago.* The University of Chicago Press, Chicago.

Knapp, P. A (1998). Spatio-temporal patterns of large grassland fires in the Intermountain West, USA. *Global Ecology and Biogeography Letters*, 7, 259-272.

Knowles, N., Dettinger, M. D., & Cayan, D. R. (2006). Trends in snowfall versus rainfall for the western United States, 1949-2004. *Journal of Climate*, *19*, 4545-4559.

Kovats, R. S., Hajat, S. S, & Wilkinson, P. (2004a). Contrasting patterns of mortality and hospital admissions during hot weather and heat waves in Greater London, UK. *Occupational & Environmental Medicine*, *61(11)*, 893-898.

Krieger, D. J. (2001). *The Economic Value of Forest Ecosystem Services: A Review*. The Wilderness Society, Washington, DC, USA, 31.

Kron, W. & Berz, G. (2007). Flood disasters and climate change: trends and options – a (re-) insurer's view. In: *Global Change: Enough Water for All?* [J. L., Lozán, H., Graßl, P., Hupfer, L. Menzel, & C. D. Schönwiese, (eds.)]. Wissenschaftliche Auswertungen/GEO, Hamburg, 268-273.

Kueppers, L. M., Snyder, M. A. & Sloan, L. C. (2007). Irrigation cooling effect: Regional climate forcing by land-use change. *Geophysical Research Letters*, *34*, L03703, doi: 10.1 029/2006GL028679.

Kundzewicz, Z. W., Mata, L. J., Arnell, N. W., Döll, P., Kabat, P., Jiménez, B., Miller, K. A., Oki, T., Sen, Z. & Shiklomanov, I. A. (2007). Freshwater resources and their management. In: *Climate Change 2007: Impacts, Adaptation and Vulnerability. Contribution of Working Group II to the Fourth Assessment Report of the Intergovernmental Panel on Climate Change* [M. L., Parry, O. F., Canziani, J. P., Palutikof, P. J. van der Linden, & C. E. Hanson, (eds.)]. Cambridge University Press, Cambridge, United Kingdom and New York, NY, USA, 173-210.

Kunkel, K. E., Easterling, D. R., Redmond, K. & Hubbard, K. (2003). Temporal variations of extreme precipitation events in the United States: 1895–2000. *Geophysical Research Letters*, *30*, 1900, doi: 10.1 029/2003GL0 18052.

Kunkel, K. E., Novak, R. J., Lampman, R. L. & Gu, W. (2006). Modeling the impact of variable climatic factors on the crossover of *Culex restauns* and *Culex pipiens* (Diptera: Culicidae), vectors of West Nile virus in Illinois. *American Journal of Tropical Medicine & Hygiene*, *74*, 168-173.

Kupfer, J. A. & Miller, J. D. (2005). Wildfire effects and post-fire responses of an invasive mesquite population: the interactive importance of grazing and non-native herbaceous species invasion. *Journal of Biogeography*, *32*, 453-466.

Larsen, P., Goldsmith, O. S., Smith, O. & Wilson, M. (2007). *A Probabilistic Model to Estimate the Value of Alaska Public Infrastructure at Risk to Climate Change*. ISER Working Paper, Institute of Social and Economic Research, University of Alaska, Anchorage, AK, USA.

Lawlor, D. W. & Mitchell, R. A. C. (2000). Crop ecosystem responses to climatic change: Wheat. Chapter 4. In: *Climate Change and Global Crop Productivity* [K. R. Reddy, & H. F. Hodges, (eds.)]. CAB International, New York, NY, 57-80.

Leakey, A. D. B., Uribelarrea, M., Ainsworth, E. A., Naidu, S. L., Rogers, A., Ort, D. R. & Long, S. P. (2006). Photosynthesis, productivity, and yield of maize are not affected by open-air elevation of CO_2 concentration in the absence of drought. *Plant Physiology*, *140*, 779- 790.

Lee, S. H., Levy, D. A., Craun, G. F. Beach, M. J. & Calderon, R. L. (2002). Surveillance for waterborne disease outbreaks – United States, 1999 – 2000. *Morbidity & Mortality Weekly Report, 51(08)*, 1-28.

Leff, B., Ramankutty, N. & Foley, J. A. (2004). Geographic distribution of major crops across the world. *Global Biogeochemical Cycles, 18*, GB 1009.

Lehodey, P., Chai, F. & Hampton, J. (2003). Modelling climate-related variability of tuna populations from a coupled ocean biogeochemical-populations dynamics model. *Fisheries Oceanography, 12*, 483-494.

Lemke, P., Ren, J., Alley, R. B., Allison, I., Carrasco, J., Flato, G., Fujii, Y., Kaser, G., Mote, P., Thomas, R. H. & Zhang, T. (2007). Observations: changes in snow, ice and frozen ground. In: *Climate Change 2007: The Physical Science Basis. Contribution of Working Group I to the Fourth Assessment Report of the Intergovernmental Panel on Climate Change* [S., Solomon, D., Qin, M., Manning, Z., Chen, M., Marquis, K. B., Averyt, M. Tignor, & H. L. Miller, (eds.)]. Cambridge University Press, Cambridge, United Kingdom and New York, NY, USA, 338-383.

Lemmen, D. S. & F. J. Warren, (eds.), (2004). *Climate Change Impacts and Adaptation: A Canadian Perspective.* Climate Change Impacts and Adaptation Directorate, Natural Resources Canada Ottawa, Ontario, 201 <environment>.

Lenton, R. (2004). Water and climate variability: development impacts and coping strategies. *Water Science and Technology, 49(7)*, 17-24.

Lettenmaier, D. P. (2003). The role of climate in water resources planning and management. In: *Water: Science, Policy, and Management.* [R., Lawford, D., Fort, H. Hartmann, & S. Eden, (eds.)]. Water Resources Monograph 16, American Geophysical Union, Washington, DC, USA, 247-266.

Lettenmaier, D. P., Major, D., Poff, L. & Running, S. (2008). Water resources. In: *The Effects of Climate Change on Agriculture, Land Resources, Water Resources, and Biodiversity.* Synthesis and Assessment Product 4.3 [P., Backlund, A., Janetos, & D. Schimel, (eds.)] by the U.S. Climate Change Science Program and the Subcommittee on Global Change Research, Washington, DC, USA.

Levitus, S., Antonov, J. L., Wang, J., Delworth, T. L., Dixon, K. W. & Broccoli, A. J. (2001). Anthropogenic warming of Earth's climate system. *Science, 292*, 267-270.

Leung, L. R., Qian, Y., Bian, X., Washington, W. M., Han, J. & Roads, J. O. (2004). Mid-century ensemble regional climate change scenarios for the western United States. *Climatic Change, 62*, 75-113.

Li, S. G., Asanuma, J., Eugster, W., Kotani, A., Liu, J. J., Urano, T., Oikawa, T., Davaa, G., Oyunbaatar, D. & Sugita, M. (2005). Net ecosystem carbon dioxide exchange over grazed steppe in central Mongolia. *Global Change Biology, 11*, 1941-1955.

Liang, J. I., Dziuban, E. J., Craun, G. F., Hill, V., Moore, M. R., Gelting, R. J., Calderon, R. L., Beach, M. J. & Roy, S. L. (2006). Surveillance for waterborne disease and outbreaks associated with drinking water and water not intended for drinking – United States, 2003 – 2004. *Morbidity & Mortality Weekly Report, 55(12)*, 32-65.

Liao, H., Chen, W. T. & Seinfeld, J. H. (2006). Role of climate change in global predictions of future tropospheric ozone and aerosols, *Journal of Geophysical Research, 111*, D12304, doi: 10.1 029/2005JD006852.

Linder, K. P. & Inglis, M. R. (1989). *The Potential Impact of Climate Change on Electric Utilities, Regional and National Estimates*. U.S. Environmental Protection Agency, Washington, DC, USA.

Lipp, E. K. & Rose, J. B. (1997). The role of seafood in foodborne diseases in the United States of America. *Revue Scientifique et Technique (Office International des Epizooties), 16(2)*, 620-640.

Lipp, E. K., Huq, A. & Colwell, R. R. (2002). Effects of global climate in infectious disease: the cholera model. *Clinical Microbiology Reviews, 15(4)*, 757-770.

Liu, L., King, J. S. & Giardina, C. P. (2005). Effects of elevated concentrations of atmospheric CO_2 and tropospheric O_3 on leaf litter production and chemistry in trembling aspen and paper birch communities. *Tree Physiology, 25*, 1511-1522.

Loaiciga, H. A., Maidment, D. R. & Valdes, J. B. (2000). Climate change impacts on a regional karst aquifer, Texas, USA. *Journal of Hydrology, 227*, 173-194.

Lobell, D.B. & G.P. Asner, 2003: Climate and management contributions to recent trends in U.S. agricultural yields. *Science, 299*, 1032.

Lobell, D. B. & Field, C. B. (2007). Global scale climate-crop yield relationships and the impact of recent warming. *Environmental Research Letters, 2*, 1-7.

Lobell, D. B., Hicke, J. A., Asner, G. P., Field, C. B., Tucker, C. J. & Los, S. O. (2002). Satellite estimates of productivity and light use efficiency in United States agriculture, 1982-98. *Global Change Biology, 8*, 722-735.

Lobell, D. B., Cahill, K. N. & Field, C. B. (2006). Historical effects of temperature and precipitation on California crop yields. *Climatic Change, 81*, 187-203.

Lofgren, B. M., Quinn, F. H., Clites, A. H., Assel, R. A., Eberhardt, A. J. & Luukkonen, C. L. (2002). Evaluation of potential impacts on Great Lakes water resources based on climate scenarios of two GCMs. *Journal of Great Lakes Research, 28*, 537-554.

Logan, J. A. & Powell, J. A. (2001). Ghost forests, global warming and the mountain pine beetle (Coleoptera: Scolytidae). *American Entomologist, 47*, 160-173.

Logan, J. A. & Powell, J. A. (2007). Ecological consequences of forest-insect disturbance altered by climate change. In: *Climate Change in Western North America. Evidence and Environmental Effects* [F. H. Wagner, (ed.)]. University of Utah Press, Salt Lake City, Utah (in press).

Logan, J. A., Régnière, J. & Powell, J. A. (2003). Assessing the impacts of global warming on forest pest dynamics. *Frontiers in Ecology and Environment, 1(3)*, 130-137.

London Climate Change Partnership, (2004). *London's Warming: A Climate Change Impacts in London Evaluation Study*, London, 293

Lonergan, S., DiFrancesco, R. & Woo, M. (1993). Climate change and transportation in northern Canada: An integrated impact assessment. *Climatic Change, 24*, 33 1-351.

Lotze, H. K., Lenihan, H. S., Bourque, B. J., Bradbury, R. H., Cooke, R. G., Kay, M. C., Kidwell, S. M., Kirby, M. X., Peterson, C. H. & Jackson, J. B. C. (2006). Depletion, degradation and recovery potential of estuaries and coastal seas. *Science, 312*, 1806-1809.

Loveland, J. E. & Brown, G. Z. (1990). *Impacts of Climate Change on the Energy Performance of Buildings in the United States*. OTA/UW/UO, Contract J3-4825.0, U.S. Congress, Office of Technology Assessment, Washington, DC, USA.

Loya, W. M., Pregitzer, K. S., Karberg, N. J., King, J. S. & Giardina, C. P. (2003). Reduction of soil carbon formation by tropospheric ozone under increased carbon dioxide levels. *Nature*, *425*, 705-707.

Lucht, W., Prentice, I. C., Myneni, R. B., Sitch, S., Friedlingstein, P., Cramer, W., Bousquet, P., Buermann, W. & Smith, B. (2002). Climatic control of the high-latitude vegetation greening trend and Pinatubo effect. *Science*, *296(5573)*, 1687-1689.

Luo, Y., Hui, D. & Zhang, D. (2006). Elevated carbon dioxide stimulates net accumulations of carbon and nitrogen in terrestrial ecosystems: A meta-analysis. *Ecology*, *87*, 53-63.

Lynch, M., Painter, J., Woodruff, R. & Braden, C. (2006). Surveillance for foodborne-disease outbreaks – United States, 1998-2002. *Morbidity & Mortality Weekly Reports*, *55(10)*, 1- 42.

Mader, T. L. (2003). Environmental stress in confined beef cattle. *Journal of Animal Science*, *81*, (electronic suppl. 2), 110-119.

Mader, T. L. & Davis, M. S. (2004). Effect of management strategies on reducing heat stress of feedlot cattle: feed and water intake. *Journal of Animal Science*, *82*, 3077-3087.

Mader, T. L., Fell, L. R. & McPhee, M. J. (1997). Behavior response of non-Brahman cattle to shade in commercial feedlots. *Proceeding 6th International Livestock Environment Symposium*. American Society of Agricultural Engineers, St. Joseph, MI, USA, 795-802.

Mader, T. L., Holt, S. M., Hahn, G. L., Davis, M. S. & Spiers, D. E. (2002). Feeding strategies for managing heat load in feedlot cattle. *Journal of Animal Science*, *80*, 2373-23 82.

Magnuson, J. J., Robertson, D. M., Benson, B. J., Wynne, R. H., Livingstone, D. M., Arai, T., Assel, R. A., Barry, R. G., Card, V., Kuusisto, E., Granin, N. C., Prowse, T. D., Stewart, K. M. & Vuglinski, V. S. (2000). Historical trends in lake and river ice cover in the Northern Hemisphere. *Science*, *289*, 1743-1746.

Malmström, C. M. & Raffa, K. F. (2000). Biotic disturbance agents in the boreal forest: considerations for vegetation change models. *Global Change Biology*, *6*, 35-48.

Mann, M. E. & Emanuel, K. A. (2006). Atlantic hurricane trends linked to climate change. *Eos: Transactions of the American Geophysical Union*, *87*, 23 3-244.

Mansur, E. T., Mendelsohn, R. & Morrison, W. (2005). *A Discrete-Continuous Choice Model of Climate Change Impacts on Energy*. SSRN Yale SOM Working Paper No. ES-43 (abstract number 738544), Yale School of Management, New Haven, CT, USA.

Mantua, N. J., Hare, R. H., Zhang, Y., Wallace, J. M. & Francis, R. C. (1997). A Pacific interdecadal climate oscillation with impacts on salmon production. *Bulletin of the American Meteorological Society*, *78*, 1069-1079.

Marshall, P. A. & Schuttenberg, H. Z. (2006). *A Reef Manager's Guide to Coral Bleaching*. Great Barrier Reef Marine Park Authority, Townsville, Australia, 163

Martinez-Navarro, F., Simon-Soria, F. & Lopez-Abente, G. (2004). Valoracion del impacto de la ola de calor del verano de 2003 sobre la mortalidad [Evaluation of the impact of the heatwave in the summer of 2003 on mortality]. *Gaceta Sanitaria*, *18*, 250-258.

Mason, S. J. (2004). Simulating climate over western North America using stochastic weather generators. *Climatic Change*, *62*, 155-187.

Mau-Crimmins, T., Schussman, H. R. & Geiger, E. L. (2006). Can the invaded range of a species be predicted sufficiently using only native-range data?: Lehmann lovegrass

(*Eragrostis lehmanniana*) in the southwestern United States. *Ecological Modelling*, *193*, 736-746.

Mauget, S.A., 2003: Multidecadal regime shifts in U.S. streamflow, precipitation, and temperature at the end of the twentieth century. *Journal of Climate*, *16*, 3905-39 16.

Maulbetsch, J. S. & DiFilippo, M. N. (2006). *Cost and Value of Water Use at Combined Cycle Power Plants*, California Energy Commission, PIER Energy-Related Environmental Research, CEC-500-2006-034, April 2006.

McBeath, J. (2003). Institutional responses to climate change: The case of the Alaska transportation system. *Mitigation and Adaptation Strategies for Global Change*, *8*, 3-28.

McCabe, G. J. & Bunnell, J. E. (2004). Precipitation and the occurrence of Lyme disease in the northeastern United States. *Vector Borne & Zoonotic Diseases*, *4(2)*, 143-148.

McCabe, G. J., Clark, M. P. & Serreze, M. C. (2001). Trends in Northern Hemisphere surface cyclone frequency and intensity. *Journal of Climate*, *14*, 2763-2768.

McCabe, G. J., Palecki, M. & Betancourt, J. L. (2004). Pacific and Atlantic Ocean influences on multi-decadal drought frequency in the United States. *Proceedings of the National Academy of Sciences*, *101*, 4136-4141.

McCarthy, H. R., Oren, R., Kim, H. S., Johnsen, K. H., Maier, C., Pritchard, S. G. & Davis, M. A. (2006). Interaction of ice storms and management practices on current carbon sequestration in forests with potential mitigation under future CO_2 atmosphere. *Journal of Geophysical Research*, *111*, 1-10, D15103, doi:10.1029/2005JD006428.

McCarthy, K., Peterson, D., Sastry, N. & Pollard, M. (2006). *The Repopulation of New Orleans after Hurricane Katrina*. Rand, Santa Monica, CA, USA.

McGeehin, M. A. & Mirabelli, M. (2001). The potential impacts of climate variability and change on temperature-related morbidity and mortality in the United States. *Environmental Health Perspectives*, *109(2)*, 185-189.

McGowan, J. E., Cayan, D. R. & Dorman, L. M. (1998). Climate-ocean variability and ecosystem response in the northeast Pacific. *Science*, *281*, 210-217.

McKenzie, D., Gedalof, Z., Peterson, D. L. & Mote, P. (2004). Climatic change, wildfire and conservation. *Conserv. Biol.*, *18*, 890-902.

McKenzie, D., Hessland, A. E. & Peterson, D. L. (2001). Recent growth of conifer species of western North America: Assessing spatial patterns of radial growth trends. *Canadian Journal of Forest Research*, *31*, 526-538.

McLaughlin, J. F., Hellmann, J. J., Boggs, C. L. & Ehrlich, P. R. (2002). Climate change hastens population extinction. *Proceedings of the National Academy of Sciences*, *99*, 6070-6074. McNulty, S.G., 2002: Hurricane impacts on U.S. forest carbon sequestration. *Environmental Pollution*, *11*, S17-S24.

McPhaden, M. J. & Zhang, D. (2002). Slowdown of the meridional overturning circulation in the upper Pacific Ocean. *Nature*, *415*, 603-608.

MEA, (2005a). *Ecosystems and Human Well-being: Biodiversity Synthesis.* Millennium Ecosystem Assessment, World Resources Institute, Washington, DC, USA, 100 MEA, 2005b: *Ecosystems and Human Well-being: Synthesis.* Millennium Ecosystem Assessment, Island Press, Washington, DC, USA, 137.

Mead, P. S., Slutsker, L., Dietz, V., McCaig, L. F., Bresee, J. S., Shapiro, C., Griffin, P. M. & Tauxe, R. V. (1999). Food-related illness and death in the United States. *Emerging Infectious Diseases*, *5(5)*, 607-625.

Mearns, L. O., Carbone, G., Doherty, R. M., Tsvetsinskaya, E. A., McCarl, B. A., Adams, R. M. & McDaniel, L. (2003). The uncertainty due to spatial scale of climate scenarios in integrated assessments: An example from U.S. agriculture. *Integrated Assessment, 4*, 225-23 5.

Medina-Ramon, M., Zanobetti, A., Cavanagh, D. P. & Schwartz, J. (2006). Extreme temperatures and mortality: assessing effect modification by personal characteristics and specific cause of death in a multi-city case-only analysis. *Environmental Health Perspectives*, *114(9)*, 133 1-1336.

Meehl, G. A. & Tebaldi, C. (2004). More intense, more frequent and longer lasting heat waves in the 21st century. *Nature*, *305*, 994-997.

Meehl, G. A., Washington, W. M., Ammann, C., Arblaster, J. M., Wigley, T. M. L. & Tebaldi, C. (2004a). Combinations of natural and anthropogenic forcings and 20th century climate. *Journal of Climate*, *17*, 3721-3727.

Meehl, G. A., Covey, C. & Latif, M. (2004b). Soliciting participation in climate model analyses leading to IPCC Fourth Assessment Report. *Eos*, *Transactions*, *American Geophysical Union*, *85(29)*, 274, doi: 10.1 029/2004EO290002.

Meehl, G. A., Covey, C., McAvaney, B., Latif, M. & Stouffer, R. J. (2005). Overview of the Coupled Model Intercomparison Project. *Bulletin of the American Meteorological Society*, *86*, 89-93.

Meehl, G. A., Stocker, T. F., Collins, W. D., Friedlingstein, P., Gaye, A. T., Gregory, J. M., Kitoh, A., Knutti, R., Murphy, J. M., Noda, A., Raper, S. C. B., Watterson, I. G., Weaver, A. J. & Zhao, Z. C. (2007). Global climate projections. In: *Climate Change 2007: The Physical Science Basis. Contribution of Working Group I to the Fourth Assessment Report of the Intergovernmental Panel on Climate Change* [S., Solomon, D., Qin, M., Manning, Z., Chen, M., Marquis, K. B., Averyt, M. Tignor, & H. L. Miller, (eds.)]. Cambridge University Press, Cambridge, United Kingdom and New York, NY, USA, 747-846.

Meites, E., Jay, M. T., Deresinski, S., Shieh, W. J., Zaki, S. R., Tomkins, L. & Smith, D. S. (2004). Reemerging leptospirosis, California. *Emerging Infectious Diseases*, *10(3)*, 406-412.

Melack, J. M., Hess, L. L. Gastil, M., Forsberg, B. R., Hamilton, S. K., Lima, I. B. T. & Novo, E. M. L. M. (2004). Regionalization of methane emissions in the Amazon Basin with microwave remote sensing. *Global Change Biology*, *10*, 530-544.

R. Mendelsohn, (ed.), (2001). *Global Warming and the American Economy: A Regional Assessment of Climate Change*. Edward Elgar Publishing, Cheltenham, UK.

Mendelsohn, R. (2003). A California model of climate change impacts on timber markets. Appendix XII. In: *Global Climate Change and California: Potential Implications for Ecosystems, Health, and the Economy*. Publication 500-03-058CF, California Energy Commission, Sacramento, CA, USA, 24 <http://www.energy -10- 3 1_500-03-058CF_A1 2.PDF>.

Menne, B. & Bertollini, R. (2000). The health impacts of desertification and drought. *Down to Earth*, *14*, 4-6.

Menzel, A., Jakobi, G., Ahas, R., Scheifinger, H. & Estrella, N. (2003). Variations of the climatological growing season (195 1-2000) in Germany compared with other countries. *International Journal of Climatology*, *23(7)*, 793-812.

Meyer, J. L., Sale, M. J., Mulholland, P. J. & Poff, N. L. (1999). Impacts of climate change on aquatic ecosystem functioning and health, *Journal of the American Water Resources Association, 35(6)*, 1373-13 86.

Mickley, L. J., Jacob, D. J., Field, B. D. & Rind, D. (2004). Effects of future climate change on regional air pollution episodes in the United States. *Geophysical Research Letters, 30*, L24103.

Middleton, K. L., Willner, J. & Simmons, K. M. (2002). Natural disasters and posttraumatic stress disorder symptom complex: evidence from the Oklahoma tornado outbreak. *International Journal of Stress Management, 9(3)*, 229-23 6.

Millerd, F. (2005). The economic impact of climate change on Canadian commercial navigation on the Great Lakes. *Canadian Water Resources Journal, 30*, 269-281.

Mills, B., Tighe, S., Andrey, J., Huen, K. & Parm, S. (2006). Climate change and the performance of pavement infrastructure in southern Canada, context and case study. In: *Proceedings of the Engineering Institute of Canada (EIC) Climate Change Technology Conference*, 9-12 May 2005, Ottawa, Ontario, Canada.

Milly, P. C. D., Dunne, K. A. & Vecchia, A. V. (2005). Global pattern of trends in streamflow and water availability in a changing climate. *Nature, 438*, 347-350.

Milly, P. C. D., Betancourt, J., Falkenmark, M., Hirsch, R. M., Kundzewicz, Z., Lettenmaier, D. P. & Stouffer, R. J. (2008). Stationarity is dead: Whither water management. *Science, 319*, 573-574.

Mimura, N., Nurse, L., McLean, R. F., Agard, J., Briguglio, L., Lefale, P., Payet, R. & Sem, G. (2007). Small islands. In: *Climate Change 2007: Impacts, Adaptation and Vulnerability. Contribution of Working Group II to the Fourth Assessment Report of the Intergovernmental Panel on Climate Change* [Parry, M.L., O.F. Canziani, J.P. Palutikof, P.J. van der Linden and C.E. Hanson (eds.)]. Cambridge University Press, Cambridge, United Kingdom and New York, NY, USA, 687-716.

Mirza, M. M. Q. (2003). Three recent extreme floods in Bangladesh: a hydrometeorological analysis. *Natural Hazards, 28*, 35-64.

Mirza, M. M. Q. (2004). *Climate Change and the Canadian Energy Sector. Report on Vulnerability and Adaptation.* Adaptation and Impacts Research Group, Atmospheric Climate Science Directorate, Meteorological Service of Canada, Downsview, Ontario, Canada, 52.

Mitchell, S. W. & Csillag, F. (2001). Assessing the stability and uncertainty of predicted vegetation growth under climatic variability: northern mixed grass prairie. *Ecological Modelling, 139*, 101-121.

Mohanti, M. (2000). Unprecedented supercyclone in the Orissa Coast of the Bay of Bengal, India. *Cogeoenvironment Newsletter. Commission on Geological Sciences for Environmental Planning of the International Union on Geological Sciences, 16*, 11-13.

Mohanty, P. & Panda, U. (2003). *Heatwave in Orissa: A Study Based on Heat Indices and Synoptic Features*. Regional Research Laboratory, Institute of Mathematics and Applications, Bubaneshwar, 15.

Monson, R. K., Sparks, J. P., Rosenstiel, T. N., Scott-Denton, L. E., Huxman, T. E., Harley, P. C., Turnipseed, A. A., Burns, S. P., Backlund, B. & Hu, J. (2005). Climatic influences on net ecosystem CO_2 exchange during the transition from wintertime carbon source to springtime carbon sink in a high-elevation, subalpine forest. *Oecologia, 146*, 130-147.

Moore, D., Copes, R., Fisk, R., Joy, R., Chan, K. & Brauer, M. (2006). Population health effects of air quality changes due to forest fires in British Columbia in 2003: Estimates from physician-visit billing data. *Canadian Journal of Public Health*, *97*, 105-108.

Morehouse, B. J., Carter, R. H. & Tschakert, P. (2002). Sensitivity of urban water resources in Phoenix, Tucson, and Sierra Vista, Arizona to severe drought. *Climate Research*, *21*, 283-297.

Morgan, J. A., Lecain, D. R., Mosier, A. R. & Milchunas, D. G. (2001). Elevated CO_2 enhances water relations and productivity and affects gas exchange in C-3 and C-4 grasses of the Colorado shortgrass steppe. *Global Change Biology*, *7*, 451-466.

Morgan, M. G., Pitelka, L. F. & Shevliakova, E. (2001). Elicitation of expert judgments of climate change impacts on forest ecosystems. *Climatic Change*, *49*, 279-307.

Morris, J. G. (2003). Cholera and other types of vibriosis: a story of human pandemics and oysters on the half shell. *Clinical Infectious Diseases*, *37*, 272-280.

Mortsch, L., Alden, M. & Scheraga, J. (2003). *Climate change and water quality in the Great Lakes Region - risks opportunities and responses*. Report prepared for the Great Lakes Water Quality Board for the International Joint Commission, 213.

Mote, P. W. (2003). Trends in snow water equivalent in the Pacific Northwest and their climatic causes. *Geophysical Research Letters*, *30*, doi: 10.1 029/2003GL0 17258.

Mote, P. W., Parson, E. A., Hamlet, A. F., Ideker, K. N., Keeton, W. S., Lettenmaier, D. P., Mantua, N. J., Miles, E. L., Peterson, D. W., Peterson, D. L., Slaughter, R. & Snover, A. K. (2003). Preparing for climatic change: the water, salmon, and forests of the Pacific Northwest. *Climatic Change*, *61*, 45-88.

Mote, P. W., Hamlet, A. F., Clark, M. P. & Lettenmaier, D. P. (2005). Declining mountain snowpack in western North America. *Bulletin of the American Meteorological Society*, *86*, 39-49.

Moulton, R. J. & Cuthbert, D. R. (2000). Cumulative impacts/risk assessment of water removal or loss from the Great Lakes St. Lawrence River system. *Canadian Water Resources Journal*, *25*, 181-208.

Muller, E. M., Rogers, C. S., Spitzack, A. S. & van Woesik, R. (2007). Bleaching increases likelihood of disease on *Acropora palmate* (Lamarck) in Hawksnest Bay, St John, U.S. Virgin Islands. *Coral Reefs*, *27(1)*, 191-195.

Muller-Karger, F. E., Varela, R., Thunell, R., Luerssen, R., Hu, C. & Walsh, J. J. (2005). The importance of continental margins in the global carbon cycle. *Geophysical Research Letters*, *32*, L0 1602, doi: 10.1 029/2004GL02 1346.

Muraca, G., MacIver, D. C., Auld, H. & Urquizo, N. (2001). The climatology of fog in Canada. In: *Proceedings of the 2nd International Conference on Fog and Fog Collection*, 15-20 July 2005, St. John's, Newfoundland.

Murdoch, P. S., Baron, J. S. & Miller, T. L. (2000). Potential effects of climate change on surface- water quality in North America. *Journal of the American Water Resources Association*, *36*, 357-366.

Nagel, J. M., Huxman, T. E., Griffin, K. L. & Smith, S. D. (2004). CO_2 enrichment reduces the energetic cost of biomass construction in an invasive desert grass. *Ecology*, *85*, 100-106.

NASA GISS, (2008a). *GISS Surface Temperature Analysis—Global Temperature Trends: 2007 Summation*. Goddard Institute for Space Studies, National Aeronautics and Space Administration, New York, NY, USA. <data.giss.nasa.gov/gistemp/2007/>.

NASA GISS, (2008b). *GISS Surface Temperature Analysis: Latest News: 2008-01-1 6.* Goddard Institute for Space Studies, National Aeronautics and Space Administration, New York, NY, USA. <http://data.giss.nasa.gov/gistemp/>.

Nash, L. L. & Gleick, P. H. (1993). *The Colorado River Basin and Climate Change: The Sensitivity of Stream Flow and Water Supply to Variations in Temperature and Precipitation.* EPA230-R-93-009, U.S. Environmental Protection Agency, Washington, DC, USA.

NAST (National Assessment Synthesis Team), (2001). *Climate Change Impacts on the United States: The Potential Consequences of Climate Variability and Change.* Report for the U.S. Global Change Research Program, Cambridge University Press, Cambridge, UK, 620.

Naumova, E. N., Jjagai, J. S., Matyas, B., DeMaria, A., MacNeill, I. B. & Griffiths, J. K. (2007). Seasonality in six enterically transmitted diseases and ambient temperature. *Epidemiology and Infection, 135(2)*, 281-292.

Nelson, E., Anisimov, O. A. & Shiklomanov, N. I. (2002). Climate change and hazard zonation in the circum-Arctic permafrost regions. *Natural Hazards, 26*, 203 -225.

Nelson, J. A., Morgan, J. A., Lecain, D. R., Mosier, A., Milchunas, D. G. & Parton, B. A. (2004). Elevated CO_2 increases soil moisture and enhances plant water relations in a long-term field study in semi-arid shortgrass steppe of Colorado. *Plant and Soil, 259*, 169-179.

Nemani, R. R., White, M. A., Cayan, D. R., Jones, G. V., Running, S. W., Coughlan, J. C. & Peterson, D. L (2001). Asymmetric warming over coastal California and its impact on the premium wine industry. *Climate Research, 19*, 25-34.

Nemani, R. R., White, M. A., Thornton, P. E., Nishida, K., Reddy, S., Jenkins, J. & Running, S. (2002). Recent trends in hydrologic balance have enhanced the terrestrial carbon sink in the United States. *Geophysical Research Letters, 29*, 1468, doi:10.1029/2002GL014867.

Newman, P. A., Nash, E. R., Kawa, S. R., Montzka, S. A. & Schauffler, S. M. (2006). When will the Antarctic ozone hole recover? *Geophysical Research Letters, 33*, doi. 10.1 029/2005GL025232.

Nicholls, K. H. (1999). Effects of temperature and other factors on summer phosphorus in the inner Bay of Quinte, Lake Ontario: Implications for climate warming. *Journal of Great Lakes Research, 25*, 250-262.

Nicholls, R. J., Wong, P. P., Burkett, V. R., Codignotto, J. O., Hay, J. E., McLean, R. F., Ragoonaden, S. & Woodroffe, C. D. (2007). Coastal systems and low-lying areas. In: *Climate Change 2007: Impacts, Adaptation and Vulnerability. Contribution of Working Group II to the Fourth Assessment Report of the Intergovernmental Panel on Climate Change* [M. L., Parry, O. F., Canziani, J. P., Palutikof, P. J., van der Linden, & C. E. Hanson, (eds.)]. Cambridge University Press, Cambridge, United Kingdom and New York, NY, USA, 3 15-357.

Nicholson, S. (2005). On the question of the "recovery" of the rains in the West African Sahel. *Journal of Arid Environments, 63*, 615-641.

Nielsen, G. D., Hansen, J. S., Lund, R. M., Bergqvist, M., Larsen, S. T., Clausen, S. K., Thygesen, P. & Poulsen, O. M. (2002). IgE-mediated asthma and rhinitis I: A role of allergen exposure? *Pharmacology & Toxicology, 90*, 231-242.

Niklaus, P. A. & Körner, C. H. (2004). Synthesis of a six-year study of calcareous grassland responses to in situ CO_2 enrichment. *Ecological Monographs*, *74*, 491-511. Nippert, J., A. Knapp, and J. Briggs, 2006: Intra-annual rainfall variability and grassland productivity: can the past predict the future? *Plant Ecology*, *184*, 75-87.

NOAA, n.d.: *United States Historical Climatology Network (USHCN): Version 1.* National Oceanic and Atmospheric Administration, National Climate Data Center. U.S. Department of Commerce. <www.ncdc.noaa.gov/oa/climate/research/ushcn/ushcn.html>.

NOAA, (2007a). *Billion Dollar Climate and Weather Disasters 1980-2006.* National Oceanic and Atmospheric Administration. <www.ncdc.noaa.gov/oa/reports/billionz.html>.

NOAA, (2007b). *The Climate of 2006.* National Oceanic and Atmospheric Administration. <www.ncdc.noaa.gov/oa/climate/research/2006/perspectives.html>.

NOAA, (2008a). *Climate of 2007 – in Historical Perspective. Annual Report.* National Oceanic and Atmospheric Administration, National Climate Data Center, U.S. Department of Commerce. 15 Jan. 2008. <www.ncdc. noaa. gov/ oa/ climate/ research/ 2007/ ann/ ann07.html>.

NOAA, (2008b). *2007 Annual Climate Review: U.S. Summary.* National Oceanic and Atmospheric Administration, National Climate Data Center, U.S. Department of Commerce. <www.ncdc.noaa.gov/oa/climate/research/2007/ann/us-summary.html>.

Norby, R. J., DeLucia, E. H., Gielen, B., Calfapietra, C., Giardina, C. P. King, J. S., Ledford, J., McCarthy, H. R., Moore, D. J. P., Ceulemans, R., De Angelis, P., Finzi, A. C., Karnosky, D. F., Kubiske, M. E., Lukac, M., Pregitzer, K. S., Scarascia-Mugnozza, G. E., Schlesinger, W. H. & Oren, R. (2005). Forest response to elevated CO_2 is conserved across a broad range of productivity. *Proceedings of the National Academy of Sciences of the United States of America*, *102*, 18052-18056.

Nordstrom, K. F. (2000). *Beaches and Dunes of Developed Coasts.* Cambridge University Press, 338.

North, C. S., Kawasaki, A., Spitznagel, E. L. & Hong, B. A. (2004). The course of PTSD, major depression, substance abuse, and somatization after a natural disaster. *The Journal of Nervous and Mental Disease*, *192(12)*, 823-829.

Novick, K. A., Stoy, Katul, G. G., Ellsworth, D. S., Siqueira, M. B. S., Juang, J. & Oren, R. (2004). Carbon dioxide and water vapor exchange in a warm temperate grassland. *Oecologia*, *138*, 259-274.

NRC, (1987). *Predicting Feed Intake of Food-Producing Animals.* National Research Council, National Academy Press, Washington, DC, USA.

NRC, (1998). *Global Water and Energy Experiment (GEWEX) Continental-Scale International Project: A Review of Progress and Opportunities*. National Academy Press, Washington, DC, USA, 93.

NRC, (2001a). *Climate Change Science: An Analysis of Some Key Questions.* Committee on the Science of Climate Change, National Research Council, National Academy Press, Washington, DC, USA, 42.

NRC, (2001b). *Under the Weather: Climate, Ecosystems, and Infectious Disease.* Committee on Climate, Ecosystems, Infectious Diseases, and Human Health, Board on Atmospheric Sciences and Climate, National Research Council, National Academy Press, Washington, DC, USA.

NRC, (2002). *Abrupt Climate Change: Inevitable Surprises.* National Research Council, National Academy Press, Washington, DC, USA, 230.

NRC, (2003). *Understanding Climate Change Feedbacks*. National Research Council, National Academy Press, Washington, DC.

NRC, (2005a). *Radiative Forcing of Climate Change: Expanding the Concept and Addressing Uncertainties*. Committee on Radiative Forcing Effects on Climate, Climate Research Committee, National Academies Press, Washington, DC, USA.

NRC, (2005b). *Thinking Strategically: The Appropriate Use of Metrics for the Climate Change Science Program*. National Research Council, National Academies Press, Washington, DC, USA.

NRC, (2006a). *Mitigating Shore Erosion along Sheltered Coasts*. National Research Council, National Academy Press, Washington, DC, USA.

NRC, (2006b). *Status of Pollinators in North America*. National Research Council, National Academies Press, Washington, DC, USA.

NRC, (2006c). *Surface Temperature Reconstructions for the Last 2,000 Years*. National Research Council, National Academy Press, Washington, DC, USA.

NRC, (2007). *Evaluating Progress of the U.S. Climate Change Science Program: Methods and Preliminary Results*. National Research Council, National Academy Press, Washington, DC, USA.

NRC, (2008). *Potential Impacts of Climate Change on U.S. Transportation: Special Report 290*. National Research Council, National Academy Press, Washington, DC, USA.

NWS CPC, (2008). *U.S. Seasonal Drought Outlook*. National Weather Service Climate Prediction Center, National Oceanic and Atmospheric Association, 21 Feb. 2008. <www.cpc.noaa.gov/products/expert_assessment>.

O'Brien, K., Sygna, L. & Haugen, J. E. (2004). Vulnerable or resilient? amulti-scale assessment of climate impacts and vulnerability in Norway. *Climatic Change*, *62*, 75-113.

Ocampo, J. A. & Martin, J. (Eds). (2003). *A Decade of Light and Shadow: Latin America and the Caribbean in the 1990s*. LC/G.2205-P/I, Economic Commission for Latin America and the Caribbean, 355.

OFCM, (2002). *Weather Information for Surface Transportation: National Needs Assessment Report*. Washington, D.C., USA, FCM-R1 8-2002, 302.

OFCM, (2004). *Urban Meteorology: Meeting Weather Needs in the Urban Community*, Washington, D.C., USA, FCM-R22-2004, 22.

OFCM, (2005). Wildland Fire in the Urban Environment. *The Federal Plan for Meteorological Services and Supporting Research Fiscal Year 2006*, Washington D.C., USA, FCM-P1- 2005, 284.

O'Meagher, B. (2005). Policy for agricultural drought in Australia: an economics perspective. In: *From Disaster Response to Risk Management: Australia's National Drought Policy*. [Botterill, L. C. & Wilhite, D. (eds.)]. Springer, Dordrecht, 139-156.

O'Reilly, C. T., Forbes, D. L. & Parkes, G. S. (2005). Defining and adapting to coastal hazards in Atlantic Canada: Facing the challenge of rising sea levels, storm surges, and shoreline erosion in a changing climate. *Ocean Yearbook*, *19*, 189-207.

Olesen, J. E. & Bindi, M. (2002). Consequences of climate change for European agricultural productivity, land use and policy. *European Journal of Agronomy*, *16*, 239-262.

O'Neill, M. S., Zanobetti, A. & Schwartz, J. (2003). Modifiers of the temperature and mortality association in seven US cities. *American Journal of Epidemiology*, *157(12)*, 1074-1082.

Orr, J. C., Fabry, V. J., Aumont, O., Bopp, L., Doney, S. C., Feely, R. A., Gnanadesikan, A., Gruber, N., Ishida, A., Joos, F., Key, R. M., Lindsay, K., Maier-Reimer, E., Matear, R., Monfray, P., Mouchet, A., Najjar, R. G., Plattner, G. K., Rodgers, K. B., Sabine, C. L., Sarmiento, J. L., Schlitzer, R., Slater, R. D., Totterdell, I. J., Weirig, M. F., Yamanaka, Y. & Yool, A. (2005). Anthropogenic ocean acidification over the twenty-first century and its impact on calcifying organisms. *Nature*, *437*, 681-686.

Ostfeld, R. S., Canham, C. D., Oggenfuss, K., Winchcombe, R. J. & Keesing, F. (2006). Climate, deer, rodents, and acorns as determinants of variation in Lyme disease risk. *PLoS Biology*, *4(6)*, e145.

Overpeck, J. T., Rind, D. & Goldberg, R. (1990). Climate-induced changes in forest disturbance and vegetation. *Nature*, *343*, 51-53.

Overpeck, J. T., Sturm, M., Francis, J. A., Perovich, D. K., Serreze, M. C., Benner, R., Carmack, E. C., Chapin III, S., Gerlach, S. C., Hamilton, L. C., Hinzman, L. D., Holland, M., Huntington, H. P., Key, J. R., Lloyd, A. H., MacDonald, G. M., McFadden, J., Noone, D., Prowse, T. D., Schlosser, P. & Vörösmarty, C. (2005). Arctic system on trajectory to new, seasonally ice- free state. *Eos*: *Transactions of the American Geophysical Union*, *86(34)*, 309, 312-313.

Pan, Z. T., Segal, M., Arritt, R. W. & Takle, E. S. (2004). On the potential change in solar radiation over the U.S. due to increases of atmospheric greenhouse gases. *Renewable Energy*, *29*, 1923-1928.

Parker, B. L., Skinner, M., Gouli, S., Ashikaga, T. & Teillon, H. B. (1999). Low lethal temperature for hemlock woolly adelgid (Homoptera: Adelgidae). *Environmental Entomology*, *28*, 1085-1091.

Parmesan, C. (1996). Climate and species' range. *Nature*, *382*, 765-766.

Parmesan, C. (2006). Ecological and evolutionary responses to recent climate change. *Annual Review of Ecology, Evolution and Systematics*, *37*, 637-669.

Parmesan, C. & Galbraith, H. (2004). *Observed Impacts of Global Climate Change in the U.S.* Pew Center on Global Climate Change, Arlington, Virginia, 67 <www.pewclimate.org/global-warming-in-depth/all_reports/observedimpacts>.

Parmesan, C. & Yohe, G. (2003). A globally coherent fingerprint of climate change impacts across natural systems. *Nature*, *421*, 37.

Parmesan, C., Ryrholm, N., Stefanescu, C., Hill, J. K., Thomas, C. D., Descimon, H., Huntley, B., Kaila, L., Kullberg, J., Tammaru, T., Tennent, W. J., Thomas, J. A. & Warren, M. (1999). Poleward shifts in geographical ranges of butterfly species associated with regional warming. *Nature*, *399*, 579-583.

Parmesan, C., Root, T. L. & Willig, M. R. (2000). Impacts of extreme weather and climate on terrestrial biota. *Bulletin of the American Meteorological Society*, *81*, 443 -450.

Parry, M. L., Canziani, O. F., Palutikof, J. P. & coauthors, (2007). Technical summary. In: *Climate Change 2007: Impacts, Adaptation and Vulnerability. Contribution of Working Group II to the Fourth Assessment Report of the Intergovernmental Panel on Climate Change* [M. L., Parry, O. F., Canziani, J. P., Palutikof, P. J. van der Linden, & C. E. Hanson, (eds.)]. Cambridge University Press, Cambridge, UK, 23-78.

Parson, E. A., Corell, R. W., Barron, E. J., Burkett, V., Janetos, A., Joyce, L., Karl, T. R., Maccracken, M.C. & coauthors, (2003). Understanding climatic impacts, vulnerabilities, and adaptation in the United States: building a capacity for assessment. *Climatic Change*, *57*, 9-42.

Patterson, D. T., Westbrook, J. K., Joyce, R. J. C., Lingren, P. D. & Rogasik, J. (1999). Weeds, insects and diseases. *Climatic Change*, *43*, 711-727.

Patz, J. A. & Balbus, J. M. (2001). Global climate change and air pollution. In: *Ecosystem Change and Public Health. A Global Perspective* [Aron, J.L. & J.A. Patz (eds.)]. The Johns Hopkins University Press, Baltimore, 379-408.

Payne, J. T., Wood, A. W., Hamlet, A. F., Palmer, R. N. & Lettenmaier, D. P. (2004). Mitigating the effects of climate change on the water resources of the Columbia River basin. *Climatic Change*, *62*, 233-256.

Pearcy, W. G. (1991). *Ocean Ecology of North Pacific Salmonids.* Washington State Sea Grant Program, The University of Washington Press, Seattle, WA, USA, 179.

Peng, S., Huang, J., Sheehy, J. E., Lanza, R. C., Visperas, R. M., Zhong, X., Centeno, G. S., Khush, G. S. & Cassman, K. G. (2004). Rice yields decline with higher night temperatures from global warming. *Proceedings of the National Academy of Sciences*, *101*, 997 1-9975.

Perez-Garcia, J., Joyce, L. A., McGuire, A. D. & Xiao, X. M. (2002). Impacts of climate change on the global forest sector. *Clim. Change*, *54*, 439-461.

Petersen, J. H. & Kitchell, J. F. (2001). Climate regimes and water temperature changes in the Columbia River: bioenergetic implications for predators of juvenile salmon. *Canadian Journal of Fisheries and Aquatic Sciences*, *58*, 183 1-1841.

Peterson, D. W. & Peterson, D. L. (2001). Mountain hemlock growth trends to climatic variability at annual and decadal time scales. *Ecology*, *82*, 3330-3345.

Peterson, D. W., Peterson, D. L. & Ettl, G. J. (2002). Growth responses of subalpine fir to climatic variability in the Pacific Northwest. *Canadian Journal of Forest Research*, *32*, 1503- 1517.

PHMSA, (2007). *Annual Reports: National Pipeline Mapping System.* Pipeline and Hazardous Materials Safety Administration, U.S. Department of Transportation, Washington, DC, USA.

Pielke, R. A., Landsea, C., Mayfield, M., Laver, J. & Pasch, R. (2005). Hurricanes and global warming. *Bulletin of the American Meteorological Society*, *86*, 1571-1575.

Pinho, O. S. & Orgaz, M. D. (2000). The urban heat island in a small city in coastal Portugal. *International Journal of Biometeorology*, *44(4)*, 198-203.

Pisano, P., Goodwin, L. & Stern, A. (2002). Surface transportation safety and operations: The impacts of weather within the context of climate change. In: *The Potential Impacts of Climate Change on Transportation: Workshop Summary and Proceedings*. Washington, DC, 20 <climate.dot.gov/publications/workshop1002/pisano.pdf>.

Plummer, D. A., Caya, D., Frigon, A., Côté, H., Giguère, M., Paquin, D., Biner, S., Harvey, R. & de Elia, R. (2006). Climate and climate change over North America as simulated by the Canadian Regional Climate Model. *Journal of Climate*, *19*, 3112-3132.

Polsky, C. & Easterling III, W. E. (2001). :Adaptation to climate variability and change in the US Great Plains: A multi-scale analysis of Ricardian climate sensitivities. *Agriculture, Ecosystems & Environment*, *85*, 133-144.

Porter, J. R. & Semenov, M. A. (2005). Crop responses to climatic variation. *Philosophical Transactions of the Royal Society B*, *360*, 202 1-2035.

Postel, S. & Richter, B. (2003). *Rivers for Life: Managing Water for People and Nature.* Island Press, Washington, District of Columbia, 220

Potter, C., Klooster, S., Hiatt, S., Fladeland, M., Genovese, V. & Gross, P. (2007). Satellite-derived estimates of potential carbon sequestration through afforestation of agricultural lands in the United States. *Climatic Change*, *80*, 323-336, doi:10.1007/s10584-006-9109-3.

Potter, J. R., Savonis, M. J. & Burkett, V. R. (2008). Executive summary. In *Impacts of Climate Change and Variability on Transportation Systems and Infrastructure: Gulf Coast Study, Phase I.* Synthesis and Assessment Product 4.7 by the U.S. Climate Change Science Program and the Subcommittee on Global Change Research, Washington, DC.

Poumadere, M., Mays, C., LeMer, S. & Blong, R. (2005). The 2003 heat wave in France: Dangerous climate change here and now. *Risk Analysis*, *25*, 1483-1494.

Pounds, J. A., Fogden, M. P. L. & Campbell, J. H. (1999). Biological response to climate change on a tropical mountain. *Nature*, *398*, 611-615.

Pounds, J. A., Bustamante, M. R., Coloma, L. A., Consuegra, J. A., Fogden, M. P. L., Foster, P. N., La Marca, E., Masters, K. L., Merino-Viteri, A., Puschendorf, R., Ron, S., Sanchez-Azofeifa, R. G. A., Still, C. J. & Young, B. E. (2006). Widespread amphibian extinctions from epidemic disease driven by global warming. *Nature*, *439*, 161-167.

Price, C. & Rind, D. (1994). The impact of a 2xCO_2 climate on lightning caused fires. *Journal of Climate*, *7*, 1484-1494.

Pulwarty, R., Jacobs, K. & Dole, R. (2005). The hardest working river: drought and critical water problems in the Colorado River Basin. In: *Drought and Water Crisis - Science, Technology and Management Issues* [Wilhite, D.A. (ed.)]. CRC Press, Boca Raton, FL, USA, 249-286.

Purse, B. V., Mellor, P. S., Rogers, D. J., Samuel, A. R., Mertens, P. P. & Baylis, M. (2005). Climate change and the recent emergence of bluetongue in Europe. *Nature Reviews Microbiology*, *3(2)*, 171-181.

Quinn, F. H. (2002). The potential impacts of climate change on Great Lakes transportation. In: *The Potential Impacts of Climate Change on Transportation: Workshop Summary and Proceedings*. Washington, DC, USA, 9

Ramstack, J. M., Fritz, S. C. & Engstrom, D. R. (2004). Twentieth century water quality trends in Minnesota lakes compared with presettlement variability. *Canadian Journal of Fisheries and Aquatic Sciences*, *61*, 561-576.

Randall, D. A., Wood, R. A., Bony, S., Colman, R., Fichefet, T., Fyfe, J., Kattsov, V., Pitman, A., Shukla, J., Srinivasan, J., Stouffer, R. J., Sumi, A. & Taylor, K. E. (2007). Climate models and their evaluation. In: *Climate Change 2007: The Physical Science Basis. Contribution of Working Group I to the Fourth Assessment Report of the Intergovernmental Panel on Climate Change* [Solomon, S., D. Qin, M. Manning, Z. Chen, M. Marquis, K.B. Averyt, M. Tignor, and H.L. Miller (eds.)]. Cambridge University Press, Cambridge, United Kingdom and New York, NY, USA.

Randerson, J. T., Liu, H., Flanner, M. G., Chambers, S. D., Jin, Y., Hess, P. G., Pfister, G., Mack, M. C., Treseder, K. K., Welp, L. R., Chapin, F. S., Harden, J. W., Goulden, M. L., Lyons, E., Neff, J. C. Schuur, E. A. G. & Zender, C. S. (2006). The impact of boreal forest fire on climate warming. *Science*, *314*, 1130-1132, doi:10.1126/science.1132075.

Ranhoff, A. H. (2000). Accidental hypothermia in the elderly. *International Journal of Circumpolar Health*, *59*, 255-259.

Raupach, M. R., Marland, G., Ciais, P., Le Quéré, C., Canadell, J. G., Klepper, G. & Field, C. B. (2007). Global and regional drivers of accelerating CO_2 emissions. *Proceedings of the National Academy of Sciences, 104(24)*, doi: 10.1 073/pnas.0700609 104.

Reading, C. J. (1998). The effect of winter temperatures on the timing of breeding activity in the common toad *Bufo bufo*. *Oecologia, 117*, 469-475.

Reale, D., McAdam, A., Boutin, S. & Berteaux, D. (2003). Genetic and plastic responses of a northern mammal to climate change. *Proceedings of the Royal Society of London B, 270*, 591-596.

Regonda, S. K., Rajagopalan, B., Clark, M. & Pitlick, J. (2005). Seasonal cycle shifts in hydroclimatology over the western United States. *Journal of Climate, 18*, 372-384.

Ren, C., Williams, G. M. & Tong, S. (2006). Does particulate matter modify the association between temperature and cardiorespiratory diseases? *Environmental Health Perspectives, 114(11)*, 1690-1696.

Reynolds, J. D., Webb, T. J. & Hawkins, L. A. (2005). Life history and ecological correlates of extinction risk in European freshwater fishes. *Canadian Journal of Fisheries and Aquatic Science, 62*, 854-862.

Richardson, D. M., Pysek, P., Rejmanek, M., Barbour, M. G., Panetta, F. D. & West, C. J. (2000). Naturalization and invasion of alien plants: concepts and definitions. *Diversity and Distributions, 6*, 93-107.

Rignot, E., (2006). Changes in ice dynamics and mass balance of the Antarctic ice sheet. *Philosophical Transactions of the Royal Societyof London, 364*, 1637-1655.

Robeson, S. (2004). Trends in time-varying percentiles of daily minimum and maximum temperature over North America. *Geophysical Research Letters, 31*, L04203, doi: 10. 1029/2003GL01 9019.

Rood, S. B., Samuelson, G. M., Weber, J. K. & Wywrot, K. A. (2005). Twentieth-century decline in streamflows from the hydrographic apex of North America. *Journal of Hydrology, 306*, 215-233.

Root, T. L., Price, J. T., Hall, K. R., Schneider, S. H., Rosenzweig, C. & Pounds, J. A. (2003). Fingerprints of global warming on wild animals and plants. *Nature, 421*, 57-60.

Root, T. L., MacMynowski, D. P., Mastrandrea, M. & Schneider, S. H. (2005). Human-modified temperatures induce species' changes: joint attribution. *Proceedings of the National Academy of Sciences, 21*, 7465-7469.

Rose, J. B., Daeschner, S., Easterling, D. R., Curriero, F. C., Lele, S. & Patz, J. A. (2000). Climate and waterborne disease outbreaks. *Journal of the American Water Works Association, 92(9)*, 77-87.

Rosenthal, D. H., Gruenspecht, H. K. & Moran, E. (1995). Effects of global warming on energy use for space heating and cooling in the United States. *Energy Journal, 16(2)*, 77-96.

Rosenzweig, C., Casassa, G., Karoly, D. J., Imeson, A., Liu, C., Menzel, A., Rawlins, S., Root, T. L., Seguin, B. & Tryjanowski, P. (2007). Assessment of observed changes and responses in natural and managed systems. In: *Climate Change 2007: Impacts, Adaptation and Vulnerability. Contribution of Working Group II to the Fourth Assessment Report of the Intergovernmental Panel on Climate Change* [Parry, M.L., O.F. Canziani, J.P. Palutikof, P.J. van der Linden, and C.E. Hanson (eds.)]. Cambridge University Press, Cambridge, United Kingdom and New York, NY, USA.

Rosenzweig, C. & Solecki, W. (eds.), (2001). *Climate Change and a Global City: The Potential Consequences of Climate Variability and Change – Metro East Coast.* Columbia Earth Institute, NY, USA.

Rosetti, M. A. (2002). Potential impacts of climate change on railroads. In: *The Potential Impacts of Climate Change on Transportation: Workshop Summary and Proceedings.* Center for Climate Change and Environmental Forecasting, Federal Research Partnership Workshop, United States Department of Transportation, Washington, DC, USA, 13 <climate.dot.gov/publications/workshop1002/>.

Rothrock, D. A., Zhang, J. & Yu, Y. (2003). The arctic ice thickness anomaly of the 1990s: A consistent view from observations and models. *Journal of Geophysical Research*, 108(C3), 3083, doi: 10. 1029/2001JC001208.

Roy, D. B. & Sparks, T. H. (2000). Phenology of British butterflies and climate change. *Global Change Biology, 6*, 407-416.

Running, S. W. (2006). Is global warming causing more larger wildfires? *Science, 313*, 927-928. Russoniello, C.V., T.K. Skalko, K. O'Brien, S.A. McGhee, D. Bingham-Alexander, and J. Beatley, (2002). Childhood posttraumatic stress disorder and efforts to cope after hurricane Floyd. *Behavioral Medicine, 28*, 61-71.

Ruth, M. & A-C Lin, (2006). Regional energy and adaptations to climate change: Methodology and application to the state of Maryland. *Energy Policy, 34*, 2820-283 3.

Ruth, M., Amato, A. & Kirshen, P. (2006). Impacts of changing temperatures on heat-related mortality in urban areas: The issues and a case study from metropolitan Boston. In: *Smart Growth and Climate Change* [Ruth, M. (ed.)]. Edward Elgar Publishers, Cheltenham, England, 364-392.

Ryan, M. G., Archer, S. R., Birdsey, R. A., Dahm, C. N., Heath, L. S., Hicke, J. A., Hollinger, D. Y., Huxman, T. E., Okin, G. S., Oren, R., Randerson, J. T. & Schlesinger, W. H. (2008). Land resources. In: *The Effects of Climate Change on Agriculture, Land Resources, Water Resources, and Biodiversity.* Synthesis and Assessment Product 4.3 by the U.S. Climate Change Science Program and the Subcommittee on Global Change Research, Washington, DC.

Rybnicek, O. & Jaeger, S. (2001). Ambrosia (ragweed) in Europe. *A CI International, 13*, 60-66.

Sabine, C. L., Heimann, M., Artaxo, P., Bakker, D. C. E., Chen, C. T. A., Field, C. B. & Gruber, N. (2004). Current status and past trends of the global carbon cycle. In: *Global Carbon Cycle: Integrating Humans, Climate, and the Natural World* [C.B. Field and M.R. Raupach, (eds.)]. Island Press, Washington, DC, USA, 17-44.

Sacks, W., Schimel, D. & Monson, R. (2007). Coupling between carbon cycling and climate in a high elevation, subalpine forest: a model-data fusion analysis. *Oecologia, 151(1)*, 54-68, doi: 10.1 007/s00442-006-0565-2.

Sailor, D. J. (2001). Relating residential and commercial sector electricity loads to climate: Evaluating state level sensitivities and vulnerabilities. *Energy, 26(7)*, 645-657.

Sailor, D. J. & Muñoz, J. R. (1997). Sensitivity of electricity and natural gas consumption to climate in the U.S: Methodology and results for eight states. *Energy, 22(10)*, 987-998.

Sailor, D. J. & Pavlova, A. A. (2003). Air conditioning market saturation and long-term response of residential cooling energy demand to climate change. *Energy, 28(9)*, 941-951.

Saito, Y. (2001). Deltas in Southeast and East Asia: their evolution and current problems. In: *Proceedings of the APN/SUR VAS/LOICZ Joint Conference on Coastal Impacts of Climate Change and Adaptation in the Asia – Pacific Region*, 14-16 November 2000, Kobe, Japan, Asia Pacific Network for Global Change Research, 185-191.

Sala, O. A., Chapin III, F. S., Armesto, J. J., Berlow, E., Bloomfield, J., Dirzo, R., Huber-Sanwald, E., Huenneke, L. F., Jackson, R. B., Kinzig, A., Leemans, R., Lodge, D. M., Mooney, H. A., Oesterheld, M., Poff, N. L., Sykes, M. T., Walker, B. H., Walker, M. & Wall, D. H. (2000). Global biodiversity scenarios for the year 2100. *Science*, *287*, 1770-1774.

Salinari, F., Giosue, S., Tubiello, F. N., Rettori, A., Rossi, V., Spanna, F., Rosenzweig, C. & Gullino, M. L. (2006). Downy mildew epidemics on grapevine under climate change. *Global Change Biology*, *12*, 1-9.

Salo, L. F. (2005). Red brome (*Bromus rubens* subsp. *madritensis*) in North America: possible modes for early introductions, subsequent spread. *Biological Invasions*, *7*, 165-180.

Sankaran, M. (2005). Fire, grazing and the dynamics of tall-grass savannas in the Kalakad-Mundanthurai Tiger Reserve, South India. *Conservation & Society*, *3*, 4-25.

Sarmiento, J. L., Wofsy, S. C. & the members of the Carbon and Climate Working Group, 1999: *A U.S. Carbon Cycle Science Plan*. U.S. Global Change Research Program, Washington, DC, USA, 69

Saunders, S., Montgomery, C., Easley, T. & Spencer, T. (2008). *Hotter and Drier: The West's Changed Climate*. The Rocky Mountain Climate Organization and Natural Resources Defense Council, New York, NY, USA, 54

Savonis, M. J. (2008). What are the key conclusions of this study? In: *Impacts of Climate Change and Variability on Transportation Systems and Infrastructure: Gulf Coast Study, Phase I.* Synthesis and Assessment Product 4.7 by the U.S. Climate Change Science Program and the Subcommittee on Global Change Research, Washington, DC, USA.

Scavia, D., Field, J. C., Boesch, D. F., Buddemeier, R. W., Burkett, V., Cayan, D. R., Fogarty, M., Harwell, M. A., Howarth, R. W., Mason, C., Reed, D. J., Royer, T. C., Sallenger, A. H. & Titus, J. G. (2002). Climate change impacts on US coastal and marine ecosystems. *Estuaries*, *25*, 149-164.

Schär, C., Vidale, P. L., Lüthi, D., Frei, C., Häberli, C., Liniger, M. A. & Appenzeller, C. (2004). The role of increasing temperature variability in European summer heatwaves. *Nature*, *427*, 332-336.

Schmidt, G. A. & 35 other authors, (2006). Present-day atmospheric simulations using GISS ModelE: Comparison to in situ, satellite, and reanalysis data. *Journal of Climate*, *19*, 153-192.

Schmittner, A. (2005). Decline of the marine ecosystem caused by a reduction in the Atlantic overturning circulation. *Nature*, *434*, 628-633.

Schneider, S. H. & Root, T. L. (1996). Ecological implications of climate change will include surprises. *Biodiversity and Conservation*, *5(9)*, 1109-1119.

Schneider, S. H., Semenov, S., Patwardhan, A., Burton, I., Magadza, C. H. D., Oppenheimer, M., Pittock, A. B., Rahman, A., Smith, J. B., Suarez, A. & Yamin, F. (2007). Assessing key vulnerabilities and the risk from climate change. In: *Climate Change 2007: Impacts, Adaptation and Vulnerability. Contribution of Working Group II to the Fourth Assessment Report of the Intergovernmental Panel on Climate Change* [Parry, M.L., O.F.

Canziani, J. P. Palutikof, P. J. van der Linden, & C.E. Hanson (eds.)]. Cambridge University Press, Cambridge, United Kingdom and New York, NY, USA.

Schoennagel, T., Veblen, T. T. & Romme, W. H. (2004). The interaction of fire, fuels, and climate across Rocky Mountain Forests. *BioScience*, *54*, 661-676.

Scholes, R. J. & Archer, S. R. (1997). Tree–grass interactions in savannas. *Annual Review of Ecology and Systematics*, *28*, 517-544.

Schwartz, J., Samet, J. M. & Patz, J. A. (2004). Hospital admissions for heart disease: The effects of temperature and humidity. *Epidemiology*, *15(6)*, 755-761.

Schwartz, M. D. & Reiter, B. E. (2000). Changes in North American spring. *International Journal of Climatology*, *20*, 929-932.

Scibek, J. & Allen, D. M. (2006). Comparing modeled responses to two high-permeability, unconfined aquifers to predicted climate change. *Global and Planetary Change*, *50*, 50-62.

Scott, M. J. & Huang, Y. J. (2007). Effects of climate change on energy use in the United States. In *Effects of Climate Change on Energy Production and Use in the United States.* Synthesis and Assessment Product 4.5 by the U.S. Climate Change Science Program and the Subcommittee on Global Change Research, Washington, DC.

Scott, M. J., Dirks, J. A. & Cort, K. A. (2005). The adaptive value of energy efficiency programs in a warmer world: Building energy efficiency offsets effects of climate change. PNNL-SA-45 118. In: *Reducing Uncertainty through Evaluation, Proceedings of the 2005 International Energy Program Evaluation Conference*, 17-19 August 2005, Brooklyn, New York.

Seager, R., Kushnir, Y., Herweijer, C., Naik, N. & Velez, J. (2005). Modeling of tropical forcing of persistent droughts and pluvials over western North America: 1856–2000. *Journal of Climate*, *18*, 4065-4088.

Seager, R., Ting, M., Held, I., Kushnir, Y., Lu, J., Vecchi, G., Huang, H. P., Harnik, N., Leetmaa, A., Lau, N. C., Li, C., Velez, J. & Naik, N. (2007). Model projections of an imminent transition to a more arid climate in southwestern North America. *Science*, *316*, 1181- 1184.

Sekercioglu, C. H., Daily, G. C. & Ehrlich, P. R. (2004). Ecosystem consequences of bird declines. *Proceedings of the National Academy of Sciences*, *101*, 18042-18047.

Semenza, J. C., McCullough, J. E., Flanders, W. D., McGeehin, M. A. & Lumpkin, J. R. (1999). Excess hospital admissions during the July 1995 heat wave in Chicago. *American Journal of Preventive Medicine*, *16(4)*, 269-277.

Sénat, (2004). France and the French face the canicule: the lessons of a crisis: appendix to the minutes of the session of February *3*, 2004, 59-62. Information report no. 195. <www.senat.fr/rap/r03-195/r03-195.html>.

Senate of Canada, (2003). *Climate Change: We are at Risk*. Final Report, Standing Senate Committee on Agriculture and Forestry, Ottawa, Canada, 123.

Setzer, C. & Domino, M. E. (2004). Medicaid outpatient utilization for waterborne pathogenic illness following Hurricane Floyd. *Public Health Reports*, *119*, 472-478.

Shaw, J., Taylor, R. B., Forbes, D. L. Ru, M. H. & Solomon, S. (1998). *Sensitivity of the Coasts of Canada to Sea Level Rise*. Bulletin *505*, Natural Resources Canada, Geological Survey of Canada, Ottawa, Ontario, 79

Sherbinin, A., Schiller, A. & Pulsipher, A. (2006). The vulnerability of global cities to climate hazards. *Environment and Urbanization*, *12(2)*, 93-102.

Sheridan, S. & Dolney, T. (2003). Heat, mortality, and level of urbanization: measuring vulnerability across Ohio, USA. *Climate Research*, *24*, 255-266.

Shindell, D. T. (2001). Climate and ozone response to increased stratospheric water vapor. *Geophysical Research Letters*, *28(8)*, 1551-1554.

Shone, S. M., Curriero, F. C., Lesser, C. R. & Glass, G. E. (2006). Characterizing population dynamics of *Aedes sollicitans* (Diptera: Culicidae) using meteorological data. *Journal of Medical Entomology*, *43(2)*, 393-402.

Shvidenko, A., Barber, C. V. & Persson, R. (2005). Forest and woodland systems. In: *Ecosystems and Human Well-being: Volume 1: Current State and Trends* [Hassan, R., R. Scholes, and N. Ash, (eds.)]. Island Press, Washington, DC, USA, 585-621.

Sillman, S. & Samson, P. J. (1995). Impact of temperature on oxidant photochemistry in urban, polluted rural, and remote environments. *Journal of Geophysical Research*, 100(D6), 11497-11508, doi:10.1029/94JD02146.

Singer, F. J. & Harter, K. (1996). Comparative effects of elk herbivory and 1988 fires on northern Yellowstone National Park grasslands. *Ecological Applications*, 6(1), 185-200. Singer, M.C. & P.R. Ehrlich, 1979. Population dynamics of the checkerspot butterfly *Euphydryas editha. Fortschritte der Zoologie*, *25*, 53-60.

Small, C. & Nicholls, R. J. (2003). A global analysis of human settlement in coastal zones. *Journal of Coastal Research*, *19*, 584-599.

Smith, S. D., Huxman, T. E., Zitzer, S. F., Charlet, T. M., Housman, D. C., Coleman, J. S., Fenstermaker, L. K., Seemann, J. R. & Nowak, R. S. (2000). Elevated CO_2 increases productivity and invasive species success in an arid ecosystem. *Nature*, *408*, 79-82.

Smith, S. J., Thomson, A. M., Rosenberg, N. J., Izaurralde, R. C., Brown, R. A. & Wigley, T. M. L. (2005). Climate change impacts for the conterminous USA: An integrated assessment: Part 1. Scenarios and context. *Climatic Change*, *69*, 7-25.

Smith, O.P. & Levasseur, G. (2002). Impacts of climate change on transportation infrastructure in Alaska. In: *The Potential Impacts of Climate Change on Transportation: Workshop Summary and Proceedings*. Washington, DC, USA, 11 <climate.dot.gov/publications/workshop1002/smith.pdf>.

Smith, T. M. & Reynolds, R. W. (2005). A global merged land and sea surface temperature reconstruction based on historical observations (1880-1997). *Journal of Climate*, *18*, 2021-2036.

SNL, (2006). *Energy and Water Research Directions – A Vision for a Reliable Energy Future*. Sandia National Laboratories, Albuquerque, NM, USA.

Soden, Brian, J., Jackson, D. L., Ramaswamy, V., Schwarzkopf, M. D & Huang, Z. (2005). The radiative signature of upper tropospheric moistening. *Science*, *310*, 841-844.

Sohngen, B. & Sedjo, R. (2005). Impacts of climate change on forest product markets: implications for North American producers. *Forestry Chronicle*, *81*, 669-674.

Solecki, W. D. & Rosenzweig, C. (2006). Climate change and the city: Observations from metropolitan New York. In: *Cities and Environmental Change* [Bai, X. (ed.)]. Yale University Press, NY, USA.

Solomon, S., Qin, D., Manning, M., Alley, R. B., Berntsen, T., Bindoff, N. L., Chen, Z., Chidthaisong, A., Gregory, J. M., Hegerl, G. C., Heimann, M., Hewitson, B., Hoskins, B. J., Joos, F., Jouzel, J., Kattsov, V., Lohmann, U., Matsuno, T., Molina, M., Nicholls, N., Overpeck, J., Raga, G., Ramaswamy, V., Ren, J., Rusticucci, M., Somerville, R., Stocker, T. F., Whetton, P., Wood, R. A. & Wratt, D. (2007). Technical summary. In: *Climate*

Change 2007: Impacts, Adaptation and Vulnerability. Contribution of Working Group I to the Fourth Assessment Report of the Intergovernmental Panel on Climate Change [Solomon, S., D. Qin, M. Manning, Z. Chen, M. Marquis, K.B. Averyt, M. Tignor, and H.L. Miller (eds.)]. Cambridge University Press, Cambridge, United Kingdom and New York, NY, USA.

Sorogin, V. P. & Co-authors, (1993). *Problemy Ohrany Zdoroviya i Socialnye Aspecty Osvoeniya Gazovyh i Neftyanyh Mestorozhdenij v Arcticheskih Regionah* [Problems of Public Health and Social Aspects of Exploration of Oil and Natural Gas Deposits in Arctic Regions]. Nadym.

Stacey, D. A. & Fellows, M. D. E. (2002). Influence of elevated CO_2 on interspecific interactions at higher trophic levels, 2002. *Global Change Biology*, *8*, 668-678.

Stanturf, J. A., Goodrick, S. L. & Outcalt, K. W. (2007). Disturbance and coastal forests: A strategic approach to forest management in hurricane impact zones. *Forest Ecology and Management*, *250*, 119-135.

Stefanescu, C., Herrando, S. & Páramo, F. (2004). Butterfly species richness in the north-west Mediterranean Basin: the role of natural and human-induced factors. *Journal of Biogeography*, *31(6)*, 905-915.

Stenseth, N. C. & Mysterud, A. (2002). Climate, changing phenology, and other life history traits: nonlinearity and match-mismatch to the environment. *Proceedings of the National Academy of Sciences*, *99*, 13379-13381.

Stevenson, D. S., et al. (2005). Impacts of climate change and variability on tropospheric ozone and its precursors. *Faraday Discuss.*, *130*, doi:10.1039/b417412g.Stewart, I.T., D.R.

Cayan, and M.D. Dettinger, 2005: Changes toward earlier streamflow timing across western North America. *Journal of Climate*, *18*, 1136-1155.

Stiger, R. W. (2001). Alaska DOT deals with permafrost thaws. *Better Roads*, June, 30-31. <obr.gcnpublishing.com/articles/brjun01c.htm>.

Stirling, I. & Parkinson, C. L. (2006). Possible effects of climate warming on selected populations of polar bears in the Canadian Arctic. *Arctic*, *59(3)*, 261-275.

Stirling, I., Lunn, N. J. & Iacozza, J. (1999). Long-term trends in the population ecology of polar bears in western Hudson Bay in relation to climate change. *Arctic*, *52*, 294-306. Stivers, L., 1999: *Crop Profile for Corn (Sweet) in New York.* <www.ipmcenters.org/CropProfiles/docs/nycorn-sweet.pdf>.

Stock, W. D., Ludwig, F., Morrow, C., Midgley, G. F., Wand, S. J. E., Allsopp, N. & Bell, T. L. (2005). Long-term effects of elevated atmospheric CO_2 on species composition and productivity of a southern African C-4 dominated grassland in the vicinity of a CO_2 exhalation. *Plant Ecology*, *178*, 211-224.

Stocks, B. J., Fosberg, M. A., Lynham, T. J., Mearns, L., Wotton, B. M., Yang, Q., Jin, J. Z., Lawrence, K., Hartley, G. R., Mason, J. A. & McKenney, D. W. (1998). Climate change and forest fire potential in Russian and Canadian boreal forests. *Climatic Change*, *38*, 1-13.

Stocks, B. J., Mason, J. A., Todd, J. B., Bosch, E. M., Wotton, B. M., Amiro, B. D., Flannigan, M. D., Hirsch, K. G., Logan, K. A., Martell, D. L. & Skinner, W. R. (2002). Large forest fires in Canada, 1959-1997. *Journal of Geophysical Research*, *107*, doi: 10.1029/2001 JD000484.

Stone, R. S., Dutton, E. G., Harris, J. M. & Longnecker, D. (2002). Earlier spring snowmelt in northern Alaska as an indicator of climate change. *Journal of Geophysical Research*, 107(D10), doi: 10. 1029/2000JD000286.

Stouffer, R. J., Yin, J., Gregory, J. M., Dixon, K. W., Spelman, M. J., Hurlin, W., Weaver, A. J., Eby, M., Flato, G. M., Hasumi, H., Hu, A., Jungclaus, J. H., Kamenkovich, I. V., Levermann, A., Montoya, M., Murakami, S., Nawrath, S., Oka, A., Peltier, W. R., Robitaille, D. Y., Sokolov, A., Vettoretti, G. & Weber, S. L. (2006a). Investigating the causes of the response of the thermohaline circulation to past and future climate changes. *Journal of Climate*, *19*, 1365-1387.

Stouffer, R. J., Delworth, T. L., Dixon, K. W., Gudgel, R., Held, I., Hemler, R., Knutson, T., Schwarzkopf, M. D., Spelman, M. J., Winton, M. W., Broccoli, A. J., Lee, H. C., Zeng, F. & Soden, B. (2006b). GFDL's CM2 global coupled climate models. Part IV: Idealized climate response. *Journal of Climate*, *19*, 723-740.

Stroeve, J. C., Serreze. M. C., Fetterer, F., Aretter, T., Meier, W., Maslanik, J. & Knowles, K. (2005). Tracking the Arctic's shrinking ice cover: Another extreme September minimum in 2004. *Geophysical Research Letters*, *32*, L04501.

Subak, S. (2003). Effects of climate on variability in Lyme disease incidence in the northeastern United States. *American Journal of Epidemiology*, *157(6)*, 53 1-538.

Sun, G., Li, C., Trettin, C. C., Lu, J. & McNulty, S. G. (2006). Simulating the biogeochemical cycles in cypress wetland-pine upland ecosystems at a landscape scale with the wetland-DNDC model. In: *Hydrology and Management of Forested Wetlands*, Proceedings of the International Conference 8-12 April 2006. American Society of Agricultural and Biological Engineers, St. Joseph, MI, USA.

Sur, D., Dutta, P., Nair, G. B. & Bhattacharya, S. K. (2000). Severe cholera outbreak following floods in a northern district of West Bengal. *Indian Journal of Medical Research*, *112*, 178-182.

Swetnam, T. W. & Betancourt, J. L. (1998). Mesoscale disturbance and ecological response to decadal climatic variability in the American Southwest. *Journal of Climate*, *11*, 3 128-3147.

Taramarcaz, P., Lambelet, B., Clot, B., Keimer, C. & Hauser, C. (2005). Ragweed (Ambrosia) progression and its health risks: will Switzerland resist this invasion? *Swiss Medical Weekly*, *135*, 538-548.

Taylor, S. W., Carroll, A. L., Alfaro, R. I. & Safranyik, L. (2006). Forest, climate, and mountain pine beetle outbreak dynamics in western Canada. In: *The Mountain Pine Beetle: A Synthesis of Biology, Management, and Impacts on Lodgepole Pine* [Safranyik, L. and Wilson, W. R. (eds)]. Natural Resources Canada, Canadian Forest Service, Pacific Forestry Centre, Victoria, British Columbia, 67-94.

Thanh, T. D., Saito, Y., Huy, D. V., Nguyen, V. L., Oanh, T. K. O. & Tateishi, M. (2004). Regimes of human and climate impacts on coastal changes in Vietnam. *Regional Environmental Change*, *4*, 49-62.

Thomas, C. D., Singer, M. C. & Boughton, D. (1996). Catastrophic extinction of population sources in a butterfly metapopulation. *American Naturalist*, *148*, 957-975.

Thomas, C. D., Cameron, A., Green, R. E., Bakkenes, M., Beaumont, L. J., Collingham, Y. C., Erasmus, B. F. N., Siqueira, M. F. d., Grainger, A., Hannah, L., Hughes, L., Huntley, B., v. Jaarsveld, A. S., Midgley, G. F., Miles, L., Ortega-Huerta, M. A., Peterson, A. T.,

Phillips, O. L. & Williams, S. E. (2004). Extinction risk from climate change. *Nature, 427*, 145-148.

Thomas, D. S. C. & Twyman, C. (2005). Equity and justice in climate change adaptation among natural-resource dependent societies. *Global Environ. Chang., 15*, 115-124.

Thomas, M. K., Charron, D. F., Waltner-Toews, D., Schuster, C., Maarouf, A. R. & Holt, J. D. (2006). A role of high impact weather events in waterborne disease outbreaks in Canada, 1975 – 2001. *International Journal of Environmental Health Research, 16(3)*, 167-180.

Thompson, A. M., Pickering, K., McNamara, D., Schoeberl, M., Hudson, R., Kim, J., Browell, E., Kirchhoff, V. & Nganga, D. (1996). Where did tropospheric ozone over southern Africa and the tropical Atlantic come from in October 1992? Insights from TOMS, GTE TRACE A, and SAFARI 1992. *Journal of Geophysical Research*, 101(D19), doi: 10.1 029/96JD0 1463.

Thomson, M. C., Doblas-Reyes, F. J., Mason, S. J., Hagedorn, R., Connor, R. J., Phindela, T., Morse, A. P. & Palmer, T. N. (2006). Malaria early warnings based on seasonal climate forecasts from multimodel ensembles. *Nature, 439*, 576-579.

Thorup, K., Tøttrup, A. P. & Rahbek, C. (2007). Patterns of phenological changes in migratory birds. *Oecologia, 151*, 697-703.

Titus, J. G., (2005). Sea-level rise effect. *Encyclopaedia of Coastal Science* [Schwart, M.L., (ed.)]. Springer, Dordrecht, 83 8-846.

Titus, J. G. & Richman, C. (2001). Maps of lands vulnerable to sea level rise: modeled elevations along the US Atlantic and Gulf Coasts. *Climate Research, 18*, 205-22 8.

Titus, J. G., et al., (2008). *Coastal Elevations and Sensitivity to Sea Level Rise*. Synthesis and Assessment Product 4.1 by the U.S. Climate Change Science Program and the Subcommittee on Global Change Research, Washington, DC, USA (peer-reviewed draft report).

Toman, E. L., Shindler, B., Absher, J. & McCaffrey, S. (2008). Post-fire communications: the influence of site visits on local support. *Journal of Forestry, 106*, 25-30.

Tran, J. K., Ylioja, T., Billings, R., Régnière, J. & Ayres. M. P. Testing a climatic model to predict populations dynamics of a forest pest, Dendroctonus frontalis (Coleptera: Scolydidae). *Ecological Applications* (in press).

Trenberth, K. E., Dai, A. G., Rasmussen, R. M. & Parsons, D. B. (2003). The changing character of precipitation. *Bulletin of the American Meteorological Society, 84*, 1205-1217.

Trenberth, K. E., Jones, P. D., Ambenje, P., Bojariu, R., Easterling, D., Klein Tank, A., Parker, D., Rahimzadeh, F., Renwick, J. A., Rusticucci, M., Soden, B. & Zhai, P. (2007). Observations: Surface and atmospheric climate change. In: *Climate Change 2007: The Physical Science Basis. Contribution of Working Group I to the Fourth Assessment Report of the Intergovernmental Panel on Climate Change* [Solomon, S., D. Qin, M. Manning, Z. Chen, M. Marquis, K.B. Averyt, M. Tignor, and H.L. Miller (eds.)]. Cambridge University Press, Cambridge, United Kingdom and New York, NY, USA.

Troyer, A. F. (2004). Background of U.S. Hybrid Corn II: Breeding, climate, and food. *Crop Science, 44*, 370-3 80.

Tubiello, F. N. (2005). Climate variability and agriculture: perspectives on current and future challenges. In: *Impact of Climate Change, Variability and Weather Fluctuations on*

Crops and Their Produce Markets [Knight, B. (ed.)]. Impact Reports, Cambridge, UK, 45-63.

U.S. Census Bureau, (2000a). *Methodology and Assumptions for the Population Projections of the United States*. Bureau of the Census, U.S. Department of Commerce, Washington, DC, USA.

U.S. Census Bureau, (2000b). *Population Projections of the United States by Age, Sex, Race, Hispanic Origin, and Nativity: 1999 to 2100*. Population Division Working Paper No. *38*, Bureau of the Census, U.S. Department of Commerce, Washington, DC, USA. U.S. Census Bureau, 2002: *Demographic Trends in the 20th Century*. CENSR-4, Bureau of the Census, U.S. Department of Commerce, Washington, DC, USA.

U.S. Census Bureau, 2005: *Florida, California and Texas to Dominate Future Population Growth, Census Bureau Reports*. Press Release, Bureau of the Census, U.S. Department of Commerce, Washington, DC, USA, 21 Apr 2005. <www.census.gov/Press-Release/www/releases/archives/population>.

U.S. Census Bureau, (2007). *Statistical Abstract of the United States: 2007* (126th Edition). Bureau of the Census, U.S. Department of Commerce, Washington, DC, USA.

Ungerer, M. J., Ayres, M. P. & Lombardero, M. A. J. (1999). Climate and the northern distribution limits of *Dendroctonus frontalis* Zimmermann (Coleoptera:Scolytidae). *Journal of Biogeography*, *26*, 1133-1145.

Urbanski, S., Barford, C., Wofsy, S., Kucharik, C., Pyle, E., Budney, J., McKain, K., Fitzjarrald, D., Czikowsky, M. & Munger, J. W. (2007). Factors controlling CO_2 exchange on timescales from hourly to decadal at Harvard Forest. *Journal of Geophysical Research*, *112*, G02020, doi: 10.1 029/2006JG000293.

USCL IVAR, (2002). *Climate Process Modeling and Science Teams (CPTs): Motivation and Concept*. Report 2002-1, Scientific Steering Committee, U.S. CLIVAR Office, Washington, DC, USA, 4

USDA Forest Service and U.S. Geological Survey, (2002). *Forest Cover Types: National Atlas of the United States*. USDA Forest Service and U.S. Geological Survey, Reston, VA, USA. <nationalatlas.gov/articles/biology/a_forest.html>.

USGS, (2004). *Estimated Use of Water in the United States in 2000*. USGS Circular 1268, U.S. Geological Survey, Washington, DC, USA.

Vaccaro, J. (1992). Sensitivity of groundwater recharge estimates to climate variability and change, Columbia Plateau, Washington. *Journal of Geophysical Research*, *97*, 2821-2833.

Valiela, I. (2006). *Global Coastal Change*. Blackwell, Oxford, 368

Van Vuuren, D., J. Weyant, & de la Chesnaye, F. (2006). Multi-gas scenarios to stabilise radiative forcing. *Energy Economics*, *28(1)*, 102-120.

van Wuijckhuise, L., Dercksen, D., Muskens, J., de Bruyn, J., Scheepers, M. & Vrouenraets, R. (2006). Bluetongue in the Netherlands; description of the first clinical cases and differential diagnosis; Common symptoms just a little different and in too many herds. *Tijdschrift Voor Diergeneeskunde*, *131*, 649-654.

Vasquez-Leon, M., West, C. T. & Finan, T. J. (2003). A comparative assessment of climate vulnerability: agriculture and ranching on both sides of the US–Mexico border. *Global Environmental Change*, *13*, 159-173.

Velicogna, I. & Wahr, J. (2005). Greenland mass balance from GRACE. *Geophysical Research Letters*, *32*, L1 8505, doi: 10.1 029/2005GRL023 955.

Velicogna, I. & Wahr, J. (2006). Measurements of time-variable gravity show mass loss in Antarctica. *Science*, *311*, 1754-1756.

Verger, P., Rotily, M., Hunault, C., Brenot, J., Baruffol, E. & Bard, D. (2003). Assessment of exposure to a flood disaster in a mental-health study. *Journal of Exposure Analysis and Environmental Epidemiology*, *13*, 436-442.

Viboud, C., Pakdaman, K., Boelle, P. Y., Wilson, M. L., Myers, M. F., Valleron, A. J. & Flahault, A. (2004). Association of influenza epidemics with global climate variability. *European Journal of Epidemiology*, *19(11)*, 1055-1059.

Vila, M., Corbin, J. D., Dukes, J. S., Pino, J. & Smith, S. D. (2007). Linking plant invasions to global environmental change. In: *Terrestrial Ecosystems in a Changing World* [Canadell, J., D. Pataki, and L. Pitelka, (eds.)]. Springer, New York.

Vincent, L. & Mekis, E. (2006). Changes in daily and extreme temperature and precipitation indices for Canada over the twentieth century. *Atmosphere-Ocean*, *44*, 177-193.

Visser, M. E. & Both, C. (2005). Shifts in phenology due to global climate change: the need for a yardstick. *Proceedings of the Royal Society of London*, *272*, 2561-2569.

Visser, M. E., Both, C. & Lambrechts, M. M. (2004). Global climate change leads to mistimed avian reproduction. *Advances in Ecological Research*, *35*, 89-110.

Visser, M. E., Holleman, L. J. M. & Gienapp, P. (2006). Shifts in caterpillar biomass phenology due to climate change and its impact on the breeding biology of an insectivorous bird. *Oecologia*, *147*, 164-172.

Vollaard, A. M., Ali, S., van Asten, H. A. G. H., Widjaja, S., Visser, L. G., Surjadi, C. & van Dissel, J. T. (2004). Risk factors for typhoid and paratyphoid fever in Jakarta, Indonesia. *Journal of the American Medical Association*, *291*, 2607-2615.

Volney, W. J. A. & Fleming, R. A. (2000). Climate change and impacts of boreal forest insects. *Agriculture*, *Ecosystems & Environment*, *82*, 283-294.

Vose, R., Karl, T., Easterling, D., Williams, C. & Menne, M. (2004). Climate (communication arising): Impact of land-use change on climate. *Nature*, *427*, 2 13-214.

Wade, T. J., Sandu, S. K., Levy, D., Lee, S., LeChevallier, M. W., Katz, L. & Colford Jr J. M., (2004). Did a severe flood in the Midwest cause an increase in the incidence of gastrointestinal symptoms? *American Journal of Epidemiology*, *159(4)*, 398-405.

Walker, R. R. (2001). Climate change assessment at a watershed scale. In: *Water and Environment Association of Ontario Conference*, Toronto, Canada, 12

Walter, M. T., Wilks, D. S., Parlange, J. Y. & Schneider, B. L. (2004). Increasing evapotranspiration from the conterminous United States. *Journal of Hydrometeorology*, 5, 405-408.

Walther, G. R., Post, E., Convey, P., Menzel, A., Parmesan, C., Beebee, T. J. C., Fromentin, J. M., Hoegh-Guldberg, O. & Bairlein, F. (2002). Ecological responses to recent climate change. *Nature*, *416*, 389-395.

Wayne, P., Foster, S., Connolly, J., Bazzaz, F. & Epstein, P. (2002). Production of allergenic pollen by ragweed (*Ambrosia artemisiifolia* L.) is increased in CO_2-enriched atmospheres. *Annals of Allergy*, *Asthma & Immunology*, *88*, 279-282.

WDR, (2003). *World Disaster Report: Focus on Ethics in Aid*. International Federation of Red Cross and Red Crescent Societies, Geneva, 240

WDR, (2004). *WorldDisaster Report: Focus on Community Resilience*. International Federation of Red Cross and Red Crescent Societies, Geneva, 240

Weber, R.W. (2002). Mother Nature strikes back: global warming, homeostasis, and implications for allergy. *Annals of Allergy, Asthma & Immunology, 88*, 251-252.

Webster, M., Forest, C., Reilly, J., Babiker, M., Kicklighter, D., Mayer, M., Prinn, R., Sarofim, M., Sokolov, A., Stone, P. & Wang, C. (2003). Uncertainty analysis of climate change and policy response. *Climatic Change, 61*, 295-320.

Wechsung, F., Becker, A. & Gräfe, P. (eds.), (2005). *Auswirkungen des globalen Wandels auf Wasser, Umwelt und Gesellschaft im Elbegebiet.* Weissensee-Verlag, Berlin, 416.

Wegbreit, J. & W. K. Reisen, (2000). Relationships among weather, mosquito abundance, and encephalitis virus activity in California: Kern County 1990-98. *Journal of the American Mosquito Control Association, 16(1)*, 22-27.

Weisler, R. H., Barbee, J. G. I. & Townsend, M. H. (2006). Mental health and recovery in the Gulf coast after hurricanes Katrina and Rita. *The Journal of the American Medical Association*, 296(5), 585-588.

Welch, C. (2006). Sweeping change reshapes Arctic. *The Seattle Times*. 1 Jan 2006. <seattletimes.nwsource.com/html/localnews/2002714404_arctic01main.html>.

Westenburg, C. L., DeMeo G. A. & Tanko, D. J. (2006). *Evaporation from Lake Mead, Arizona and Nevada, 1997–99.* U.S. Geological Survey Scientific Investigations Report 2006- 5252, 24 p.

Westerling, A. L., Hidalgo, H. G., Cayan, D. R. & Swetnam, T. W. (2006). Warming and earlier spring increase western U.S. forest wildfire activity. *Science, 313*, 940-943.

Wheaton, E., Wittrock, V., Kulshretha, S., Koshida, G., Grant, C., Chipanshi, A. & Bonsal, B. (2005). *Lessons Learned from the Canadian Drought Years of 2001 and 2002: Synthesis Report.* Saskatchewan Research Council Publication No. 11602-46E03, Saskatoon, Saskatchewan, 38 <www.agr.gc.ca/pfra/drought>.

Wheeler, T. R., Crauford, P. Q., Ellis, R. H., Porter, J. R. & Vara Prasad, P. V. (2000). Temperature variability and the yield of annual crops. *Agriculture, Ecosystems & Environment, 82*, 159-167.

White, M. A., Diffenbaugh, N. S., Jones, G. V., Pal, J. S. & Giorgi, F. (2006). Extreme heat reduces and shifts United States premium wine production in the 21 stcentury. *Proceedings of the National Academy of Sciences, 103*, 11217-11222, doi:10.1073/pnas.0603230103.

White, N., Sutherst, R. W., Hall, N. & Whish-Wilson, P. (2003). The vulnerability of the Australian beef industry to impacts of the cattle tick (*Boophilus microplus*) under climate change. *Climatic Change, 61*, 157-190.

Whitney, S., Whalen, J., VanGessel, M. & Mulrooney, B. (2000). *Crop profiles for Corn (Sweet) in Delaware.* <www.ipmcenters.org/CropProfiles/docs/DEcorn-sweet.html>.

WHO, (2002). *World Health Report 2002: Reducing Risks, Promoting Healthy Life*. World Health Organization, Geneva, 268

Wielgolaski, F. E. & Inouye, D. W. (2003). High latitude climates. In: *Phenology: an Integrative Environmental Science* [Schwartz, M.D. (ed.)]. Kluwer Academic Publishers, Dordrecht, The Netherlands, 175-194.

Wijffels, S. J., Willis, C., Domingues, C., Barker, P., White, N., Gronell, A., Ridgway, K. & Church, J. (2008). Changing expendable bathythermograph fall-rates and their impact on estimates of thermosteric sea level rise. *Journal of Climate* (in press).

Wilbanks, T. J. (2003). Integrating climate change and sustainable development in a place-based context. *Clim. Policy, 3*, S147-S154.

Wilbanks, T. J., Romero Lankao, P., Bao, M., Berkhout, F., Cairncross, S., Ceron, J. P., Kapshe, M., Muir-Wood, R. & Zapata-Marti, R. (2007a). Industry, settlement and society. In: *Climate Change 2007: Impacts, Adaptation and Vulnerability. Contribution of Working Group II to the Fourth Assessment Report of the Intergovernmental Panel on Climate Change* [Parry, M.L., O.F. Canziani, J.P. Palutikof, P.J. van der Linden and C.E. Hanson (eds.)]. Cambridge University Press, Cambridge, United Kingdom and New York, NY, USA.

Wilbanks, T. J., et al., (2007b). Executive Summary. In *Effects of Climate Change on Energy Production and Use in the United States.* Synthesis and Assessment Product 4.5 by the U.S. Climate Change Science Program and the Subcommittee on Global Change Research, Washington, DC, USA.

Wilbanks, T. J. et al., (2008). (in press): Effects of global change on human settlements. Ch. 3 in: *Analyses of the Effects of Global Change on Human Health and Welfare and Human Systems.* Synthesis and Assessment Product 4.6 by the U.S. Climate Change Science Program and the Subcommittee on Global Change Research, Washington, DC.

Wiley, J. W. & Wunderle J J. M.r., (1994). The effects of hurricanes on birds, with special reference to Caribbean islands. *Bird Conservation International*, *3*, 319-349.

Wilkinson, P., Pattenden, S., Armstrong, B., Fletcher, A., Kovats, R. S., Mangtani, P. et al., (2004). Vulnerability to winter mortality in elderly people in Britain: population based study. *British Medical Journal*, *329(7467)*, 647.

Williams, A. A. J., Karoly, D. J. & Tapper, N. (2001). The sensitivity of Australian fire danger to climate change. *Climatic Change*, *49*, 171-191.

Williams, D. G. & Baruch, Z. (2000). African grass invasion in the Americas: ecosystem consequences and the role of ecophysiology. *Biological Invasions*, *2*, 123-140.

Williams, D. W. & Liebhold, A. M. (2002). Climate change and the outbreak ranges of two North American bark beetles. *Agricultural and Forest Meteorology*, *4*, 87-99.

Wilson, R. J., Gutiérrez, D., Gutiérrez, J. & Monserrat, V. J. (2007). An elevational shift in butterfly species richness and composition accompanying recent climate change. *Global Change Biology*, *13*, 1873-1887.

Winkler, D. W., Dunn, P. O. & McCulloch, C. E. (2002). Predicting the effects of climate change on avian life-history traits. *Proceedings of the National Academy of Sciences*, *99*, 13595-13 599.

Wisdom, M. J., Rowland, M. M. & Suring, L. H. (eds.), (2005). *Habitat Threats in the Sagebrush Ecosystem: Methods of Regional Assessment and Applications in the Great Basin*. Allen Press/Alliance Communication Group Publishing, Lawrence, KS, USA.

Wittenberg, A. T., Rosati, A., Lau, N. C. & Ploshay, J. J. (2006). GFDL's CM2 global coupled climate models. Part III: Tropical Pacific climate and ENSO. *Journal of Climate*, *19*, 698-722.

WMO, (2007). *Scientific Assessment of Ozone Depletion: 2006*. Global Ozone Research and Monitoring Project Report No. *50*, World Meteorological Organization, Geneva, 572

Wolfe, D. W., Schwartz, M. D., Lakso, A. N., Otsuki, Y., Pool, R. M. & Shaulis, N. (2005). Climate change and shifts in spring phenology of three horticultural woody perennials in northeastern USA. *International Journal of Biometeorology*, *49(5)*, 303-309.

Wollenweber, B., Porter, J. R. & Schellberg, J. (2003). Lack of interaction between extreme high- temperature events at vegetative and reproductive growth stages in wheat. *Journal of Agronomy and Crop Science*, *189*, 142-150.

Wood, A. W., Leung, L. R., Sridhar, V. & Lettenmaier, D. P. (2004). Hydrologic implications of dynamical and statistical approaches to downscaling climate model outputs. *Climatic Change*, *62*, 189 -216

Woodhouse, C. A. & Overpeck, J. T. (1998). 2000 years of drought variability in the central United States. *Bulletin of the American Meteorological Society*, *79*, 2693-2714.

Woodward, F. I. & Lomas, M. R. (2004a). Simulating vegetation processes along the Kalahari transect. *Global Change Biology*, *10*, 383-392.

Woodward, F. I. & Lomas, M. R. (2004b). Vegetation dynamics - Simulating responses to climatic change. *Biological Reviews*, *79*, 643-370.

Xu, H. Q. & Chen, B. Q. (2004). Remote sensing of the urban heat island and its changes in Xiamen City of SE China. *Journal of Environmental Sciences*, *16(2)*, 276-281.

Yu, H., Kaufman, Y. J., Chin, M., Feingold, G., Remer, L. A., Anderson, T. L., Balkanski, Y., Bellouin, N., Boucher, O., Christopher, S., DeCola, P., Kahn, R., Koch, D., Loeb, N., Reddy, M. S., Schultz, M., Takemura, T. & Zhou, M. (2006). A review of measurement-based assessments of the aerosol direct radiative effect and forcing. *Atmospheric Chemistry and Physics*, *6*, 613-666.

Zavaleta, E. S. & Hulvey, K. B. (2004). Realistic species losses disproportionately reduce grassland resistance to biological invaders. *Science*, *306*, 1175-1177.

Zervas, C. E. (2001). *Sea Level Variations of the United States: 1854-1999*. NOAA Technical Report NOS CO-OPS *36*, National Ocean Service, National Oceanic and Atmospheric Administration, Silver Spring, Maryland, 201 <tidesandcurrents. noaa. gov/ publications/techrpt36doc.pdf>.

Zha, Y., Gao, J. & Zhang, Y. (2005). Grassland productivity in an alpine environment in response to climate change. *Area*, *37*, 332-340.

Zhang, K. Q., Douglas, B. C. & Leatherman, S. P. (2000). Twentieth-century storm activity along the U.S. east coast. *Journal of Climate*, *13*, 1748-1761.

Zhang, K. Q., Douglas, B. C. & Leatherman, S. P. (2004). Global warming and coastal erosion. *Climatic Change*, *64*, 41-58.

Zhou, G., Liu, S., Li, Z., Zhang, D., Tang, X., Zhou, C., Yan, J. & Mo, J. (2006). Old-growth forests can accumulate carbon in soils. *Science*, *314*, 1417, doi:10.1126/science.1130168.

Zhou, L. M., Tucker, C. J., Kaufmann, R. K., Slayback, D., Shabanov, N. V. & Myneni, R. B. (2001). Variations in northern vegetation activity inferred from satellite data of vegetation index during 1981 to 1999. *Journal of Geophysical Research*, *106*(D17), 20069-20083.

Zhu, Z. I. & Evans, D. L. (1994). United States forest types and predicted percent forest cover from AVHRR data. *Photogrammetric Engineering and Remote Sensing*, *60*, 525-531.

Zimmerman, R. (2002). Global climate change and transportation infrastructure: lessons from the New York area. In: *The Potential Impacts of Climate Change on Transportation: Workshop Summary and Proceedings*. Washington District of Columbia, 11 <climate.dot.gov/publications/workshop1002/zimmermanrch.pdf>.

Zimov, S. A., Schuur, E. A. G. & Chapin III, F. S. (2006). Permafrost and the global carbon budget, *Science*, *312*, 1612-1613.

Zishka, K. M. & Smith, P. J. (1980). The climatology of cyclones and anticyclones over North America and surrounding ocean environs for January and July, 1950–1977. *Monthly Weather Review*, *108*, 387-401.

Ziska, L. H. (2003). Evaluation of the growth response of six invasive species to past, present and future carbon dioxide concentrations. *Journal of Experimental Botany*, *54*, 395-404.

Ziska, L.H. & George, K. (2004). Rising carbon dioxide and invasive, noxious plants: potential threats and consequences. *World Resource Review*, *16*, 427-447.

Ziska, L.H. &Runion, G.B. (2006). Future weed, pest and disease problems for plants. Chapter 11. In: *Agroecosystems in a Changing Climate* [Newton, P., A. Carman, G. Edwards, and P. Niklaus (eds.)]. CRC, NY, USA, 262-287.

Ziska, L. H., Teasdale, J. R. & Bunce, J. A. (1999). Future atmospheric carbon dioxide may increase tolerance to glyphosate. *Weed Science*, *47*, 608-615.

Ziska, L. H., Gebhard, D. E., Frenz, D. A., Faulkner, S., Singer, B. D. & Straka, J. G. (2003). Cities as harbingers of climate change: Common ragweed, urbanization, and public health. *Journal of Allergy and Clinical Immunology*, *111*, 290-295.

Zolbrod, A. N. & Peterson, D. L. (1999). Response of high-elevation forests in the Olympic Mountains to climatic change. *Canadian Journal of Forest Research*, *29*, 1966-1978.

Zvereva, E. L. & Kozlov, M. V. (2006). Consequences of simultaneous elevation of carbon dioxide and temperature for plant–herbivore interactions: a meta-analysis. *Global Change Biology*, *12*, 27-41.

End Notes

[1] 15 USC Chapter 56A

[2] Definitions for terms used in statements of confidence and likelihood can be found in Section II.3: Characterization of Uncertainty.

[3] The report section(s) denoted in square brackets are where information supporting a particular statement can be found.

[4] Non-italicized terms and phrases such as "may," "are expected," and "are projected" are used to indicate the possibility of the occurrence of an event or phenomenon without assignment of a formal level of likelihood.

[5] A scenario is a coherent, internally consistent and plausible description of a possible future state of the world. It is not a forecast; rather, each scenario is one alternative image of how the future can unfold. The scenarios cited in this document do not include greenhouse gas emission reduction initiatives.

[6] 15 USC Chapter 56A

[7] The U.S. Global Change Research Program is incorporated within CCSP.

[8] The "Council" refers to the Federal Coordinating Council on Science, Engineering, and Technology, which has been supplanted by the National Science and Technology Council. The "Committee" refers to the Committee on Earth and Environmental Sciences, which has been supplanted by the Committee on Environment and Natural Resources under the council.

[9] 15 USC Chapter 56A, Subchapter 1, Section 2936

[10] See <www.ipcc.ch>.

[11] See <www.climatescience.gov>.

[12] This approach is described in *Guidance Notes for Lead Authors of the IPCC Fourth Assessment Report on Addressing Uncertainties*, which can be found at <www.ipcc.ch/pdf/supporting-material/uncertainty->.

[13] The date of the *Our Changing Planet* reports refers to the fiscal year for which they were prepared, not the year in which they were published.

[14] See <www.narccap.ucar.edu>.

[15] See <www-pcmdi.llnl.gov>.

[16] See <www.ispe.arizona.edu/climas/conferences/NSAW>.

[17] See <www.cara.psu.edu>.

[18] See <www.ccsr.columbia.edu/cig/taskforce>.

[19] See <www.eia.doe.gov/oiaf/1605/frntvrgg.html>.

[20] Definitions for terms used in statements of confidence and likelihood can be found in Section II.3: Characterization of Uncertainty.

[21] Definitions for terms used in statements of confidence and likelihood can be found in Section II.3: Characterization of Uncertainty.

[22] The last nine 5-year periods referred to here are: 2003–2007, 2002–2006, 2001–2005, 2000–2004, 1999–2003, 1998–2002, 1997–2001, 1996–2000, 1995–1999.
[23] Deep, narrow holes drilled vertically on land.
[24] Definitions for terms used in statements of confidence and likelihood can be found in Section II.3: Characterization of Uncertainty.
[25] See <nsidc.org/news/press/2007_seaiceminimum/20070810_index.html>.
[26] See <www.pol.ac.uk/psmsl/>.
[27] Fingerprint studies use rigorous statistical methods to compare the patterns of observed temperature changes with model expectations and determine whether or not similarities could have occurred by chance.
[28] Linear trend comparisons are less powerful than fingerprint analyses for studying cause–effect relationships, but can highlight important differences and similarities between models and observations (as in Figures IV.10 and IV.1 1).
[29] Annular modes are preferred patterns of change in atmospheric circulation corresponding to changes in the zonally averaged mid-latitude westerly winds. The Northern Annular Mode has a bias to the North Atlantic and has a large correlation with the North Atlantic Oscillation. The Southern Annular Mode occurs in the Southern Hemisphere.
[30] 1 gigaton (Gt) = 1 billion metric tons.
[31] SAP 3.3 (Karl et al., 2008) does not use discrete quantitative bounds to characterize the likelihood of events, in contrast to the IPCC (Meehl et al., 2007). The characterization of likelihood is, however, qualitatively similar between the two assessments.
[32] According to the NRC (2002), changes in Earth's orbit occur too slowly to be prime movers of abrupt change but might determine the timing of events. Abrupt climate changes of the past were especially prominent when orbital processes were forcing the climate to change during the cooling leading into and warming leading out of ice ages (NRC, 2002).
[33] ENSO describes the full range of the Southern Oscillation (seesaw of atmospheric mass or pressure between the Pacific and Indo-Australian areas) that includes both sea surface temperature increases as well as sea surface temperature decreases when compared to a long-term average. It has sometimes been used by scientists to relate only to the broader view of El Niño or the warm events–the warming of sea surface temperatures in the central and eastern equatorial Pacific. The acronym, ENS O, is composed of El Niño–Southern Oscillation, where El Niño is the oceanic component and the Southern Oscillation is the atmospheric component of the phenomenon.
[34] The North Atlantic Oscillation (NAO) is the dominant mode of winter climate variability in the North Atlantic region ranging from central North America to Europe and into Northern Asia. The NAO is a large-scale seesaw in atmospheric mass or pressure between the subtropical high and the polar low. Similarly, the Arctic Oscillation (AO) refers to opposing atmospheric pressure patterns in northern middle and high latitudes. The NAO and AO are different ways of describing the same general phenomenon.
[35] A planetary wave is a large horizontal atmospheric undulation that is associated with the polar-front jet stream and separates cold, polar air from warm, tropical air.
[36] Definitions for terms used in statements of confidence and likelihood can be found in Section II.3: Characterization of Uncertainty.
[37] Definitions for terms used in statements of confidence and likelihood can be found in Section II.3: Characterization of Uncertainty.
[38] Definitions for terms used in statements of confidence and likelihood can be found in Section II.3: Characterization of Uncertainty.
[39] See <www.nsidc.org>.
[40] A keystone species is defined as a species that has a disproportionate effect on its environment relative to its abundance or total biomass. Typically, ecosystems experience dramatic changes with the removal of such a species.
[41] Ecological cascades are defined as sequential chains of ecological effects, including starvation and death, beginning at the bottom levels of the food chain and ascending to higher levels, including apex predators.
[42] C_3 and C_4 refer to different carbon fixation pathways in plants during photosynthesis. C_3 is the most common pathway, and C_3 crops (e.g., wheat, soybeans, and rice) are more responsive than C_4 crops such as maize to increases in CO_2.
[43] See <www.bnl.gov/face/>.
[44] High Plains Regional Climate Center < http://www.hprcc.unl.edu/>
[45] 1 Pg = 1 ,000,000,000,000,000 grams = 1 Gt.
[46] The Palmer Drought Severity Index is used by the National Oceanic and Atmospheric Administration and uses a formula that includes temperature and rainfall to determine dryness. It is most effective in determining long-term drought.
[47] Redox potential is defined as the tendency of a chemical species to acquire electrons and therefore be reduced.
[48] A freshwater lens is defined as a relatively thin layer of freshwater within island aquifer systems that floats on an underlying mass of denser seawater. Numerous factors control the shape and thickness of the lens, including

the rate of recharge from precipitation, island geometry, and geologic features such as the permeability of soil layers.

[49] The daily maximum 8-hour ozone concentration is a measurement of air quality based on averaging the fourth-highest daily maximum ozone concentrations over an 8-hour period. The standard is set at 0.08 ppm.

[50] Hypolimnetic water is the layer of water in a thermally stratified lake that lies below the thermocline, is noncirculating, and remains perpetually cold.

In: Assessing Climate Change Impacts on the United States ISBN: 978-1-60876-160-9
Editor: Alex B. McNeill © 2010 Nova Science Publishers, Inc.

Chapter 2

POTENTIAL IMPACTS OF CLIMATE CHANGE IN THE UNITED STATES*

Congressional Budget Office

Introduction

Human activities around the world—primarily fossil fuel use, forestry, and agriculture—are producing growing quantities of emissions of greenhouse gases, other gases, and particulates and are also greatly altering the Earth's vegetative cover. A strong consensus has developed in the expert community that if allowed to continue unabated, the accumulation of those substances in the atmosphere and oceans, coupled with widespread changes in patterns of land use, will have extensive, highly uncertain, but potentially serious and costly impacts on regional climate and ocean conditions throughout the world.

This paper summarizes the current state of scientific understanding of the potential effects of projected changes in climate and related developments. The paper describes the wide range of potential impacts, including changes in: seasonal weather patterns; the amount and type of precipitation; storms and sea level; regular climate fluctuations; ocean acidity; ecosystems and biodiversity; agriculture, forestry, and fishing; water supply and other infrastructure; and human health. The discussion focuses mainly on projections of impacts in the United States but also refers to impacts elsewhere that could be particularly severe and could indirectly affect the United States. The paper draws from various primary and secondary sources, particularly the *Fourth Assessment Report of the Intergovernmental Panel on Climate Change* (IPCC), a major national assessment recently released by the National Science and Technology Council in the Executive Office of the President, and a number of surveys published by the U.S. Climate Change Science Program, which integrates executive-branch research on climate and global change.[1]

---

* This is an edited, reformatted and augmented version of a Congressional Budget Office publication dated May 2009.

The paper emphasizes the extensive uncertainty about the future growth of greenhouse-gas emissions and resulting climate-related developments, and the implications of that uncertainty for climate policy. Uncertainty arises from several sources, including limitations in current data, imperfect understanding of physical processes, and the inherent unpredictability of economic activity, technological innovation, and many aspects of the interacting components (land, air, water and ice, and life) that make up the Earth's climate system.

Uncertainty does not imply that nothing is known about future developments, but rather that projections of future changes in climate and of the resulting impacts should be considered in terms of ranges or probability distributions.[2] For example, some recent research suggests that the median increase in average global temperature during the 21st century will be in the vicinity of 9° Fahrenheit (F) if no actions are taken to reduce the growth of greenhouse-gas emissions. However, warming could be much less or much greater than that median level, depending on the growth of emissions and the response of the climate system to those emissions.

The uncertainties surrounding the extent of climate change and its effects on human welfare complicate policy choices because any given emissions scenario could result in much more or much less dramatic changes in climate (and consequent impacts) than are expected. A full evaluation of any policy requires an assessment not only of its most likely effect on climate impacts but also of how it might shift the probabilities of impacts that are unlikely but would be very severe if they occurred. Because of that uncertainty, policymakers might wish to consider taking more action as a hedge against those severe outcomes than they would just to address the expected or most likely outcome.

THE LINK BETWEEN EMISSIONS AND CLIMATE

Over the past century, researchers have developed an increasingly sophisticated understanding of the climate system through direct observations of the system, statistical analyses of those observations, and, more recently, simulations of the system using computer models. (See Box 1 for a discussion of climate modeling.) According to current understanding, as energy from the sun is absorbed by the Earth's climate system and radiated back into space, greenhouse gases increase the amount of energy temporarily held in the lower atmosphere, keeping the Earth's surface warmer than it would otherwise be.[3] Those gases include carbon dioxide (CO_2), methane (CH_4), nitrous oxide (N_2O), and several man-made gases containing fluorine and chlorine. Soot in the atmosphere also absorbs incoming energy and tends to warm the lower atmosphere, whereas aerosols tend to exert offsetting cooling effects.[4]

Since the onset of the Industrial Revolution about two centuries ago, the growth of emissions from human activities has resulted in rising atmospheric concentrations of greenhouse gases, other gases, and aerosols.[5] The accumulation of greenhouse gases and other warming substances has been the dominant influence contributing to climate change, triggering an irregular but accelerating warming of the Earth's surface and various consequent changes.[6] By current estimates, the average global temperature (currently about 58.0°F, or 14.4° Celsius) has most probably risen by about 1.4°F since the mid-19th century.[7] However, that estimate involves substantial uncertainty. There is roughly a 5 percent chance that the warming has been less than 1.0°F and an equal chance that it has been more than 1.7°F.

In spite of extensive uncertainties, both in the data and in the projections based on that data, researchers are increasingly confident about their ability to decipher the relationship between past activities and recent warming, to distinguish the effect of rising concentrations of greenhouse gases and changing land-use patterns from natural variability and other influences on climate, and to develop projections of the pace and ultimate magnitude and distribution of future warming and related changes.[8] The great majority of experts conclude that they cannot explain observed patterns of warming and related changes without considering emissions from human activities and that it is very likely that most of the warming is due to human activities.[9] Those experts also conclude that ongoing emissions at current or rising levels will continue to raise atmospheric concentrations and temperatures indefinitely.[10] Even immediate, dramatic reductions in emissions would not necessarily halt changes to which past emissions have already committed the climate: If concentrations were stabilized today by cutting emissions immediately to a small fraction of current levels, the average global temperature would gradually continue to rise—increasing by another 0.5°F to 1.6°F above recent levels by the end of this century, according to one study.[11]

The accumulation of gases in the atmosphere, changes in land use, and the resulting changes in climate are having further cascading effects. The oceans are absorbing CO_2 and heat from the atmosphere and will continue to do so for centuries, even if emissions are curtailed, changing the chemistry of the oceans and raising sea level.[12] Glaciers and ice caps are gradually melting in response to rising temperatures, also contributing to a rise in sea level.[13] Forests and soils are responding to elevated CO_2 concentrations and temperatures as well.[14] Moreover, the effects of greenhouse gases interact with and can be enhanced or offset in particular regions by the direct and indirect effects of emissions of different types of aerosols as well as by changes in land cover.[15]

The climate system's response to emissions and to changes in land use is thus a complex, gradually unfolding process involving both rapid changes and interactions among several slowly responding components that adjust at differing and uncertain rates. Once perturbed by a large increase in greenhouse-gas concentrations and extensive changes in land cover, the system will take centuries to respond fully, as the oceans warm, ice melts, and other parts of the system adjust.[16] The responses will not be uniform but will vary across regions and seasons. Furthermore, the response need not be entirely smooth or evenly paced but could involve rather abrupt shifts long after the processes have been triggered.[17]

Temperature trends can be masked to some extent by inherent short-term variability in global and regional temperatures—a phenomenon replicated by climate models. For example, between 1980 and 2007, the average annual global temperature grew at an average rate of about 0.03°F per year, but in one year (1997) it rose by 0.42°F, and in another (1999) it fell by even more.[18] Some recent projections suggest that annual temperatures may cool slightly over the next decade in Europe and North America, while the average global temperature remains fairly steady.[19] Aerosols from volcanic eruptions and other sources could temporarily result in regional or even global cooling. Such fluctuations are not inconsistent with an ongoing long-term warming trend brought about by human activities.

Box 1. The Current State of the Art in Climate Modeling

Climate models have gradually but steadily improved in detail, in the range of phenomena that they include, and in their ability to replicate characteristics of the Earth's climate system.[1] Studies that measure and compare the ability of the current generation of models to simulate recent climate conditions—such as the extensive comparison of 23 of the most complex models in the *Fourth Assessment Report of the Intergovernmental Panel on Climate Change*—show that nearly all of them have improved in most respects. The most advanced models typically include linked representations of the atmosphere, oceans, sea ice, and land surface; most types of greenhouse gases and other relevant components of atmospheric chemistry; and characteristics of the carbon cycle.

The models replicate seasonal and large-scale regional variations in temperature and, to a lesser extent, precipitation; they also replicate large-scale ocean currents, large-scale ocean and climate oscillations, and storms and jet streams in the middle latitudes. However, for some phenomena, such as the dynamics of glaciers, the models remain in early stages of development. Moreover, because the global models have relatively coarse spatial resolutions, regional models of higher resolution are used to "downscale" the results of large models to analyze smaller-scale phenomena. The quality of such downscaling exercises is constrained by the limitations of and uncertainties in the global models.

The models plausibly replicate 20th-century climate trends when they are run with historical emissions of greenhouse gases, other types of emissions, and variations in natural forces, such as volcanic eruptions and fluctuations in solar energy. No model replicates those climate trends through variations in natural forces alone.

Because research groups vary in the way they represent uncertain aspects of the Earth's climate system, the models produce a range of results for many important climate indicators. Studies comparing the models find that the average of the models' simulations (referred to as the "ensemble-mean model") generally replicates features of the system better than does any single model. As a result, researchers have focused on ensemble-mean projections of climate change under various scenarios as being a type of "best guess" of likely changes, taking the range of model results as representing, to an extent, the uncertainty in researchers' current understanding of likely developments.[2]

However, that approach understates uncertainty because it overlooks the fact that each model incorporates its builders' best guesses for uncertain parameters (guesses that do not reflect the full range of uncertainty about them) and that no model includes all human influences on regional climates. To better analyze the full extent of uncertainty, researchers turn to simpler models that can be "tuned" to replicate the results of larger models but that can also be systematically varied to simulate other possible combinations of parameters.

[1] This discussion draws primarily from David C. Bader and others, *Climate Models: An Assessment of Strengths and Limitations. Synthesis and Assessment Product 3.1. Report by the U.S. Climate Change Science Program and the Subcommittee on Global Change Research* (Washington, D.C.: U.S. Climate Change Science Program, 2008); and D. A. Randall and others, "Climate Models and Their Evaluation," in Susan Solomon and others, eds., *Climate Change 2007: The Physical Science Basis. Contribution of Working Group I to the Fourth Assessment Report of the Intergovernmental Panel on Climate Change* (Cambridge, U.K.: Cambridge University Press, 2007).

[2] See Gerald A. Meehl and others, "Global Climate Projections," in Susan Solomon and others, eds., *Climate Change 2007: The Physical Science Basis. Contribution of Working Group I to the Fourth Assessment Report of the Intergovernmental Panel on Climate Change* (Cambridge, U.K.: Cambridge University Press, 2007), pp. 797–801.

Longer-term scenarios involve considerable uncertainties that become more extensive the further into the future researchers attempt to project. Important uncertainties include the following:

- *How population growth, technological developments, and economic change will influence land cover and the growth of emissions.* Projections of long-term changes in population, output, emissions, and land use are necessarily quite uncertain. For example, one analysis concludes that the global population will probably be around 8.4 billion at century's end (compared with nearly 6.8 billion today), but the study also projects a 10 percent chance that the population will be less than 5.6 billion and a 10 percent chance that it will be more than 12.1 billion.[20] Wide ranges of uncertainty also apply to projections of long-term changes in productivity, advances in energy technology, and the demand for fossil fuel and land use.[21] As a result, cumulative emissions of greenhouse gases and changes in land use over the course of the 21st century and beyond can be projected only as the extremely uncertain outcome of several very uncertain long-term trends.

- *How rapidly the climate system will respond to accumulating greenhouse gases and other changes, and how much warming will ultimately occur (what is referred to as the climate's sensitivity).* The response to rising concentrations of greenhouse gases and other perturbations involves a range of feedbacks that can begin to develop immediately but, in some important cases, can take centuries or even millennia to fully unfold.[22] With respect to the short-term response, the most important uncertainties include the influence of clouds (which can enhance or offset warming, depending on where and when they form) and the effects of aerosols (which can not only offset warming directly but also influence cloud formation).[23] Longer-term uncertainties involve the uptake of CO_2 by the oceans and forests, the absorption of heat by the oceans, and the effect of warming on the continental ice sheets of Greenland and Antarctica.[24] Reasonable differences in how such processes are treated in climate models lead to large differences in the projected magnitude and distribution of warming and other changes, even when the models are all run with the same emissions scenario. Taking such uncertainties into account, the IPCC concluded that for any specific scenario, the change in average global temperature in 2100 was likely to be within a range of 40 percent below to 60 percent above the models' average.[25] That is, if all of the models were run with a specific emissions scenario, and the warming in 2100 (averaged across all models) was 5°F, then there are about two chances in three that the actual warming under such a scenario would be between 3°F and 8°F—and one chance in three that it could be higher or lower. There is substantial uncertainty about the amount of further warming over following centuries as well, because it will depend on how rapidly natural processes remove greenhouse gases from the atmosphere and how rapidly the system responds to the remaining gases.[26]

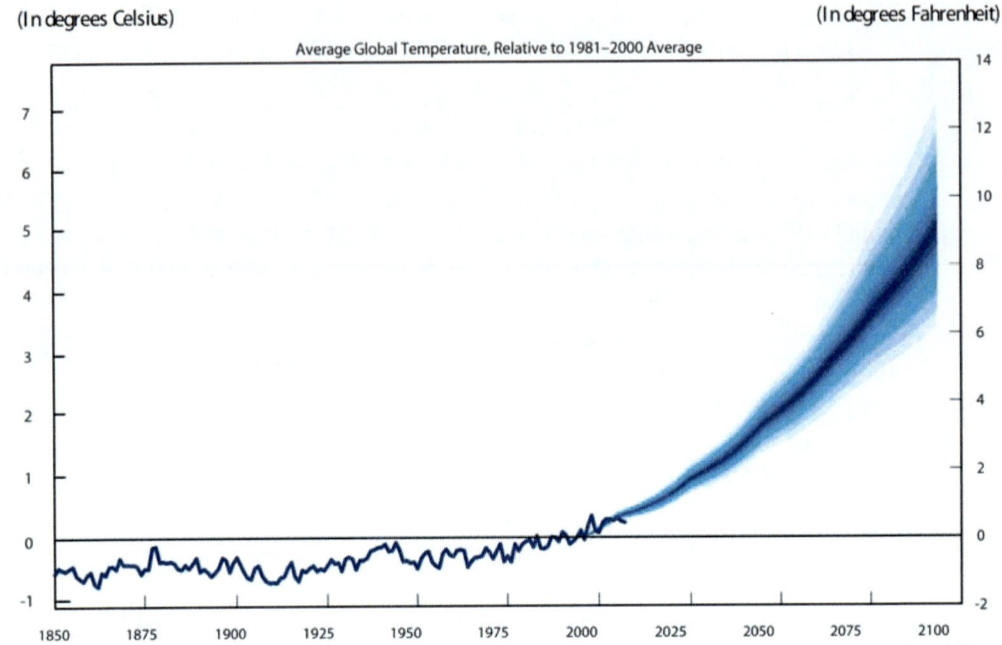

Source: Congressional Budget Office. Historical data are from the Hadley Centre for Climate Prediction and Research, http://hadobs. metoffice.com/hadcrut3/diagnostics/global/nh+sh/annual, and described primarily in P. Brohan and others, "Uncertainty Estimates in Regional and Global Observed Temperature Changes: A New Dataset from 1850," *Journal of Geophysical Research,* vol. 111 (June 24, 2006). The projection is based on data provided by Henry Jacoby, Massachusetts Institute of Technology, in a personal communication to CBO, December 22, 2008; the results are discussed in A.P. Sokolov and others, *Probabilistic Forecast for 21st Century Climate Based on Uncertainties in Emissions (Without Policy) and Climate Parameters,* Report No. 169 (Cambridge, Mass.: MIT Joint Program on the Science and Policy of Global Change, 2009), http://globalchange.mit.edu/files/document/ MITJPSPGC_Rpt169.pdf.

Note: The projection, which is interpolated from decadal averages beginning in 1995, shows the possible distribution of changes in average global temperature as a result of human influence, relative to the 1981–2000 average and given current understanding of the climate. Under the Sokolov study's assumptions, the probability is 10 percent that the actual global temperature will fall in the darkest area and 90 percent that it will fall within the whole shaded area. However, actual temperatures could be affected by factors that were not addressed in the study (such as volcanic activity and the variability of solar radiation) and whose effects are not included in the figure.

Figure 1. Historical and Projected Climate Change

Uncertainties regarding the amount of future emissions and the climate system's response appear to make roughly comparable contributions to the overall uncertainty about the impact of warming over the 21st century.[27] Taken together, those uncertainties are sufficiently large that many experts have been reluctant to project a likely range of long-term changes in the global climate, and no firm consensus on such a range exists. Nevertheless, an analysis by one research group that has undertaken a particularly detailed treatment of both economic and scientific uncertainties provides a useful illustration of the range of possible outcomes over the 21st century. That group recently revised its projections upward substantially, concluding that if no actions are taken to reduce the growth of emissions, the median increase in the average annual global temperature over the course of the 21st century would be around 9°F, with a two-in-three chance that warming will be between 7°F and 11°F and a 9-in10 chance that it will be between 6°F and 13°F.[28] By way of comparison, the transition from the depths of the last ice age (the Last Glacial Maximum, about 21,000 years ago) to the present climate involved a long-term increase in temperature of roughly 7°F to 13°F.[29] Even if global emissions were nearly eliminated over the next hundred years, one study indicates that temperatures would still rise by between 1°F and 5°F above recent levels.[30] (See Figure 1 for an illustration of historical and projected warming.)

POTENTIAL IMPACTS OF CLIMATE CHANGE

Rising concentrations of greenhouse and other gases, changes in land use, and the resulting shifts in climate will result in many different kinds of impacts on widely differing scales and developing over widely varying periods of time. A number of such changes are already evident in the data, although, in many cases, the effects of human influences on the climate are very difficult to distinguish from other contributing causes.[31]

- A changing climate will involve changes in typical patterns of regional and seasonal temperature, rainfall, and snowfall, as well as changes in the frequency and severity of extreme events, such as heat waves, cold snaps, droughts, storms, and floods.[32] (See Figure 2 for an illustration of changes in averages and extremes of temperature and precipitation.) Regional climates in the United States are expected to become more variable, with more intense and more frequent extremes of high temperature and rainfall. In general, extreme events tend to have disproportionately greater effects: A small percentage increase in hurricane wind speeds, for example, can greatly increase the potential damage.[33] Unfortunately, changes in the frequency and intensity of extreme events—especially precipitation—are also more difficult to simulate and project.[34]
- Some effects, including the melting of ice caps, a rise in sea level, and increasing acidity of the oceans, will unfold relatively gradually. Other effects could appear comparatively abruptly. Some extreme, abrupt changes—such as major shifts in ocean currents and regional patterns of rainfall—could occur unexpectedly, even centuries after emissions have been curtailed and concentrations have been stabilized.[35] A further complication is that changes, both gradual and extreme, will unfold at varying paces that, in some cases, are very difficult to project.
- In some cases, interactions among effects can amplify them in unpredictable ways. For example, rising sea level exacerbates coastal flooding from more intense storms.[36] Similarly, agricultural impacts are driven not only by changes in temperature and rainfall, but also by the enhancement of plant growth by higher atmospheric concentrations of CO_2 (the carbon fertilization effect), by the destructive

effects of more ground- level ozone, and by more subtle effects, such as changes in susceptibility to pests and fire.[37]

- Yet another complicating factor is that the effects of climate change will occur along with demographic, economic, and environmental developments that are also difficult to project, that will have an impact on many of the same activities affected by climate change, and that, in many cases, will interact with (and mitigate or exacerbate) the effects of climate change in ways that are particularly difficult to untangle or project.[38]

Some changes, such as gradual shifts in average regional weather conditions, are likely to be relatively easy for a modern economy to anticipate and adapt to. Others, such as abrupt and unexpected changes in regional patterns of rainfall, could present much greater difficulties and impose much higher costs. Consideration of such complexities has gradually led researchers to shift from trying to gauge impacts precisely and to focus instead on characterizing vulnerabilities—the extent to which systems may be unable to cope with or adapt to adverse changes from multiple stresses.[39]

Effects on the Physical Environment

In general, projections of regional and local changes in climate are increasingly uncertain, the smaller the scale. Nevertheless, researchers find agreement among current climate models in consistently projecting certain types of changes in climate patterns over large geographic areas.[40] That agreement suggests that those projected changes are indeed likely to occur.

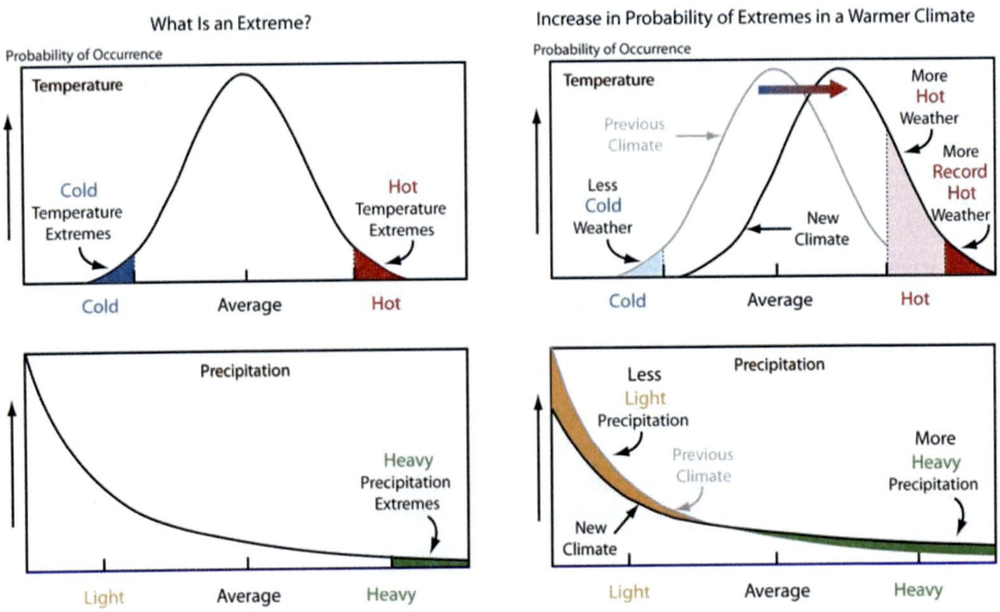

Source: Thomas R. Karl and others, eds., *Weather and Climate Extremes in a Changing Climate. Regions of Focus: North America, Hawaii, Caribbean, and U.S. Pacific Islands. A Report by the U.S. Climate Change Science Program and the Subcommittee on Global Change Research*

(Washington, D.C.: Department of Commerce, National Oceanic and Atmospheric Administration, National Climate Data Center, June 2008), p. 2.

Figure 2. Illustration of Changes in Averages and Extremes in Temperature and Precipitation

Temperature

Models suggest that as concentrations of greenhouse gases increase, warming will tend to be relatively greater at high latitudes in the Northern Hemisphere.[41] Warming will also tend to be greater over the continents than over the oceans and greater in the interiors of continents than near the coasts.[42] Average annual warming in the continental United States is likely to exceed the global average by roughly 25 percent to 35 percent—and by roughly 70 percent in Alaska.[43] Thus, if warming over the course of the 21st century is in the middle of the likely range indicated by the study discussed in the previous section (that is, about 9°F), average annual temperatures would rise by roughly 12°F in the continental United States and by nearly 16°F in Alaska. Further warming can be anticipated over the longer term even if emissions are curtailed by the end of this century.

In the central United States, the range of warming can be very roughly illustrated in terms of moving south—or, conversely, of having temperature patterns move north. At the low end of the study's likely range of 21st century warming (an increase of 7°F in average global temperatures), average annual temperatures would shift about two states north; that is, North Dakota's temperature range would become more like Nebraska's is today. At the high end of that likely range (that is, an increase of 11°F in average global temperatures), temperatures would shift about three states north over the 21st century: North Dakota would become almost as warm as Kansas is today, while central Kansas would become as warm as south Texas is today. (See Figure 3 for historical patterns of temperature and precipitation in the continental United States.)

Warming will also cause seasonal shifts throughout the world. It is very likely to result in shorter and milder winters, fewer and warmer cold days and nights, and fewer frost days; and it is also likely to yield longer and hotter summer conditions over most land areas, longer growing seasons, more and hotter warm days and nights, and more and hotter heat waves.[44] Northerly regions of the United States will tend to warm most in the winter, largely because declining snow and ice cover will increase the amount of heat absorbed by the land surface, while regions in the Southwest will tend to warm most in the summer.[45] Nevertheless, even northerly regions of the contiguous United States could warm substantially during the summer.[46]

Precipitation

Although projections of regional patterns of rainfall and snowfall are generally more uncertain than those for temperature, a warmer world is virtually certain also to experience more precipitation because warmer air can hold more moisture.[47] Indeed, precipitation has increased in most of the United States over the past halfcentury.[48] According to the results from current climate models, however, the increase in average precipitation over the course of the coming century will be unevenly distributed: Many regions and seasons that already have greater precipitation will tend to get more, while drier regions in the northern subtropics and midlatitudes will tend to get less.[49] Polar regions are likely to be an exception to that rule: They will tend to experience more precipitation despite being relatively arid at present.[50]

Projections of precipitation are particularly uncertain across a broad swath of northern Mexico and the continental United States. In those regions, a transition zone between relatively dry and wet climates moves north and south with the seasons, and the response of precipitation in that zone to a particular amount of warming is unclear.[51] Nevertheless, most models project that almost any amount of warming will cause much of the Southwest (as well as neighboring countries in much of Central America) to become more arid over the course of the 21st century—and at the high end of the range of possible warming, considerably more arid.[52]

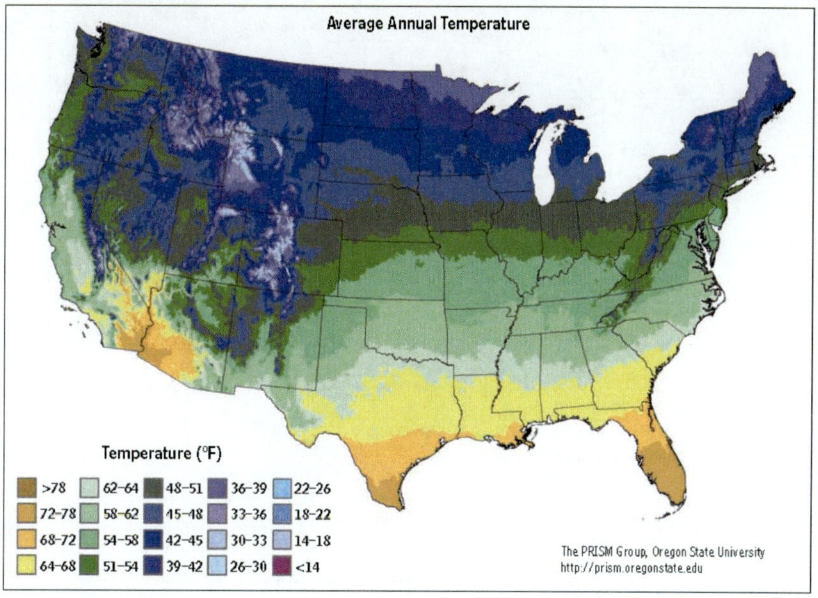
Average Annual Temperature
Temperature (°F)
>78
72-78
68-72
64-68
62-64
58-62
54-58
51-54
48-51
45-48
42-45
39-42
36-39
33-36
30-33
26-30
22-26
18-22
14-18
<14
The PRISM Group, Oregon State University
http://prism.oregonstate.edu

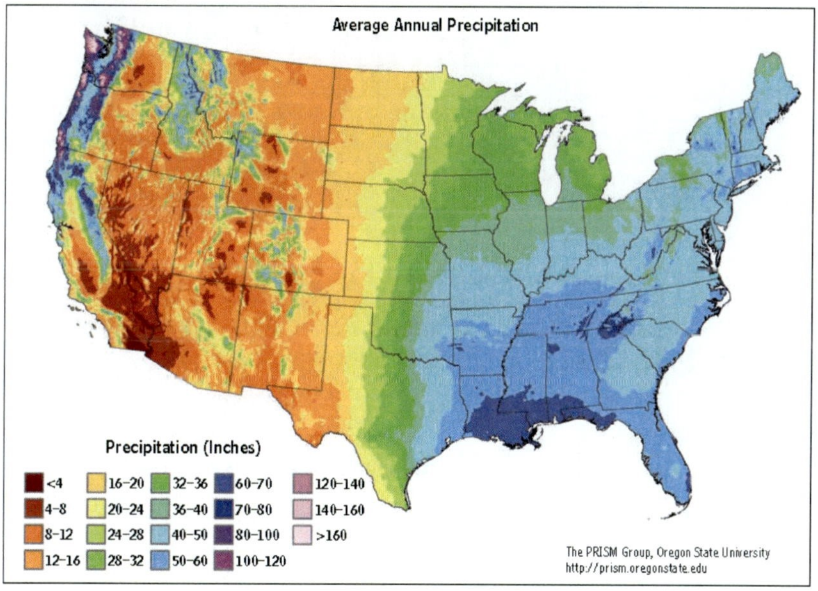
Average Annual Precipitation
Precipitation (Inches)
<4
4-8
8-12
12-16
16-20
20-24
24-28
28-32
32-36
36-40
40-50
50-60
60-70
70-80
80-100
100-120
120-140
140-160
>160
The PRISM Group, Oregon State University
http://prism.oregonstate.edu

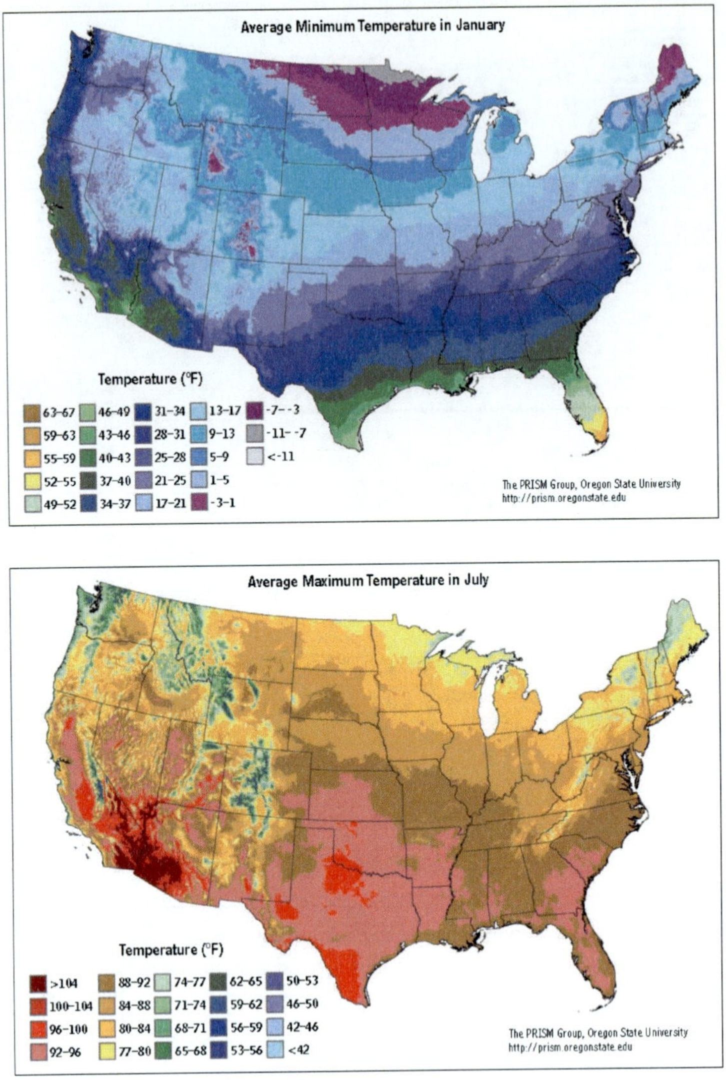

Source: The PRISM Group, Oregon State University; http://prism.oregonstate.edu.

Figure 3. Historical Climate Patterns in the Continental United States, 1971 to 2000

Changes in seasonal weather patterns are more uncertain than annual averages, but as a rule, models show agreement in predicting that warmer winters will bring more precipitation in northerly regions of the United States, while summers will tend to become drier in southerly regions.[53] With warmer temperatures, less of the precipitation will fall as snow, and the snow will tend to melt earlier in the spring.[54] Warming even in the lower half of the projected range could dramatically shorten the snow season in much of the lower 48 states by the end of this century.[55] As discussed below, less snowfall and earlier melting could greatly reduce the supply of water during dry summer months in large parts of the western United States.

Patterns of rain and snowfall in specific locations also are likely to become more variable.[56] Severe summer thunderstorms are likely to become more frequent in large areas of the country.[57] In wet and dry regions alike, more intense and heavy rainfall will tend to be inter-spersed with longer relatively dry periods.[58] That pattern implies, somewhat paradoxically, that in a wetter, warmer world, many regions will experience greater risks both of flooding and of drought.

Regional patterns of precipitation also could be affected by changes in patterns of land use, by emissions and deposition of aerosols (and by their influence on cloud formation), and by the response of vegetation to changes in temperature and CO_2 concentrations.[59] Such regional responses are particularly difficult to project.

Cyclones, Typhoons, and Hurricanes

Many researchers have concluded that, on balance, warming over the 21st century will also tend to increase the overall size and intensity of tropical cyclones (called typhoons in much of the Pacific Ocean but called hurricanes in the Atlantic and eastern Pacific Oceans) by raising surface temperatures of the water in regions where such storms form.[60] A recent economic assessment concludes that the stronger hurricanes that would result from a temperature increase of 4.5°F—below the low end of the projected range of warming—would increase average annual U.S. hurricane damages by about $8 billion (in constant 2005 dollars), or about 0.06 percent of current gross domestic product.[61] However, future trends in the frequency and intensity of hurricanes remain very uncertain, especially in the North Atlantic, where nearly all hurricanes that reach the United States form.[62] An accurate assessment will require a better understanding of how large climate fluctuations (discussed below), which also influence the formation and persistence of hurricanes, are likely to change in a warming world.[63] Large storms outside of the tropics, such as nor'easters, may become less frequent but are likely to become more intense.[64]

Ocean currents

Warming also could influence the Earth's major ocean currents, which are driven in part by wind patterns and which, in turn, strongly affect many regional patterns of climate. At present, warming appears to be strengthening the Antarctic Circumpolar Current, a strong westerly circulation around Antarctica in the Southern Ocean that is believed to account for the bulk of the wind-driven currents of the world's oceans.[65] In contrast, warming during the coming century is expected to slow the Atlantic meridional overturning circulation (MOC), a key component of the ocean circulation system that drives warm tropical waters far into the North Atlantic, into the deep ocean, and back to the south. That circulation is regulated in part by processes that increase the salinity and density of surface waters in the north Atlantic, causing them to sink into the deep ocean. Increasing rainfall and decreasing formation of sea ice could reduce surface salinity in the region, slowing the circulation process. No consensus currently exists regarding the likelihood that the MOC could shut down completely or the temperature increase that might trigger such an event.[66]

Climate oscillations

Warming could be expected to have significant impacts on the large semiregular climate oscillations that dramatically affect weather patterns throughout the world, such as the El Niño/Southern Oscillation (ENSO), which results from interactions between ocean currents and wind patterns across the Pacific Ocean and causes significant variations in regional weather patterns from southern Africa to northeastern North America on a timescale of several years. Although researchers conclude that the overall global climate is likely to become more like that of a typical El Niño year, they currently cannot determine whether even dramatic warming would cause any significant change in the frequency or intensity of ENSO events.[67] Some evidence suggests that warming will tend to shift the typical effects of ENSO events in the United States—such as intense rainfall in the Southwest—northward and eastward over the 21st century.[68]

Sea level

Warming will gradually raise sea level for many centuries after concentrations of greenhouse gases have been stabilized, partly by causing the oceans to expand, and partly by melting glaciers, mountain ice caps, and parts of the Greenland and Antarctic ice sheets.[69] Warming of the oceans alone is projected eventually to raise sea level by roughly 4 to 13 inches per degree Fahrenheit of warming.[70] Moreover, even warming at the very low end of the projected range could result in melting of ice sheets that generates an unexpectedly rapid rise in sea level. The rate of that rise is very poorly understood, however, and constitutes one of the crucial scientific uncertainties surrounding climate change—so much so that the IPCC was unwilling to estimate a probable range for the contribution from melting ice sheets.[71]

Rising sea level will cause a gradual, progressive inundation and more rapid erosion of shorelines, coastal wetlands, and coastal infrastructure.[72] In the United States, most of the land vulnerable to inundation is in coastal Louisiana and east Texas, south Florida, the Pamlico- Albemarle Peninsula of North Carolina, and the Eastern Shore of Maryland.[73] Rising sea level will also tend to exacerbate the impacts of storm surges and floods in tidal zones: By one estimate, a three-foot rise in sea level would roughly double the incremental damages resulting from warming-induced increases in the intensity of hurricanes.[74] Because most large North American cities are on tidewater or rivers, they will face a gradual increase in the probability of what are currently unusually large storm surges. One study, for example, concludes that in the middle of the range of projected warming, New York City could experience what are now considered 100-year floods every three to four years by the end of the 21st century.[75]

Ocean acidification

The world's oceans have partly offset the accumulation of CO_2 in the atmosphere by absorbing about one-third of historical emissions from human activities.[76] However, the absorption of CO_2 is also making ocean water more acidic. Ongoing emissions will continue to influence ocean chemistry: Even if the amount of emissions is at the low end of the projected range, seawater in some regions is expected to become more acidic during the 21st century than it has been in millions of years, with potentially serious effects on ocean eco-systems.[77]

Consequences for Biological Systems

By shifting climate zones in the United States about 50 miles poleward (and mountain climates about 300 feet higher) per degree Fahrenheit of warming, by shifting patterns of precipitation, and by changing ocean conditions, climate change will shift the types of vegetation and animals that thrive in specific locations.[78] Such environmental shifts are likely to affect unmanaged ecosystems, agriculture, commercial forestry, and fisheries.[79]

Ecosystems and Biodiversity

Ecosystems in the United States and around the world have already been measurably affected by rising concentrations of greenhouse gases, land-use changes, and changes in climate.[80] Under projected rates of warming over the next century, ongoing shifts in climate zones are very likely to further affect ecosystems. The impact on unmanaged ecosystems, including those in many of the world's regions of unusually rich biodiversity, could be particularly harmful.[81] In addition, shellfish, plankton, and corals face a highly uncertain threat from acidification of the world's oceans, which is making it more difficult for them to form and maintain their shells and skeletons.[82] In some regions, the pace of change in conditions on land and in the oceans could exceed species' capacity to respond by adapting or migrating.[83]

Warming is thus likely to place a substantial but very uncertain share of the Earth's species at risk of extinction—a share that is likely to grow with the rate and ultimate amount of emissions and warming.[84] Further uncertainties involve the combined effects on plant growth from shifts in climate, greater concentrations of CO_2, and higher levels of ozone; the pace at which various species will be able to disperse in the face of a shift in climate that, in geologic terms, is very rapid; and the simultaneous stresses caused by other widespread types of human disruption of natural habitats.[85] Given that about 2 percent of all known species of plants and animals have disappeared in the past few centuries and that the bulk of human population growth and resource use has occurred only since the mid-20th century, increasing pressures on biodiversity unrelated to climate are likely to interact with and exacerbate any stresses associated with climate change.[86]

Agriculture

Studies suggest that, in contrast to the difficulties for unmanaged ecosystems, the risks to agriculture in the United States are comparatively modest. Farmers in most regions of the United States can largely mitigate adverse impacts during the 21st century by adapting crops and techniques to changing climate conditions— for instance, by shifting northward the cultivation of specific crops and altering the timing of planting and harvesting in response to the earlier onset of warm weather and drier summer conditions.[87] If warming remains in the lower part of the projected range and if conventional air pollution controls mitigate the potential for damage from ozone, aggregate crop yields could even increase in the United States from the combined impact of changing regional climates and the carbon fertilization effect.[88]

Nevertheless, agriculture in some regions of the United States is likely to be adversely affected by climate change during the 21st century, mainly because of changes in rainfall rather than changes in temperature, and the effects are likely to be more adverse, the greater the amount of warming.[89] Although northerly regions of the country (as well as Canada) are likely to benefit from longer growing seasons and greater precipitation, more southerly regions tend to be much more sensitive to higher temperatures and would benefit less from increased precipitation.[90] Declining water supply could further affect agriculture in Western states. In other agricultural regions, greater and more intense precipitation would probably increase damage to crops from flooding.[91] Some effects of a changing climate—increasing variability of local climates as well as unpredictable abrupt climate shifts—could impair farmers' ability to determine what crops to cultivate and where and when to cultivate them. Over the coming century, the United States may become largely unsuitable for the cultivation of some crops, such as premium wine grapes, even as some areas of the country may become more suitable for others.[92]

Forestry

Studies generally conclude that warming and higher atmospheric concentrations of carbon dioxide will boost global forest growth and timber production over the course of the 21st century, although the effects on timber supply in specific regions will vary.[93] Warming and higher CO_2 concentrations have very likely already enhanced the growth of forests in regions of the United States where water is not a limiting factor.[94] Over the next century, changes in patterns of temperature and precipitation will shift the distribution of forest types throughout the country. The area of arid shrubland and steppe is likely to expand in the Southwest; in temperate regions, the ranges of forests that are mainly coniferous (dominated by evergreen trees such as spruce, fir, and pine) or deciduous (dominated by trees that shed their leaves each year) will expand; and the ranges of mixed forests in the northernmost temperate regions and in colder, more northerly boreal regions will shift northward and cover less land.[95] Even below the lower end of the projected range of warming, the tree line could shift northward by up to 250 miles in the Arctic over the course of the 21st century, dramatically increasing the Arctic forest area, shrinking the tundra, and releasing potentially large quantities of methane from melting tundra and from wetlands that would form on currently frozen ground.[96] Such shifts, in turn, could affect regional climates. Forests throughout the country are likely to be adversely affected by the northerly spread of insects and diseases, and, most importantly throughout much of the West, considerably greater risk of wildfires.[97] (Changes in CO_2 concentrations and climate have very likely already contributed to such developments, although their effects are very difficult to distinguish from those of other causes.)[98]

Fisheries

In rivers, reservoirs, lakes, estuaries, and oceans throughout the United States and its territories, rising temperatures will shift the geographic ranges of many fish northward, upriver, or deeper. Warm-water species will generally prosper, but even in the lower half of the likely range of warming, cold-water species such as salmon are likely to disappear from all but the deepest lakes in the continental United States over the course of the 21st century, while cod could disappear from Georges Bank off Cape Cod.[99]

Impacts on the Economy and Human Health

In addition to their effects on agriculture, forestry, and fisheries, changing climate conditions will affect a few industries and various types of infrastructure, such as water supply, and are also expected to affect human health in a variety of ways. Although those impacts could be significant on a regional scale, studies of the overall impact of climate change on measured economic activity in the United States have generally yielded estimates of economic costs that are modest in relation to overall economic activity, at least so long as warming is in the lower half of the projected range.

Water Supply

Changing precipitation patterns have important implications for water supply, especially in the arid Western states, where snowmelt supplies water through much of the dry summer season. During the 21st century, less winter snow accumulation and earlier spring melt will tend to decrease the amount of water available in summer months, significantly affecting the water supply in much of that region.[100] Under existing institutional arrangements, Western states will find it increasingly difficult to meet current levels of demand for water—not to mention the likely higher future levels needed by a larger population and a growing economy. The Colorado River flow is already declining, and recent research indicates that major water shortages are likely to occur in the Colorado River system within a matter of years.[101] Even in areas of the country that will experience a stable or increasing water supply overall, more intense rainstorms and other changes will tend to flush more materials into water supplies, while saltwater will tend to intrude into groundwater in coastal regions, increasing the cost of maintaining water quality.[102]

Infrastructure

While water supply appears to be the most important challenge, particularly in the West, various other types of infrastructure in the United States could be adversely affected by climate change in ways that would require significant expenditures for adaptation. The increased likelihood of flooding from a rising sea level, hurricanes, and intense precipitation may require investment in flood-control infrastructure along coasts and major rivers.[103] In Alaska, progressively thawing permafrost is expected to damage structures and roads.[104] However, warming is likely to reduce energy demand, all else being equal: In most American cities, warming is likely to raise the demand for energy for summer cooling but lower the demand for winter heating, which accounts for roughly twice as much energy use as cooling.[105]

Human Health

Warming is likely to affect human health in various ways. The most comprehensive national reviews of effects of climate change in the United States concluded that projected changes are likely to affect risks of morbidity (the incidence of disease) and mortality for several climate-sensitive aspects of human health, but that the overall net impact is uncertain.[106] For example, although heat waves are likely to become increasingly intense and potentially lethal, extreme cold snaps are likely to become less frequent. Recent research suggests that taken together, those changes in extremes are likely to yield only a very small increase in mortality over the course of the 21st century, especially once the potential for adaptation is taken into account.[107] Warmer climates could encourage the spread of a variety of insect-borne infectious diseases, however.[108] Warmer conditions could also be associated with higher levels of conventional air pollution in summer months but lower levels in winter, though the most recent national assessment concluded that projections of such changes were "somewhat speculative."[109] In general, most of the potential impacts on human health projected for the United States over the coming century appear to be modest, on net.

Aggregate Economic Impacts

Many of the projected impacts of climate change (such as damage from storms, hurricanes, and floods) will affect agriculture, forestry, and fishing; energy demand; and infrastructure. Because prices in those sectors are readily measured, evaluating the potential economic costs of specific effects of climate change is relatively straightforward. Thus, the significant uncertainty in estimates of economic costs arises more from uncertainty about the magnitude of future climate change and its physical impacts than from uncertainty about the economic costs of given physical impacts.

Despite the wide range of projected impacts of climate change over the course of the 21st century, published estimates of the economic costs of direct impacts in the United States tend to be modest.[110] Most of the economy involves activities that are not likely to be directly affected by changes in climate. Moreover, researchers generally expect the U.S. economy to grow dramatically over the coming century, mainly in sectors (such as information technology and medical care) that are relatively insulated from climate effects. Damages are therefore likely to be a smaller share of the future economy than they would be if they occurred today. As a consequence, a relatively pessimistic estimate for the loss in projected real (inflation- adjusted) U.S. gross domestic product is about 3 percent for warming of about 7°F by 2100.[111]

However, such estimates tend to mask larger losses in subsectors of the economy. Some sectors in certain regions are likely to bear sizable costs requiring significant adjustments and adaptations, and a few sectors in a few regions may be eliminated altogether. Even at the low end of the projected range of warming, for example, changing winter conditions would cut the Western ski season by up to four months and would virtually eliminate the Eastern snowmobiling season.[112]

In addition, most of the published studies do not include all of the potential costs of climate change to the country over the coming century and beyond. Most important, there are few detailed estimates of the costs of warming in the upper half of the projected range of 6°F to 13°F of warming during the 21st century.

Moreover, even for the levels of warming that have been examined, most of the estimates discussed above cover only a portion of the potential costs, for several reasons:

- *Nonmarket impacts.* Some types of impacts are very difficult to evaluate in monetary terms because they do not directly involve products that are traded in markets.[113] Although such difficulties apply to effects on human health and quality of life, they are particularly significant for biological impacts, such as loss of species' habitat, biodiversity, and ecosystem services (the various resources and processes that are supplied by natural ecosystems). Experts in such issues generally believe that those nonmarket impacts are much more likely to be negative than positive and could be large. It is particularly difficult to estimate how future generations, which are likely to be wealthier than current ones, will value such impacts relative to the much higher income that they are projected to have.

- *The potential for abrupt changes.* Experts believe that there is a small possibility that even relatively modest warming could trigger abrupt and unforeseen effects during the 21st century that could be associated with quite large economic costs in the United States. Two examples of such effects are shifts in ocean currents that could change weather patterns and affect agriculture over large areas, and rapid disintegration of ice sheets, which could dramatically raise sea level. The sources and nature of such abrupt changes, their likelihood, and their potential impacts remain very poorly understood.

- *Impacts outside the United States.* Most experts agree that populations in other countries—especially poor countries near the equator and bordering on desert zones—are likely to suffer the bulk of the damage from climate change during the 21st century.[114] Some countries could experience significant losses even at relatively low levels of warming—losses that would constitute a very small fraction of global economic output but that would nonetheless be catastrophic for the countries concerned. Some experts conclude that warming could result in "multiple chronic conditions" in some regions—for instance, simultaneous flooding from a rise in sea level, a reduced water supply, agricultural losses, and the spread of disease. Such conditions could seriously harm living standards that are already marginal in regions of Asia, Africa, and the Middle East, perhaps contributing to widespread political instability, with potentially large but very uncertain implications—including significant national security consequences—for the United States.[115]

The most comprehensive published study includes estimates of nonmarket damages as well as costs arising from the risk of catastrophic outcomes associated with about 11°F of warming by 2100. That study projects a loss equivalent to about 5 percent of U.S. output and, with substantially larger losses in a number of other countries, a loss of about 10 percent of global output.[116]

POLICY IMPLICATIONS OF UNCERTAINTY

The uncertainties surrounding future emissions of greenhouse and other gases and land-use changes, the climate system's response to those developments, and the resulting impacts greatly complicate the crafting of a policy response. They make the climate outcome of any particular policy very difficult to determine and, conversely, make the appropriate policy to achieve any desired climate outcome very hard to predict.[117] Given current knowledge:

- *Any particular amount of emissions could result in much more or much less warming and associated impacts than expected.* Emissions in the middle of the range of projections discussed previously are most likely to result in 9°F of warming. However, that scenario has about a one-in-six chance of yielding less than 5°F of warming and about an equal chance of yielding more than 14°F. A policy that limited emissions with the goal of stabilizing atmospheric concentrations of greenhouse gases at roughly 650 parts per million of CO_2 equivalent (roughly 2.3 times the preindustrial CO_2 concentration of 280 parts per million) would significantly reduce warming, compared with a midrange scenario, but the policy could result in warming of anywhere between 3°F and nearly 6°F over the course of the 21st century (see Figure 4). Even if concentrations were stabilized at today's levels by dramatically reducing emissions, there is a small chance that a long-term temperature increase of nearly 4°F over the preindustrial level would nevertheless occur. Conversely, a fourfold increase in atmospheric CO_2 concentrations would have a small probability of yielding long-term warming of only about 6°F.

- *The amount of emissions that is compatible with a given climate goal is extremely uncertain.* For instance, suppose that policymakers wanted to limit total warming during the 21st century to no more than 5°F. If the climate's sensitivity to rising concentrations of greenhouse gases is at the low end of the likely range, policymakers could allow concentrations to more than double over their current level before stabilizing them. Little if any action would be required in the near term to meet that goal. However, if the climate is particularly sensitive, policymakers would have to sharply limit the growth of emissions over the next few decades to meet the same temperature goal.

Therefore, policies that target emissions or concentrations of greenhouse gases cannot guarantee specific climate outcomes; they can only shift the odds of those outcomes. Keeping concentrations to less than twice their preindustrial level—a commonly discussed target—would leave a more than 50 percent chance of ultimately exceeding a 5°F increase in average global temperature and a small chance even of exceeding 8°F.[118] Limiting concentrations further would reduce the chance of exceeding 5°F, but concentrations would have to be kept under about 420 parts per million of CO_2 equivalent to nearly eliminate that chance altogether. (With only the long-lived greenhouse gases taken into account, the current concentration is roughly 455 parts per million of CO_2 equivalent. However, with the offsetting effects of aerosols and other gases taken into account, the net current concentration is roughly 375 parts per million of CO_2 equivalent, with a very wide range of uncertainty around that estimate. See Figure 5 for approximate distributions of temperature under different scenarios for CO_2 concentrations.)

Given those uncertainties, crafting a policy response to climate change involves balancing two types of risks: the risks of limiting emissions to reach a temperature target and experiencing much more warming and much greater impacts than expected versus the risks of incurring costs to limit emissions when warming and its impacts would, in any event, have been less severe than anticipated. Climate policies thus have a strong element of risk management: Depending on the costs of doing so, society may find it economically sensible to invest in reducing the risk of the most severe possible impacts from climate change even if their likelihood is relatively remote.[119] In particular, the potential for unexpectedly severe and even catastrophic outcomes, even if unlikely, would justify more stringent policies than would result from simply balancing the costs of reducing emissions against the benefits associated with the expected or most likely resulting degree of warming.

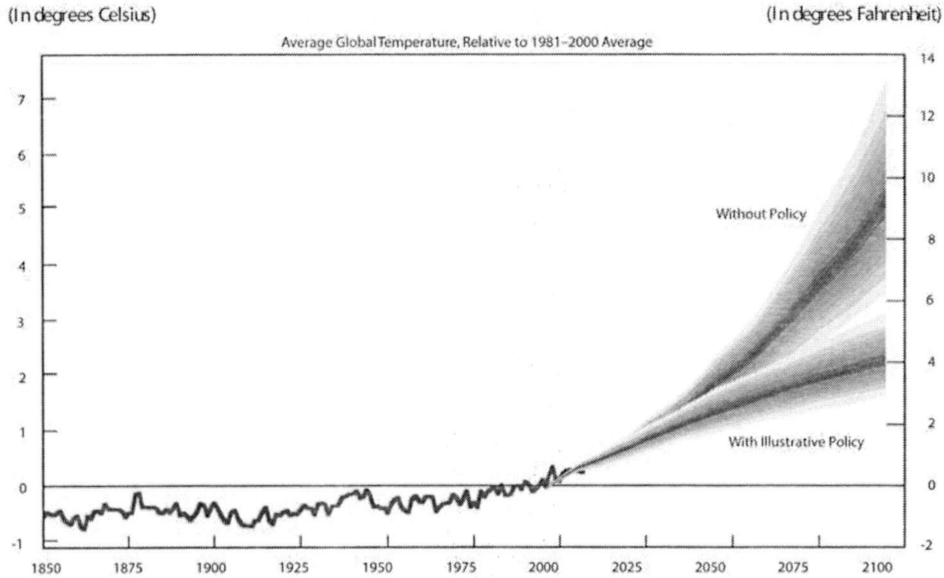

Source: Congressional Budget Office. Historical data are from the Hadley Centre for Climate Prediction and Research, http://hadobs.met office.com/hadcrut3/diagnostics/global/nh+sh/annual, and described primarily in P. Brohan and others, "Uncertainty Estimates in Regional and Global Observed Temperature Changes: A New Dataset from 1850," *Journal of Geophysical Research,* vol. 111 (June 24, 2006). The projections are based on data provided by Henry Jacoby, Massachusetts Institute of Technology, in a personal communication to CBO, December 22, 2008. The projection without policy is presented in A.P. Sokolov and others, *Probabilistic Forecast for 21st Century Climate Based on Uncertainties in Emissions (Without Policy) and Climate Parameters,* Report No. 169 (Cambridge, Mass.: MIT Joint Program on the Science and Policy of Global Change, 2009), http://globalchange.mit.edu/files/ document/MITJPSPGC_Rpt169.pdf (in press in the *Journal of Climate*). The projection with an illustrative policy is the "Level 2" scenario discussed in L. Clarke and others, *Scenarios of Greenhouse Gas Emissions and Atmospheric Concentrations: Subreport 2.1A of Synthesis and Assessment Report 2.1 by the U.S. Climate Change Science Program and the Subcommittee on Global Change Research* (Washington, D.C.: Department of Energy, Office of Biological and Environmental Research, 2007).

Note: The projections, which are interpolated from decadal averages beginning in 1995, show the possible distribution of changes in average global temperature as a result of human influence, relative to the 1981–2000 average and given current understanding of the climate. The projection without policy assumes that no actions are taken to limit the growth of emissions. The projection with policy simulates the results over time of a cap on emissions that is designed to ensure that atmospheric concentrations of carbon dioxide remain below 550 parts per million. Under such a policy, cumulative global carbon dioxide emissions from human activities are assumed to be limited to 667 billion tons of carbon over the 21st century. The policy constrains emissions, not concentrations, so the resulting carbon dioxide concentrations over the 21st century grow to a median of 557 parts per million, with a 95 percent probability interval of 508 to 602 parts per million; concentrations of all greenhouse gases, summed using Global Warming Potential indices, have a 95 percent range of 599 to 714 parts per million of carbon dioxide equivalent. For each scenario, the probability is 10 percent that the actual global temperature will fall in the darkest area and 90 percent that it will fall within the whole shaded area. However, actual temperatures could be affected by factors that were not addressed in either scenario (such as volcanic activity and the variability of solar radiation) and whose effects are not included in the figure.

Figure 4. Historical and Projected Climate Change, With and Without an Illustrative Policy Applied

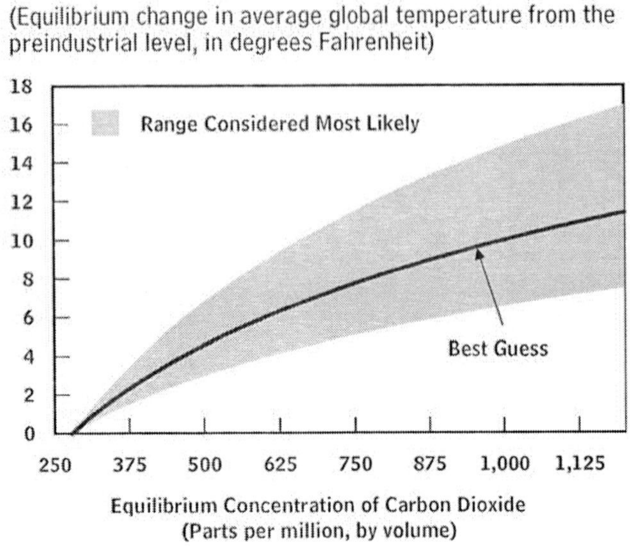

Source: Congressional Budget Office based on Table 10.8 in Gerald A. Meehl and others, "Global Climate Projections," in Susan Solomon and others, eds., *Climate Change 2007: The Physical Science Basis. Contribution of Working Group I to the Fourth Assessment Report of the Intergovernmental Panel on Climate Change* (Cambridge, U.K.: Cambridge University Press, 2007), p. 826.

Note: The figure shows the "best guess" and "most likely" range for the ultimate long-term amounts of warming that would result if, all else being equal, atmospheric concentrations of carbon dioxide were stabilized at specified levels. For example, if concentrations were stabilized at 750 parts per million by volume, the best guess is that the average global temperature would increase by nearly 8 degrees Fahrenheit above the preindustrial level, and there are about two chances in three (the "most likely" range) that the increase would be between 5 degrees and 11 degrees.

Figure 5. Uncertainty in the Climate's Response to Rising Atmospheric Concentrations of Carbon Dioxide

Those insights have spurred some researchers who are particularly worried about low-probability but high-impact outcomes to call for limiting long-term warming to no more than 3°F to 5°F with a high degree of certainty. However, since about 1.4°F of warming has already occurred, and past emissions have made a substantial amount of further warming inevitable, limiting long-term warming to such levels with a substantial degree of certainty would probably require very dramatic and potentially very expensive curtailment of expected future emissions. There is a large difference in costs between a policy that leaves a 50 percent risk of warming exceeding 5°F and a policy that virtually eliminates that risk. In moving along the continuum of risk from the former to the latter, each increment of risk reduction is likely to come at an increasing price.

Although the possibility of very serious impacts from warming may justify more stringent near-term policies than might be justified by balancing expected costs and benefits alone, that rationale may shift over time as researchers learn more about the likely range of warming and related risks in the future. On the one hand, researchers may gradually learn that the climate's response to rising concentrations is much larger or the impacts much more severe than expected, requiring even more stringent policies than were initially adopted. On the other hand, researchers may discover that the climate's response or the impacts of its changes are relatively mild, in which case policies might be justifiably left as originally set.

The uncertainties in the link between emissions and climate change, coupled with the potential for improvements in future understanding of the climate problem, imply that policies that gradually raise the price of emissions with the expectation of meeting a cumulative target for emissions are likely to yield greater long-term net economic benefits than policies that impose increasingly stringent quantitative limits on emissions with the same cumulative target. Neither type of policy is likely to achieve a specific warming target, and policies of either type, if adopted, would be very likely to require adjustments and refinements as better information became available. Uncertainties may thus justify flexible mechanisms even though they may simultaneously justify relatively stringent policies.

End Notes

[1] See Intergovernmental Panel on Climate Change, *Climate Change 2007: Fourth Assessment Report of the Intergovernmental Panel on Climate Change,* vols. I, II, and III (Geneva, Switzerland: 2008); National Science and Technology Council, *Scientific Assessment of the Effects of Global Change on the United States* (May 2008); and the online library of the U.S. Climate Change Science Program, www.climatescience.gov/Library/default.htm.

[2] Much of the research presented in the *Fourth Assessment Report* and other recent studies is concerned with documenting uncertainties in an increasingly standardized framework. For example, events that are considered to have a greater than 99 percent probability are referred to as "virtually certain"; "very likely" events are more than 90 percent probable; "likely" ones are greater than 66 percent probable; events that are "about as likely as not" have a probability of 33 percent to 66 percent; "unlikely" events have less than a 33 percent likelihood; "very unlikely" ones have less than a 10 percent probability; and "exceptionally unlikely" outcomes have less than 1 percent probability. This paper applies that framework where appropriate. See Susan Solomon and others, "Technical Summary," in Solomon and others, eds., *Climate Change 2007: The Physical Science Basis. Contribution of Working Group I to the Fourth Assessment Report of the Intergovernmental Panel on Climate Change* (Cambridge, U.K.: Cambridge University Press, 2007), pp. 22–23.

[3] For more background information on climate change, see Congressional Budget Office, *The Economics of Climate Change: A Primer.* (April 2003).

[4] See Solomon and others, "Technical Summary," p. 29.

[5] Ibid., pp. 24–35.

[6] Ibid., pp. 31–36.

[7] Ibid., pp. 36–37. Temperatures and changes in temperature are presented here in degrees Fahrenheit (°F), but most scientific documents present them in degrees Celsius (°C). The formula for conversion of levels is °C = (0.56 * °F) - 32; the formula for changes in temperature is Δ°C = 0.56 * Δ °F.

[8] Ibid., pp. 58–66.

[9] Ibid., pp. 60–62.

[10] Ibid., pp. 68–69.

[11] Ibid., p. 70.

[12] Ibid., pp. 26–27, 47–48, 77–79; and James Hansen and others, "Earth's Energy Imbalance: Confirmation and Implications," *Science*, vol. 308, no. 5727 (June 3, 2005), pp. 143 1–1435.

[13] See Solomon and others, "Technical Summary," pp. 42–46.

[14] Ibid., pp. 26–27, 43–48.

[15] See National Research Council, Climate Research Committee, Committee on Radiative Forcing Effects on Climate, *Radiative Forcing of Climate Change: Expanding the Concept and Addressing Uncertainties* (Washington, D.C.: National Academies Press, 2005); Daniel Rosenfeld and others, "Flood or Drought: How Do Aerosols Affect Precipitation?" *Science*, vol. 321, no. 5894 (June 2, 2008), pp. 1309–1313; and Gregg Marland and others, "The Climatic Impacts of Land Surface Change and Carbon Management, and the Implications of Climate-Change Mitigation Policy," *Climate Policy,* vol. 3 (2003), pp. 149–157.

[16] Ibid., pp. 79–80.

[17] See P.U. Clark and others, *Abrupt Climate Change, A report by the U.S. Climate Change Science Program and the Subcommittee on Global Change Research* (Reston, Va.: U.S. Geological Survey, 2008); R. B. Alley and others, "Abrupt Climate Change," *Science*, vol. 299, no. 5615 (March 28, 2003), pp. 2005–2010; and José Rial and others, "Nonlinearities, Feedbacks, and Critical Thresholds Within the Earth's Climate System," *Climatic Change*, vol. 65 (2004), pp. 11–38.

[18] Calculated using data from the Met Office Hadley Centre for Climate Change, http://hadobs.metoffice.com/hadcrut3/diagnostics/global/nh+sh/annual.

[19] See N. S. Keenlyside and others, "Advancing Decadal-Scale Climate Prediction in the North Atlantic Sector," *Nature*, vol. 453 (May 1, 2008), pp. 84–88.

[20] Wolfgang Lutz, Warren C. Sanderson, and Sergei Scherbov, eds., *The End of World Population Growth in the 21st Century: New Challenges for Human Capital Formation and Sustainable Development* (London: Earthscan, 2004), p. 40. The estimate of current world population is from the U.S. Census Bureau's World Population Clock, www.census.gov/main/www/popclock.html.

[21] For example, see Nebojša Nakićenović and Rob Swart, eds., *Emission Scenarios* (Cambridge, U.K.: Cambridge University Press, 2000).

[22] See Solomon and others, "Technical Summary," p. 60; and K. L. Denman and others, "Couplings Between Changes in the Climate System and Biogeochemistry," in Susan Solomon and others, eds., *Climate Change 2007: The Physical Science Basis,* p. 534.

[23] See Solomon and others, "Technical Summary," pp. 3 1–35; and D. A. Randall and others, "Climate Models and Their Evaluation," in Susan Solomon and others, eds., *Climate Change 2007: The Physical Science Basis,* p. 636.

[24] See Gerald A. Meehl and others, "Global Climate Projections," in Susan Solomon and others, eds., *Climate Change 2007: The Physical Science Basis,* pp. 822–831.

[25] Ibid., p. 810; and R. Knutti and others, "A Review of Uncertainties in Global Temperature Projections over the Twenty-First Century," *Journal of Climate,* vol. 21, no. 11 (June 1, 2008), pp. 2651–2663.

[26] See Susan Solomon and others, "Irreversible Climate Change Due to Carbon Dioxide Emissions," *Proceedings of the National Academy of Sciences,* vol. 106, no. 6 (February 10, 2009), pp. 1704–1709.

[27] For example, note that the ratio of highest to lowest best estimates of warming from different emissions scenarios presented by the IPCC is roughly comparable to the ratio of upper to lower bounds of the likely range of outcomes for any particular scenario. See Solomon and others, "Technical Summary," p. 70. Also see A. P. Sokolov and others, *Probabilistic Forecast for 21st Century Climate Based on Uncertainties in Emissions (Without Policy) and Climate Parameters*, Report No. 169 (MIT Joint Program on the Science and Policy of Global Change, 2009), p.26, http://globalchange.mit.edu/files/document/MITJPSPGC_Rpt169.pdf.

[28] See Sokolov and others, *Probabilistic Forecast for 21st Century Climate*. These estimates represent a substantial upward revision from the Joint Program's previous estimates, which CBO cited in *The Economics of Climate Change: A Primer*. The revisions result mainly from a reappraisal of scientific uncertainties, particularly about the rate at which oceans will absorb heat from the atmosphere. The median level of projected warming is substantially greater than the warming expected to result from the IPCC's A1B emissions scenario, which is used extensively in the *Fourth Assessment Report* to illustrate the current state of scientific understanding of sectoral and regional impacts of warming. Another recent study that develops probabilistic projections of emissions reports a median warming of about 7°F during the 21st century, with a 9-in-10 probability that warming will be between 5°F and 9°F. However, that study does not consider emissions of any greenhouse gases except CO_2. See David von Below and Torsten Persson, *Uncertainty, Climate Change and the Global Economy*, Working Paper No. 14426 (Cambridge, Mass.: National Bureau of Economic Research, October 2008), www.nber.org/ papers/w14426.

[29] See E. Jansen and others, "Palaeoclimate," in Susan Solomon and others, eds., *Climate Change 2007: The Physical Science Basis*, pp. 435, 447.

[30] See D. P. Van Vuuren and others, "Temperature Increase of 21st Century Mitigation Scenarios," *Proceedings of the National Academy of Sciences*, vol. 105, no. 40 (October 7, 2008), pp. 15258–15262.

[31] See, for example, Solomon and others, "Technical Summary," pp. 35–54; and Peter Backlund, Anthony Janetos, and David Schimel, *The Effects of Climate Change on Agriculture, Land Resources, Water Resources, and Biodiversity in the United States. A Report by the U.S. Climate Change Science Program and the Subcommittee on Global Change Research* (Washington, D.C.: U.S. Environmental Protection Agency, 2008), p. 3.

[32] For a discussion of the importance of the distinction between changes in average conditions and changes in extreme events, see Stanley A. Changnon, "Economic Impacts of Climate Conditions in the United States: Past, Present, and Future," *Climatic Change*, vol. 68 (2005), pp. 1–9.

[33] See William D. Nordhaus, *The Economics of Hurricanes in the United States,* Working Paper No. 12813 (Cambridge, Mass.: National Bureau of Economic Research, December 2006), www.nber.org/papers/w12813.

[34] See Randall and others, "Climate Models and Their Evaluation," pp. 627–629.

[35] See Alley and others, "Abrupt Climate Change," p. 2008.

36 See C. B. Field and others, "North America," in M. L. Parry and others, eds., *Climate Change 2007: Impacts, Adaptation and Vulnerability. Contribution of Working Group II to the Fourth Assessment Report of the Intergovernmental Panel on Climate Change* (Cambridge, U.K.: Cambridge University Press, 2007), p. 630.

37 See Randall and others, "Climate Models and Their Evaluation," p. 631.

38 See T. R. Carter and others, "New Assessment Methods and the Characterisation of Future Conditions," in M. L. Parry and others, eds., *Climate Change 2007: Impacts, Adaptation and Vulnerability*, pp. 144–160; Field and others, "North America," p. 635; and Backlund, Janetos, and Schimel, *The Effects of Climate Change*, p. 3.

39 Hans-Martin Füssel and Richard J. T. Klein, "Climate Change Vulnerability Assessments: An Evolution of Conceptual Thinking," *Climatic Change*, vol. 75 (2006), pp. 30 1–329.

40 See Solomon and others, "Technical Summary," p. 71.

41 Ibid.

42 Ibid.

43 J. H. Christensen and others, "Regional Climate Projections," in Susan Solomon and others, eds., *Climate Change 2007: The Physical Science Basis*, pp. 852–856.

44 Ibid., p. 862; Thomas R. Karl and Gerald A. Meehl, "Executive Summary," in Thomas R. Karl and others, eds., *Weather and Climate Extremes in a Changing Climate. Regions of Focus: North America, Hawaii, Caribbean, and U.S. Pacific Islands. A Report by the U.S. Climate Change Science Program and the Subcommittee on Global Change Research* (Washington, D.C.: Department of Commerce, National Oceanic and Atmospheric Administration, National Climatic Data Center, 2008), pp. 4, 8; and Claudia Tebaldi and others, "Going to the Extremes: An Intercomparison of Model-Simulated Historical and Future Changes in Extreme Events," *Climatic Change*, vol. 79 (2007), p. 22. Such changes are already evident in most of North America and are considered likely to be linked to human influences on the climate.

45 Christensen and others, "Regional Climate Projections," p. 887.

46 Ibid., p. 890.

47 See Randall and others, "Climate Models and Their Evaluation," p. 591; and Meehl and others, "Global Climate Projections," pp. 750, 762, 768. Consistent with that projection, global precipitation over land has increased by about 2 percent over the past century. See Backlund, Janetos, and Schimel, *The Effects of Climate Change,* p. 14.

48 See Backlund, Janetos, and Schimel, *The Effects of Climate Change*, p. 149.

49 See Solomon and others, "Technical Summary," p. 75; and Christensen and others, "Regional Climate Projections," p. 890.

50 See Meehl and others, "Global Climate Projections," p. 750.

51 See Solomon and others, "Technical Summary," p. 75.

52 Ibid.; and Richard Seager and others, "Model Projections of an Imminent Transition to a More Arid Climate in Southwestern North America," *Science*, vol. 316, no. 5828 (May 25, 2007), pp. 1181–1184

53 Christensen and others, "Regional Climate Projections," p. 891.

54 Ibid.

55 See Field and others, "North America," p. 634.

56 See Meehl and others, "Global Climate Projections," pp. 750–751.

57 Robert J. Trapp and others, "Changes in Severe Thunderstorm Environment Frequency During the 21st Century Caused by Anthropogenically Enhanced Global Radiative Forcing," *Proceedings of the National Academy of Sciences*, vol. 104, no. 50 (2007), pp. 19719–19723, www.pnas.org/cgi/doi/10.1073/pnas.0705494104.

58 See Meehl and others, "Global Climate Projections," pp. 750, 782; and Tebaldi and others, "Going to the Extremes," pp. 185–211.

59 See National Research Council, *Radiative Forcing of Climate Change*, pp. 5–6; Rosenfeld and others, "Flood or Drought," pp. 1309–1313; and R. A. Pielke Sr. and others, "An Overview of Regional Land-Use and Land-Cover Impacts on Rainfall," *Tellus*, vol. 59B (2007), pp. 587–601.

60 See Meehl and others, "Global Climate Projections," p. 751; and Karl and Meehl, "Executive Summary," p. 6.

61 See Nordhaus, *The Economics of Hurricanes in the United States.*

62 See Meehl and others, "Global Climate Projections," p. 788.

63 See Jeffrey P. Donnelly and Jonathan D. Woodruff, "Intense Hurricane Activity over the Past 5,000 Years Controlled by El Niño and the West African Monsoon," *Nature*, vol. 447, no. 7143 (May 24, 2007), pp. 465–468; and Gabriel A. Vecchi and Brian J. Soden, "Increased Tropical Atlantic Wind Shear in Model Projections of Global Warming," *Geophysical Research Letters*, vol. 34 (April 18, 2007).

64 See Meehl and others, "Global Climate Projections," pp. 788–789.

65 See J. R. Toggweiler and Joellen Russell, "Ocean Circulation in a Warming Climate," *Nature*, vol. 451 (January 17, 2008), pp. 286–28 8.

66 See Solomon and others, "Technical Summary," p. 72; Karl and Meehl, "Executive Summary," p. 6; and Kirsten Zickfeld and others, "Expert Judgments on the Response of the Atlantic Meridional Overturning Circulation to Climate Change," *Climatic Change*, vol. 82, no. 3-4 (June 2007), pp. 235–265.

67 See Solomon and others, "Technical Summary," p. 73; and Meehl and others, "Global Climate Projections," p. 779.

[68] See Gerald A. Meehl and others, "Current and Future U.S. Weather Extremes and El Niño," *Geophysical Research Letters*, vol. 34 (October 19, 2007).

[69] See Solomon and others, "Technical Summary," pp. 70–71, 80; and National Science and Technology Council, *Scientific Assessment of the Effects of Global Change on the United States*, p. 6.

[70] See Susan Solomon and others, "Irreversible Climate Change Due to Carbon Dioxide Emissions," *Proceedings of the National Academy of Sciences*, vol. 106, no. 6 (February 10, 2009), pp. 1704–1709.

[71] Solomon and others, "Technical Summary," p. 71.

[72] M. L. Parry and others, "Technical Summary," in Parry and others, eds., *Climate Change 2007: Impacts, Adaptation and Vulnerability*, pp. 40–41, 44–45; and Field and others, "North America," p. 630.

[73] See James G. Titus and Charlie Richman, "Maps of Lands Vulnerable to Sea Level Rise: Modeled Elevations Along the U.S. Atlantic and Gulf Coasts," *Climate Research*, vol. 18 (2001), pp. 205–228.

[74] See National Science and Technology Council, *Scientific Assessment of the Effects of Global Change on the United States*, pp. 6–7; and Nordhaus, *The Economics of Hurricanes in the United States*.

[75] See Field and others, "North America," p. 633.

[76] Estimate based on an extrapolation from Christopher L. Sabine and others, "The Oceanic Sink for Anthropogenic CO_2," *Science*, vol. 305, no. 5682 (July 16, 2004), pp. 367–371.

[77] See J. A. Raven and others, *Ocean Acidification Due to Increasing Atmospheric Carbon Dioxide* (London: Royal Society, 2005); and Andy Ridgwell and Richard E. Zeebe, "The Role of the Global Carbonate Cycle in the Regulation and Evolution of the Earth System," *Earth and Planetary Science Letters*, vol. 234, no. 3-4 (June 15, 2005), pp. 299–315.

[78] See Wilfried Thuiller, "Climate Change and the Ecologist," *Nature*, vol. 448 (August 2, 2007), pp. 5 50–552.

[79] See Parry and others, "Technical Summary," pp. 38–40, 44–45.

[80] See Backlund, Janetos, and Schimel, *The Effects of Climate Change*, pp. 178–181.

[81] See Jay R. Malcolm and others, "Global Warming and Extinctions of Endemic Species from Biodiversity Hotspots," *Conservation Biology*, vol. 20, no. 2 (April 2006), pp. 538–548. Thirty-four hotspots around the world are thought to contain more than 50 percent of all plant species and 42 percent of all terrestrial vertebrate species on 2.3 percent of the Earth's land area. The only hotspots in the United States are the California Floristic Province, comprising the California coast and central valley, and the northernmost Madrean Pine-Oak Woodlands in the Southwest. For further information on hotspots of biodiversity, see www.biodiversity-hotspots.org/Pages/default.aspx.

[82] See A. Fischlin and others, "Ecosystems, Their Properties, Goods, and Services," in Parry and others, eds., *Climate Change 2007: Impacts, Adaptation and Vulnerability*, p. 213; Raven and others, *Ocean Acidification Due to Increasing Atmospheric Carbon Dioxide*; Ridgwell and Zeebe, "The Role of the Global Carbonate Cycle in the Regulation and Evolution of the Earth System"; Victoria J. Fabry, "Marine Calcifiers in a High-CO_2 Ocean," *Science*, vol. 320, no. 5879 (May 23, 2008), pp. 1020–1022; and Kent E. Carpenter and others, "One-Third of Reef-Building Corals Face Elevated Extinction Risk from Climate Change and Local Impacts," *Science*, vol. 321, no. 5888 (July 23, 2008), pp. 560–563.

[83] See John W. Williams, Stephen T. Jackson, and John E. Kutzbach, "Projected Distributions of Novel and Disappearing Climates by 2100 AD," *Proceedings of the National Academy of Sciences*, vol. 104, no. 14 (April 3, 2007), pp. 5738–5742.

[84] See Fischlin and others, "Ecosystems, Their Properties, Goods, and Services," p. 213.

[85] Parry and others, "Technical Summary," pp. 37–38; Inger Greve Alsos and others, "Frequent Long-Distance Plant Colonization in the Changing Arctic," *Science*, vol. 316, no. 5831 (June 15, 2007), pp. 1606–1609; and Ronald P. Neilson and others, "Forecasting Regional to Global Plant Migration in Response to Climate Change," *BioScience*, vol. 55, no. 9 (September 2005), pp. 749–759.

[86] Documentation on extinction from the International Union for Conservation of Nature is available at www.iucnredlist.org/info/stats.

[87] See, for example, John Reilly and others, "U.S. Agriculture and Climate Change: New Results," *Climatic Change*, vol. 57 (2003), pp. 43–69.

[88] See John Reilly and others, *Global Economic Effects of Changes in Crops, Pasture, and Forests Due to Changing Climate, Carbon Dioxide, and Ozone*, Report No. 149 (MIT Joint Program on the Science and Policy of Global Change, 2007), http://globalchange.mit.edu/files/document/MITJPSPGC_Rpt149.pdf.

[89] See Allison M. Thomson and others, "Climate Change Impacts for the Conterminous USA: An Integrated Assessment," *Climatic Change*, vol. 69, no. 1 (March 2005), "Part 3: Dryland Production of Grain and Forage Crops," pp. 43–65, and "Part 5: Irrigated Agriculture and National Grain Crop Production," pp. 89–105.

[90] See Raymond P. Motha and Wolfgang Baier, "Impacts of Present and Future Climate Change and Climate Variability on Agriculture in the Temperate Regions: North America," *Climatic Change*, vol. 70, no. 1–2 (May 2005), pp. 137–164; and Robert Mendelsohn and Michelle Reinsborough, "A Ricardian Analysis of U.S. and Canadian Farmland," *Climatic Change*, vol. 81. no. 1 (March 2007), pp. 9–17.

[91] Cynthia Rosenzweig and others, "Increased Crop Damage in the US from Excess Precipitation Under Climate Change," *Global Environmental Change*, vol. 12, no. 3 (October 2002), pp. 197–202.

[92] See Katharine Hayhoe and others, "Regional Assessment of Climate Impacts on California Under Alternative Emission Scenarios—Key Findings and Implications for Stabilisation," Chapter 24 in *Avoiding Dangerous Climate Change* (Cambridge, U.K.: Cambridge University Press, 2006); and M. A. White and others, "Extreme Heat Reduces and Shifts United States Premium Wine Production in the 21st Century," *Proceedings of the National Academy of Sciences*, vol. 103, no. 30 (July 25, 2006), pp. 11217–11222.

[93] See M. G. Ryan and S. R. Archer, "Land Resources: Forests and Arid Lands," in Backlund, Janetos, and Schimel, *The Effects of Climate Change*, pp. 79–80; and W. E. Easterling and others, "Food, Fibre, and Forest Products," in Parry and others, eds., *Climate Change 2007: Impacts, Adaptation and Vulnerability*, pp. 288–290.

[94] See Backlund, Janetos, and Schimel, *The Effects of Climate Change*, p. 119.

[95] See R. César Izaurralde and others, "Climate Change Impacts for the Conterminous U.S.A.: An Integrated Assessment—Part 6: Distribution and Productivity of Unmanaged Ecosystems," *Climatic Change*, vol. 69, no. 1 (March 2005), pp. 107–126.

[96] See Fischlin and others, "Ecosystems, Their Properties, Goods, and Services," pp. 230–232; Jed O. Kaplan and Mark New, "Arctic Climate Change with a 2°C Global Warming: Timing, Climate Patterns, and Vegetation Change," *Climatic Change*, vol. 79, no. 3–4 (December 2006), pp. 213–241; and Clark and others, *Abrupt Climate Change*, p. 16.

[97] See Field and others, "North America," p. 631; and Backlund, Janetos, and Schimel, *The Effects of Climate Change*, pp. 7–8.

[98] See Ryan and Archer, "Land Resources," pp. 78, 118.

[99] See Field and others, "North America," pp. 63 1–632; Michael Fogarty and others, "Potential Climate Change Impacts on Atlantic Cod (Gadus Morhua) off the Northeastern U.S.A.," *Mitigation and Adaptation Strategies for Global Change*, vol. 13, no. 5–6 (June 2008), pp. 453–466; and Heinz G. Stefan, Xing Fang, and John G. Eaton, "Simulated Fish Habitat Changes in North American Lakes in Response to Projected Climate Warming," *Transactions of the American Fisheries Society*, vol. 130, no. 3 (May 2001), pp. 459–477.

[100] See Field and others, "North America," p. 627; and National Science and Technology Council, *Scientific Assessment of the Effects of Global Change on the United States* (May 2008), pp. 12–13.

[101] See Tim P. Barnett and David W. Pierce, "When Will Lake Mead Go Dry?" *Water Resources Research*, vol. 44, no. 3 (March 2008).

[102] See Field and others, "North America," p. 629; and National Science and Technology Council, *Scientific Assessment of the Effects of Global Change on the United States.*

[103] Field and others, "North America," pp. 632–633.

[104] Ibid., p. 632; and Larry D. Hinzman and others, "Evidence and Implications of Recent Climate Change in Northern Alaska and Other Arctic Regions," *Climatic Change*, vol. 72, no. 3 (October 2005), pp. 251–298.

[105] See National Science and Technology Council, *Scientific Assessment of the Effects of Global Change on the United States*, pp. 16–17; Olivier Deschênes and Michael Greenstone, *Climate Change, Mortality, and Adaptation: Evidence from Annual Fluctuations in Weather in the US*, Working Paper No. 13178 (National Bureau of Economic Research, June 2007), www.nber.org/papers/w13178; and Department of Energy, Energy Information Administration, *Annual Energy Outlook 2008*, DOE/EIA-0383 (2008), pp. 123, 125, www.eia.doe.gov/oiaf/aeo/pdf/0383(2008).pdf.

[106] See J. L. Gamble and others, "Executive Summary," in Gamble and others, *Analyses of the Effects of Global Change on Human Health and Welfare and Human Systems. A Report by the U.S. Climate Change Science Program and the Subcommittee on Global Change Research* (Washington, D.C.: U.S. Environmental Protection Agency, 2008), pp. ES 5-7; National Science and Technology Council, *Scientific Assessment of the Effects of Global Change on the United States*, pp. 14–16; National Assessment Synthesis Team, *Climate Change Impacts on the United States: The Potential Consequences of Climate Variability and Change* (Cambridge, U.K.: Cambridge University Press, 2001), pp. 102–107; and Kristie L. Ebi and others, "Climate Change and Human Health Impacts in the United States: An Update on the Results of the U.S. National Assessment," *Environmental Health Perspectives*, vol. 114, no. 9 (2006), pp. 1318–1324.

[107] See Mercedes Medina-Ramón and Joel Schwartz, "Temperature, Temperature Extremes, and Mortality: A Study of Acclimatisation and Effect Modification in 50 United States Cities," *Occupational and Environmental Medicine*, vol. 64 (2007), pp. 827–833; and Deschênes and Greenstone, *Climate Change, Mortality, and Adaptation.*

[108] See Field and others, "North America," p. 632.

[109] Ibid.; and National Science and Technology Council, *Scientific Assessment of the Effects of Global Change on the United States*, p. 15.

[110] See Dale W. Jorgenson and others, *U.S. Market Consequences of Global Climate Change* (Arlington, Va.: Pew Center on Global Climate Change, 2004); Robert Mendelsohn and James E. Neumann, "Synthesis and Conclusions," in Robert Mendelsohn and James E. Neumann, eds., *The Impact of Climate Change on the United States Economy* (Cambridge, U.K.: Cambridge University Press, 2004), pp. 315–331; William D. Nordhaus and Joseph Boyer, *Warming the World: Economic Models of Global Warming* (Cambridge, Mass.: MIT Press,

2000); and Richard S. J. Tol, "Estimates of the Damage Costs of Climate Change—Part II: Dynamic Estimates," *Environmental and Resource Economics*, vol. 21, no. 2 (February 2002), pp. 135–160.

[111] See Jorgenson and others, p. 36.

[112] See Field and others, "North America," p. 634.

[113] See Backlund, Janetos, and Schimel, *The Effects of Climate Change,* pp. 3–4.

[114] See Parry and others, "Technical Summary," pp. 59–63.

[115] See, for example, Kurt M. Campbell and others, *The Age of Consequences: The Foreign Policy and National Security Implications of Global Climate Change* (joint publication of the Center for Strategic and International Studies and the Center for a New American Security, 2007), www.csis.org/media 071 105_ageofconsequences.pdf; and CNA Corp., *National Security and the Threat of Climate Change* (Alexandria, Va., 2007), http://securityandclimate.cna.org/report/SecurityandClimate_Final.pdf.

[116] See Nordhaus and Boyer, *Warming the World*, pp. 9 5–96.

[117] For more background information on the policy implications of uncertainty, see Congressional Budget Office, *Uncertainty in Analyzing Climate Change: Policy Implications* (January 2005).

[118] Different greenhouse gases persist in the atmosphere for different periods of time and have different warming characteristics. To simplify matters, researchers commonly refer to emissions and atmospheric concentrations of different greenhouse gases in terms of the amount of CO_2 that would cause an equivalent amount of warming over an arbitrary period of time—typically 100 years. Researchers often discuss concentration goals with reference to the preindustrial level of CO_2 (about 280 parts per million, or ppm). For example, a commonly mentioned goal is to limit the increase in concentrations to 550 ppm of CO_2 equivalent (CO_2e), or about twice the preindustrial level of CO_2, even though there were other greenhouse gases in the atmosphere before industrialization as well. The doubling limit would most likely constrain the long-term increase in average global temperature to slightly more than 5°F, but the likely outcome could be anywhere between 3°F and 8°F.

[119] The risk management aspect of climate policy is more analogous to buying air bags than buying insurance, for two reasons. One is that climate policy involves making investments to reduce the risk of adverse outcomes rather than to ensure compensation in the event of such outcomes. The other reason is that unlike the case of automobile insurance, in which millions of drivers can pool their risks to compensate those who actually have accidents, climate change involves only one outcome on one planet. In that respect, no risk pooling is feasible. Thus, for society as a whole, expenditures to avert the greatest risks associated with warming could not be used to compensate anyone if those risks ultimately proved to be relatively mild.

INDEX

A

B

C

D

E

F

G

H

I

J

K

L

M

N

O

P

Q

R

S

T

U

V

W

Y

Z